Design Parameters of Electrical Network Grounding Systems

Osama El-Sayed Gouda
Cairo University, Egypt

A volume in the Advances in Computer and Electrical Engineering (ACEE) Book Series

IGI Global
DISSEMINATOR OF KNOWLEDGE

Published in the United States of America by
IGI Global
Engineering Science Reference (an imprint of IGI Global)
701 E. Chocolate Avenue
Hershey PA, USA 17033
Tel: 717-533-8845
Fax: 717-533-8661
E-mail: cust@igi-global.com
Web site: http://www.igi-global.com

Library of Congress Cataloging-in-Publication Data

Names: Gouda, Osama El-Sayed, 1951- author.
Title: Design parameters of electrical network grounding systems / by Osama El-Sayed Gouda.
Description: Hershey, PA : Engineering Science Reference, [2018] | Includes bibliographical references.
Identifiers: LCCN 2017022951| ISBN 9781522538530 (hardcover) | ISBN 9781522538547 (ebook)
Subjects: LCSH: Electric currents--Grounding. | Electric networks--Safety measures.
Classification: LCC TK3227 .G68 2018 | DDC 621.31/7--dc23 LC record available at https://lccn.loc.gov/2017022951

This book is published in the IGI Global book series Advances in Computer and Electrical Engineering (ACEE) (ISSN: 2327-039X; eISSN: 2327-0403)

British Cataloguing in Publication Data
A Cataloguing in Publication record for this book is available from the British Library.

Advances in Computer and Electrical Engineering (ACEE) Book Series

ISSN:2327-039X
EISSN:2327-0403

Editor-in-Chief: Srikanta Patnaik, SOA University, India

MISSION

The fields of computer engineering and electrical engineering encompass a broad range of interdisciplinary topics allowing for expansive research developments across multiple fields. Research in these areas continues to develop and become increasingly important as computer and electrical systems have become an integral part of everyday life.

The **Advances in Computer and Electrical Engineering (ACEE) Book Series** aims to publish research on diverse topics pertaining to computer engineering and electrical engineering. **ACEE** encourages scholarly discourse on the latest applications, tools, and methodologies being implemented in the field for the design and development of computer and electrical systems.

COVERAGE

- Optical Electronics
- VLSI Design
- Circuit Analysis
- Algorithms
- Sensor Technologies
- Computer Architecture
- VLSI Fabrication
- Programming
- Digital Electronics
- Power Electronics

IGI Global is currently accepting manuscripts for publication within this series. To submit a proposal for a volume in this series, please contact our Acquisition Editors at Acquisitions@igi-global.com or visit: http://www.igi-global.com/publish/.

The Advances in Computer and Electrical Engineering (ACEE) Book Series (ISSN 2327-039X) is published by IGI Global, 701 E. Chocolate Avenue, Hershey, PA 17033-1240, USA, www.igi-global.com. This series is composed of titles available for purchase individually; each title is edited to be contextually exclusive from any other title within the series. For pricing and ordering information please visit http://www.igi-global.com/book-series/advances-computer-electrical-engineering/73675. Postmaster: Send all address changes to above address.

Titles in this Series

For a list of additional titles in this series, please visit:
https://www.igi-global.com/book-series/advances-computer-electrical-engineering/73675

Design and Use of Virtualization Technology in Cloud Computing
Prashanta Kumar Das (Government Industrial Training Institute Dhansiri, India) and Ganesh Chandra Deka (Government of Indi, India)
Engineering Science Reference • ©2018 • 315pp • H/C (ISBN: 9781522527855) • US $235.00

Smart Grid Test Bed Using OPNET and Power Line Comunication
Jun-Ho Huh (Catholic University of Pusan, South Korea)
Engineering Science Reference • ©2018 • 425pp • H/C (ISBN: 9781522527763) • US $225.00

Transport of Information-Carriers in Semiconductors and Nanodevices
Muhammad El-Saba (Ain-Shams University, Egypt)
Engineering Science Reference • ©2017 • 677pp • H/C (ISBN: 9781522523123) • US $225.00

Accelerating the Discovery of New Dielectric Properties in Polymer Insulation
Boxue Du (Tianjin University, China)
Engineering Science Reference • ©2017 • 388pp • H/C (ISBN: 9781522523093) • US $210.00

Handbook of Research on Nanoelectronic Sensor Modeling and Applications
Mohammad Taghi Ahmadi (Urmia University, Iran) Razali Ismail (Universiti Teknologi Malaysia, Malaysia) and Sohail Anwar (Penn State University, USA)
Engineering Science Reference • ©2017 • 579pp • H/C (ISBN: 9781522507369) • US $245.00

Field-Programmable Gate Array (FPGA) Technologies for High Performance Instrumentation
Julio Daniel Dondo Gazzano (University of Castilla-La Mancha, Spain) Maria Liz Crespo (International Centre for Theoretical Physics, Italy) Andres Cicuttin (International Centre for Theoretical Physics, Italy) and Fernando Rincon Calle (University of Castilla-La Mancha, Spain)
Engineering Science Reference • ©2016 • 306pp • H/C (ISBN: 9781522502999) • US $185.00

Design and Modeling of Low Power VLSI Systems
Manoj Sharma (BVC, India) Ruchi Gautam (MyResearch Labs, Gr Noida, India) and Mohammad Ayoub Khan (Sharda University, India)
Engineering Science Reference • ©2016 • 386pp • H/C (ISBN: 9781522501909) • US $205.00

For an enitre list of titles in this series, please visit:
https://www.igi-global.com/book-series/advances-computer-electrical-engineering/73675

IGI Global
DISSEMINATOR OF KNOWLEDGE

701 East Chocolate Avenue, Hershey, PA 17033, USA
Tel: 717-533-8845 x100 • Fax: 717-533-8661
E-Mail: cust@igi-global.com • www.igi-global.com

Table of Contents

Preface

The present book is focused to evaluate design parameters of ground system for the predetermined grounding resistance value and soil resistivity without need of complex calculations. A general introduction for the purpose of grounding, requirements of satisfactory ground system and grounding or earthing principles and definitions are presented in this book. Objectives of safe grounding design, advantages of grounding, design considerations of grounding system and grounding resistance values are also presented. The following points are discussed in detail through the chapters of the book:

1. The purpose and requirements of satisfactory ground system and phenomena affecting ground resistance
2. Evaluation of soil resistivity using different mathematical expressions and comparison with the experimental measurements are investigated, also the factors affecting the apparent soil resistivity of multilayer soil is investigated.
3. Different mathematical expressions and methods to compute the grounding resistance have been introduces and several computer programs based on some of these methods have been developed to design charts for grounding system design by using rods and/or grids connected together.
4. A detailed analysis of potential profile at a surface above energized unequally spaced grounding grid has been carried out, for soils with multiple layers.
5. The corrosion theory of grounding electrodes and basic electrochemistry in corrosion reactions. It contains also the forms of substation grounding grid corrosion (uniform corrosion, pitting corrosion, galvanic corrosion, microbial influenced corrosion, survey on corrosion rate of substation grounding grid, copper and steel corrosion rates, corrosion protection methods (coating, cathodic protection (CP)). The book contains also the methods of applying cathodic protection in grounding grids, anode

selection, anode spacing, and impressed current in the grounding grid cathodic protection. Finally, it contains the required information for design grounding system cathodic protection and sacrificial anode (galvanic) cathodic protection system design steps.

6. Performance of grounding system under transient conditions has been studied.

Finally, the information of this book is valuable for grounding system engineers, students of electrical power departments and technicians of that field. The book is interesting also for the researchers in the grounding system development.

Introduction

WHY GROUNDING?

By grounding or earthing we mean making a connection to the general mass of earth. Practically every point of the supply network, from the generators to the apparatus on the consumers' earth electrodes will be found. Some standards discriminate between grounding and earthling, it define erthing as connecting to earth, but grounding is defined as common reference to the network. In this book grounding and earthing will be used as the same meaning. The purposes of grounding are:

1. Increases the reliability of the supply service
2. It helps to provide stability of voltage conditions, preventing excessive voltage sparks during disturbances
3. It provides protection against lightning
4. It creates path for the protection of electrical networks against earth faults
5. It ensures safety for human and animal's bodies by discharging the electrical energy to the earth
6. To provide means to carry electric currents into the earth under normal and fault conditions without exceeding any operating and equipment limits or adversely affecting continuity of service.
7. To assure that a person in the vicinity of grounded facilities is not exposing to the danger of critical electric shock.

GROUNDING OR EARTHING PRINCIPLES DEFINITIONS

Grounding or Earthing

It is defined as a connection to general mass of earth by means of an earth electrode. An object is earthed when it is electrically connected to an earth electrode by a conductor. Practically the earthing or grounding can be classified to neutral or power grounding in which the neutral point is connected to the earth and safety grounding in which the metallic cover or sheath of the equipment is connected to the earth.

Solidly Grounding

When the object is electrically connected to an earth electrode without a fuse, switch circuit breaker, resistance or impedance in the earth connection, the object grounding is define as solid grounding.

Earth Electrode

It is defined as a metal plate, rod, grid or other conductors buried in or driven into the ground and used for grounding metal work of electrical network.

Grounding Lead

The final conductor by which the connection to the earth electrode is made

Live Object

An object is said to be live when:

1. A difference of potential exists between it and earth
2. It is connected to common return, or neutral of a supply system in which such common return or neutral is not solidly earthed.

Earth Leakage Switch

It is an automatic device used to disconnect faulty appliances and is operated by a small leakage current flowing as result of the connection of the metal work to an earth electrode through the operating coil of the switch.

Core Balance Relay

An automatic device to disconnect faulty appliances as result of the out of balance currents of the supply circuit to the leakage of part of the current to earth through an earth electrode connected to the frame.

Grounding Resistance

The ohmic resistance between an earth electrode system and general mass of earth is defined as grounding resistance. In the case of consumer grounding, the resistance is the sum of the resistance of the consumer earth continuity conductor and grounding rod, and the resistance of the earth electrode to the general mass of earth. Measurements show that 90% of the total resistance around an earth rod is within a radius of three meters.

Earth Resistivity

It could be in ohms per cubic centimeter of a sample of earth, or in ohm. Meter

Ground Potential Rise (GPR)

It is the maximum electrical potential that a substation grounding grid may attain relative to a distant grounding point assumed to be at the potential of remote earth. This voltage, (GPR), equals the maximum grid current times the grid resistance as reported by the IEEE 80 standard (2000).

Mesh Voltage

It is the maximum touch voltage within a mesh of a ground grid as explained by the IEEE 80 standard (2000)

Step Voltage

It is defined as the difference in surface potential experienced by a person bridging a distance of 1 m with the feet without contacting any other grounded object.

The IEEE 80 standard (2000) uses the maximum mesh voltage as the touch voltage, and this usually exists at the corner mesh. UK practice (2011) defines the touch voltage differently. In practice, the voltage at the surface of the ground is a maximum adjacent to a corner of a grid. UK practice is to define touch voltage as the sum of the step voltage plus the voltage difference between the ground surface adjacent to a corner and the grid beneath. Although the mesh voltage is used as the defining touch voltage in American practice, the maximum permitted touch voltage used is less than that used in British

Standards. In practice, compliance with American usage thus also ensures the arrangement will comply with UK requirements.

Touch Voltage

It is defined as the potential difference between the ground potential rise (GPR) and the surface potential at the point where a person is standing while at the same time having a hand in contact with a grounded structure. This is reported by the IEEE 80 standard (2000)

Transferred Voltage

It is considered as a special case of the touch and step voltages where the voltage is transferred into or out of the substation from or to a remote point external to the substation site

Grounding

A conducting connection, whether intentional or accidental, between an electrical circuit or equipment and the earth, or to some conducting body that serves in place of the earth, See (Figure 1)

Earthing

The conductive mass of the earth, whose electric potential at any point is conventionally taken as equal to zero. (In some countries, the term "ground" is used instead of "earth."), see (Figure 2)

Figure 1. Grounding

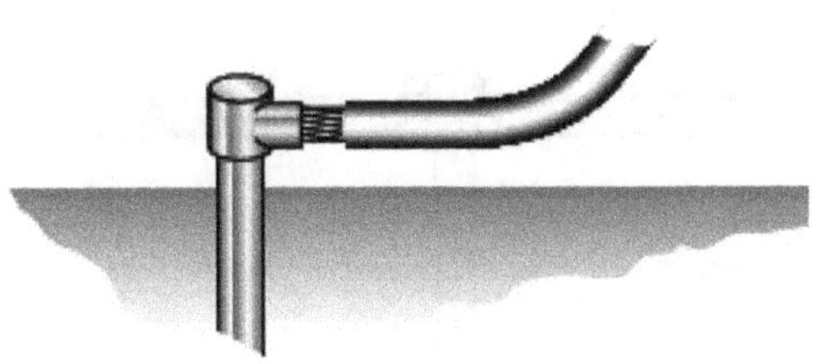

Figure 2. Earthing

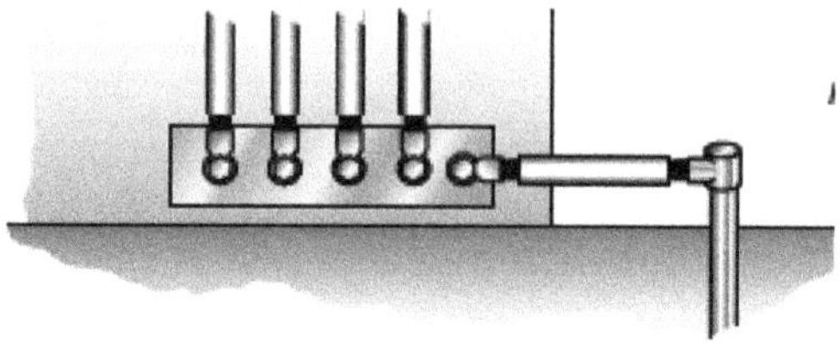

Grounding Grid

A system of horizontal ground electrodes that consists of a number of interconnected, bare conductors buried in the earth, providing a common ground for electrical devices or metallic structures, usually in one specific location.

Grounding Mat

A solid metallic plate or a system of closely spaced bare conductors that are connected to and often placed in shallow depths above a ground grid or elsewhere at the earth surface, in order to obtain an extra protective measure minimizing the danger of the exposure to high step or touch voltages in a critical operating area or places that are frequently used by people. Grounded metal gratings placed on or above the soil surface or wire mesh placed directly under the crushed rock, are common forms of a ground mat.

Bonding

It is defined as the permanent joining of metallic parts to form an electrically conductive path that will ensure electrical continuity and the capacity to conduct safely any current likely to be imposed. See (Figure 3)

Figure 3. Bonding

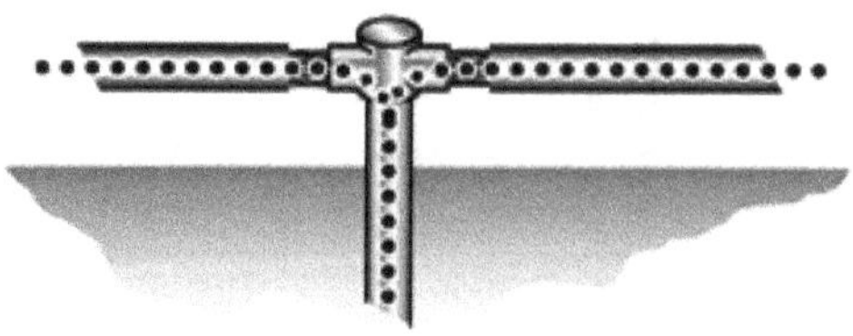

Ungrounded System

A power system not having any intentional connection to ground is referred to as an ungrounded system. However, because of the capacitive coupling between the phase conductors and ground, an ungrounded system is in reality grounded through the distributed capacitance of the system conductors to ground. A line to ground fault on an ungrounded system cause a very small ground fault current to flow through the capacitance of cables, transformers, and other electrical equipment on the system. Thus, ungrounded systems are designed to operate such that phase over current devices do not trip for the first phase to ground fault. If a second ground fault occurs on another phase, the result is a line to line fault with high fault current. The relatively high current of the L-L fault should trip phase over current devices, as it would with the high resistance grounded system.

OBJECTIVES OF SAFE GROUNDING DESIGN

The good grounding design of the electrical networks must achieve two main goals:

1. To provide means to carry electric currents into the earth under normal and fault conditions without exceeding any operating and equipment limits or adversely affecting continuity of service.
2. To assure that a person in the vicinity of grounded facilities is not exposed to the danger of critical electric shock.

A good grounding system should satisfy the following criteria:

- To provide a low impedance return path for surge current which is necessary for the timely operation of the over current/voltage protection system.
- To reduce the risks of breakdown hazard to electrical systems or electronic equipment.
- To reduce electric shock hazard to personnel.
- To minimize the cost of the grounding system.

For the electrical/electronic systems connected to the grounding structure, the voltage/ (voltage difference) due to the surge current should not be larger than the breakdown value of the system. It is well known that different systems have different breakdown voltages. In order to meet the above said criteria, there are several basic rules that should be followed:

1. The size of the grounding system should be large enough to reduce the maximum potential rise when the surge current enters it.
2. The spacing between the grounding wires should be so arranged that the touch and step voltages will be smaller than the safe value for the personnel.
3. The downward conductor should be connected to the grounding system at such points in order to reduce the ground potential rise, for example, at the mid-point of the grounding system.
4. For different soil structures, the grounding system should be laid in such a way that it can take advantage of the low resistivity part of the soil to reduce the ground potential rise as much as possible.
5. The effective length/area should be considered when one tries to minimize the cost.
6. Consequently, a good grounding system can dissipate the fault/lightning current into the soil as fast as possible, thereby, reduce the potential rise and potential difference within the different parts of the system and finally, there will be minimum damage to inter connected electronic devices and injury to the personnel working on site.

ADVANTAGES OF GROUNDING

The following are the grounded network merits:

1. The star point is always at earth potential to that when an earth fault occurs on one line. The potential difference between the healthy line and earth cannot exceed the phase voltage.
2. The maintenance and operation expenses are less.
3. Simple protection system based on the direction of the earth leakage current.
4. The system provides greater safety to person and equipment.
5. An arc in ground fault cannot occur.
6. The system provides a reliable service.

DESIGN CONSIDERATION OF GROUNDING SYSTEM

Some parameters have to be considered in designing the grounding system of electrical networks and apparatus. These parameters include:

1. Purpose of facility
2. Design life of facility
3. Soil resistivity at three different depths
4. Corrosive nature of soil
5. Shape and available area of facility site
6. Existing structures and their grounding systems
7. Seasonal variations in moisture and temperature for facility site
8. Public access and personnel use
9. Adjacent facilities and electrical systems
10. Future uses, additions, equipment for facility

For proper operation of over current devices, it is important to have a low DC ohmic resistance to remote earth. In many instances, this is best achieved by installing a deep ground electrode on site. It should be driven deep enough to reach the permanent water table. For dissipation of direct or indirect lightning currents, it is better to have many horizontal ground conductors in the soil, preferably in a radial array. This provides a low impedance path of dissipation to the high frequency component of the lightning energy. For safety of personnel, particularly where people congregate or where equipment operators will be located, it is important to have a grid system or other equipotential plane to reduce "step potential" and have equipment and metal structures bonded to the ground system to reduce "touch potential".

A proper facility grounding system incorporates these necessities in the most cost-effective manner that will last for the design life of the facility.

GROUNDING RESISTANCE VALUES

There is a good deal of confusion as to what constitutes a good ground and what the ground resistance value needs to be. Ideally a ground should be of zero ohms resistance. The NEC has stated that "A single electrode consisting of a rod, pipe, or plate which does not have a resistance to ground of 25 ohms or less shall be augmented by one additional electrode." Once you have added the supplemental ground you have met the requirement for the NEC.

This does not mean that the value of the ground now has to be 25 ohms or less. The ground resistance values objectives vary from industry to industry. Telecommunications industry has often used 5 ohms or less as their value for grounding and bonding. The goal in ground resistance values is to achieve the lowest ground resistance value possible that makes sense economically and physically. Generally, the following values are recommended by the author to be reference for different purposes of grounding systems:

- **Generator Substations:** 0.5 ohms or less
- **Transformer Substations Over 100 kV:** 1.0 ohms or less
- **Transformers Substations Over 33 kV:** 2.0 ohms or less
- **Transformers Substations Over 11 kV:** 5.0 ohms or less
- **Pole Mounted Type:** 5 ohms or less
- **Earth Wire of Transmission Lines:** 0.2 ohms
- **Electronic and Communication Devices:** 0.2 ohms

TRANSIENT RESPONCE OF GROUNDING SYSTEMS

This section explains models used to simulate lightning current. Also, a comparison between different approaches used for simulating frequency dependent of the soil has been mentioned. Transmission line approach, which is used to simulate the grounding system behavior under lighting condition, has been explained.

Lightning Current Impulse Model

Two lightning current waveforms corresponding to first and subsequent lightning stroke are usually used. Heidler lightning current function (HF) is obtained by Heidler and Cvetic (2002) and it represents the current waveform as follows:

$$i(t) = \frac{I_0}{\eta} \frac{\left(\frac{t}{\tau_1}\right)^{n'}}{1 + \frac{t}{\tau_1}^{n'}} e\left(-t/\tau_2\right) \tag{1}$$

Table 1. Parameters for the return stroke currents

Parameters	I_0	τ_1	τ_2	n'	I_0	τ_1	τ_2	n'
First stroke	28	1.8	95	2	--	--	--	--
Subsequent stroke	10.7	0.25	2.5	2	6.5	2	230	2

$$\eta = e^{-\left(\tau_1/\tau_2\right)\left(n'\left(\tau_2/\tau_1\right)\right)^{1/n'}} \tag{2}$$

where

I_0 = the amplitude of the current pulse.
τ_1 = the front time constant.
τ_2 = the decay time constant.
n' = exponent having values between 2 to 10.
η = the amplitude correction factor.

The first stroke waveform is reproduced by one and the subsequent stroke is reproduced by the sum of two Heidler's functions with the parameters given

Figure 4. Typical waveforms of lightning first and subsequent return stroke
Adapted by Heidler and Cvetic (2002).

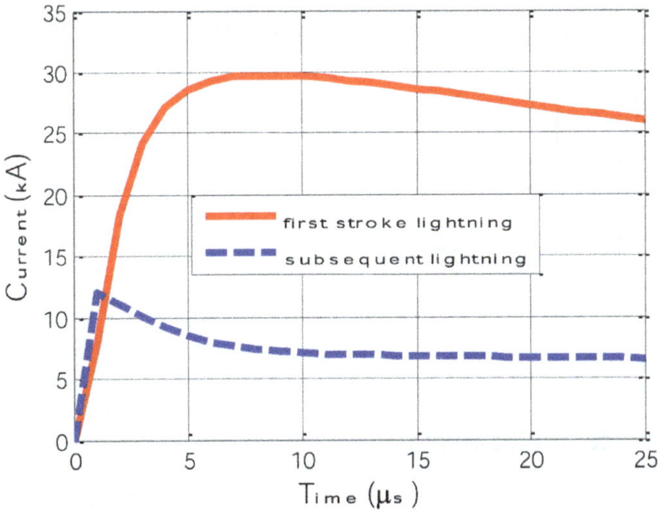

in Table 1. (Figure 4) shows lightning current wave of first and subsequent return stroke according to equations 1 and 2.

Modeling of Grounding System Under Lightning

There are several numerical methods used to study the performance of grounding systems under lightning current stroke such as:

1. Circuit approach as reported by Leonid Grcev (2009)
2. Transmission line approach as discussed by L. Grcev and S. Grceva, (2009).
3. Electromagnetic field approach as reported by L.Grcev and F. Dawalibi, (1990).
4. Hybrid approach as reported by P. Yuttagowith, A. ametani, N. Nagaoka, and Y. Baba, (2011).

Transmission line method is usually used by the authors to simulate the grounding grids as transmission line. In this method, any transmission line is divided into N segment each segment consists of R_g, C_g, and L_g. The formula of R_g, C_g, and L_g are as follows:

Resistance for vertical electrode is

$$R_g = \frac{\rho}{2\pi} \ln\left(\frac{4l}{a} - 1\right) \tag{3}$$

And the resistance for the horizontal electrode is

$$R_g = \frac{\rho}{\pi} \ln\left(\frac{2l}{\sqrt{2ad}} - 1\right) \tag{4}$$

The grounding capacitance is

$$C_g = \frac{\rho\varepsilon}{R} \tag{5}$$

The inductance of electrode is

$$L_g = \frac{\mu}{2\pi}\left[\ln\frac{2l}{a} - 1\right] \tag{6}$$

where μ is permeability of soil (H/m)

After calculating the parameter of the transmission line, the transient behavior of grounding systems can be simulated by ATP-EMTP, ATP rule book, Can/Am EMTP User Group, (1995). In ATP-EMTP, each electrically small segment of the transmission line is represented by lumped resistance, inductance, and capacitance. The number of the segments required for an accurate simulation is dependent on the highest frequency of the injected impulse, the higher the frequency, the larger the number of sections is required. Another solution to simulate transient behavior of grounding systems is simple analytical equation. These simplified equations are used to calculate the harmonic impedance, transient voltage, impulse impedance, impulse coefficient, effective length and decrement factor.

1. **Harmonic Impedance:** One of the most important properties in the high-frequency analysis of grounding system is harmonic impedance to ground $Z(jw)$. It can be computed as the input impedance of an open transmission line in frequency domain as reported by Leonid Grcev (2000) and (2007). This is discussed also by Grcev, Markovski, Arnautovski-Toseva et al. (2012).

$$Z(jw) = Z_o \coth(\gamma l) \tag{7}$$

$$Z_0 = \sqrt{\frac{j\omega L_g}{\left(1/R_g + j\omega C_g\right)}} \tag{8}$$

$$\gamma = \sqrt{j\omega L_g\left(1/R_g + j\omega C_g\right)} \tag{9}$$

From above equations, it is noticed that the harmonic impedance depends on only geometry, electromagnetic properties of the ground system and electric soil parameter.

2. **Transient Potential:** After solving equation 7 and obtaining $Z(jw)$, it enables evaluation of the time functions of the potential $v(t)$ as response to an arbitrary current pulse $i(t)$ by using equation 10 which is suggested by Grcev and Grceva, (2009).

$$v(t) = F^{-1}\left\{F\left[i(t)\right]Z(jw)\right\} \tag{10}$$

where F and F^{-1} denote Fourier and inverse Fourier transforms, respectively.

The transient voltage is parameter of great practical important since it is able to reveal the maximum voltage which the grounding is submitted when it is subjected to impulsive phenomena. Moreover, as it can also evaluate the time that the ground is subjected to certain levels of voltage transient.

3. **Impulse Impedance:** The impulse impedance Z_p is defined as the ratio between the peak voltage and peak current at point of injection and its concept is originally in the time domain.

$$Z_p = \frac{V_p}{I_p} \tag{11}$$

Impulse impedance is a very important parameter as it allows to quickly estimating the maximum voltage at the point of injection just multiplying the value of peak current by Z_p.

4. **Effective Length:** The effective length is defined as the maximum length of the electrode from which increases in size do not cause significant decrease in the value of the impedance impulsive.
5. **Decrement Factor:** The decrement factor is defined as the ratio of maximum ground potential rise for frequency dependent soil parameters and maximum ground potential rise for constant soil parameter as reported by Majed Akbari, Keyhan Sheshyekani, and Mohamad Reza Alemi, (2013).

$$DF = \frac{\max\left\{GPRforfrequencydependentsoilparameter\right\}}{\max\left\{GPRforconstantsoilparameter\right\}} \tag{12}$$

Frequency Dependent Soil Models

Five models have been proposed by the researchers for representation of soil electrical parameters (the electrical conductivity and permittivity). All models based on laboratory and field measurements.

1. **Scott (S) Expressions:** According to laboratory and field measurements carried out using variable frequency test voltage source on many soil samples with assorted water content, Scott et al. (1966) proposed plots of conductivity, permittivity, and permeability as a function of frequency and water content. The following equation for calculating the relative permittivity (dielectric constant) and conductivity as a function of frequency are given by Scott, Carroll and Cunningham (1964).

$$\varepsilon_r\left(f\right) = 10^D \tag{13}$$

$$\begin{aligned} D = {} & 5.491 + 0.946\log_{10}\left(\sigma_{100Hz}\right) - 1.097\log_{10}\left(f\right) + 0.069\log_{10}^2\left(\sigma_{100Hz}\right) \\ & -0.1141.097\log_{10}\left(f\right)\log_{10}\left(\sigma_{100Hz}\right) + 0.067\log_{10}^2\left(f\right) \end{aligned} \tag{14}$$

$$\sigma\left(f\right) = 10^K \text{ [mS/m]} \tag{15}$$

$$\begin{aligned} K = {} & 0.028 + 1.098\log_{10}\left(\sigma_{100Hz}\right) - 0.068\log_{10}\left(f\right) + 0.036\log_{10}^2\left(\sigma_{100Hz}\right) \\ & -0.046\log_{10}\left(f\right)\log_{10}\left(\sigma_{100Hz}\right) + 0.0180.067\log_{10}^2\left(f\right) \end{aligned} \tag{16}$$

where, f is frequency in [Hz] and σ_{100Hz} is the conductivity at 100 Hz in [m S/m].

2. **Smith and Longmire (SL) Expressions:** Smith and Longmire (1975) developed Scott's model for electrical soil based on experimental data.

They proposed curve-fit model for frequency using experimental data and proposed the universal soil Model. The next equations are used for calculating soil electrical parameters (relative permittivity and conductivity) as a function of frequency proposed by Longmire and Longley. (1973).

$$\varepsilon_r(f) = \varepsilon_\infty + 2\pi\varepsilon_0 \sum_{i=1}^{13} \frac{a_i}{1+\left(\frac{f}{f_i}\right)^2} \quad (17)$$

$$\sigma(f) = \sigma_{DC} + 2\pi\varepsilon_0 \sum_{i=1}^{13} a_i f_i \frac{\left(\frac{f}{f_i}\right)^2}{1+\left(\frac{f}{f_i}\right)^2} \quad (18)$$

where, σ_{DC} is the dc soil conductivity, ε_∞ is permittivity of high frequency limit of the dielectric constant and is set to 5, and f_i is defined as:

$$\boldsymbol{f_i} = \boldsymbol{F}_{(\tilde{A}_{DC})} \cdot 10^{i-1}\,[\mathrm{Hz}] \quad (19)$$

$$F_{(\sigma_{DC})} = (125\sigma_{DC})^{0.8312} \quad (20)$$

Finally, the coefficient a_i is given in Table 2.

3. **Messier (M) Expression:** Another formula of the electrical soil parameters based on Scott's data was developed by M. A. Messier (1980). According to Messier, the frequency dependence of the soil parameters can be estimated by equations 21 and 22, the high-frequency limit $\boldsymbol{\mu}_\infty$ of the dielectric is constant and equals to 8. Also, the dc conductivity σ_{DC} is the measured conductivity at low frequency:

Table 2. Coefficients a_i for universal soil

i	a_i
1	3.4×10^6
2	2.74×10^5
3	2.58×10^4
4	3.38×10^3
5	5.26×10^2
6	1.33×10^2
7	2.72×10^1
8	1.25×10^1
9	4.8×10^0
10	2.17×10^0
11	9.78×10^{-1}
12	3.92×10^{-1}
13	1.73×10^{-1}

Reported by Cavka, Mora and Rachidi (2014).

$$\varepsilon_r\left(f\right) = \frac{\varepsilon_\infty}{\varepsilon_0}\left(1 + \sqrt{\frac{\sigma_{DC}}{\pi f \varepsilon_\infty}}\right) \quad (21)$$

$$\sigma\left(f\right) = \sigma_{DC}\left(1 + \sqrt{\frac{4\pi f \varepsilon_\infty}{\sigma_{DC}}}\right) \text{[S/m]} \quad (22)$$

4. **Visacor and Portola (VP) Expressions:** Based on the Laboratory measurements of many soil samples in the frequency range 40 Hz to 2 MHz Visacor and Portela (2007) developed empirical equations for soil permittivity and conductivity as function of frequency as follows

$$\varepsilon_r\left(f\right) = 2.34 \times 10^6 \left(\frac{1}{\sigma_{100\,Hz}}\right)^{-0.535} \cdot f^{-0.597} \quad (23)$$

$$\sigma(f) = \sigma_{100Hz}\left(\frac{f}{100}\right)^{0.072} \tag{24}$$

5. **Visacor and Alipio (VA) Expression:** A new expression based on experimental data to determine the frequency variation of soil resistivity and permittivity is suggested by Visacor and Alipio (2012). The following equations are formed by the authors based on measurements in field condition:

$$\varepsilon_r(f) = 7.6\times10^3 f^{-0.4} + 1.3f \tag{26}$$

$$\sigma(f) = \sigma_{100Hz}\times\left\{1+\left[1.2\times10^{-6}\left(\frac{1}{\sigma_{100Hz}}\right)^{0.73}\right](f-100)^{0.63}\right\} \tag{27}$$

The frequency range suggested by above models is 10 KHz to 4 MHz for relative permittivity, while it is 100 KHz to 4MHz for conductivity.

REFERENCES

Akbari, M., Sheshyekani, K., & Alemi, M. R. (2013). The effect of frequency dependence of soil electrical parameters on the lightning performance of grounding systems. *IEEE Trans. Electromagnetic Compatibility*, *55*(4), 739–746.

Damir, C., Nicolas, M., & Farhad, R. (2014). A Comparison of Frequency-Dependent Soil Models: Application to the Analysis of Grounding Systems. *IEEE Transactions on Electromagnetic Compatibility*, *56*(1), 177–187. doi:10.1109/TEMC.2013.2271913

Grcev, L. (2007). Impulse efficiency of grounding electrode arrangements. In *Proceedings of the 18th International Conference Zurich symposium on EMC*, Munich /Germany (pp. 325-328).

Grcev, L., & Grceva, S. (2009). On high-frequency circuit of horizontal grounding electrodes. *IEEE Transactions on Electromagnetic Compatibility*, *51*(3), 873–875. doi:10.1109/TEMC.2009.2023330

Grcev, L., Markovski, B., Arnautovski-Toseva, V., & Drissi, K. E. K. (2012). Transient analysis of grounding systems without computer. In *Proceedings of the International Conference on Lightning Protection*, Vienna, Austria.

Heidler, F., & Cvetic, J. (2002). A Class of analytical functions to study the lightning effects associated with the current front. *Electrical Energy Systems*, *12*(2), 141–150.

IEEE. (2000). *Std. 80*. IEEE Guide for Safety in AC Substation Grounding.

IEEE. (2011). Code of Practice for Protective Earthing of Electrical Installations, BS7430 IEEE 80, 2000 BS7430.

Leonid, G. (2000). High-frequency performance of ground rods in highly resistive soil. In *Proceedings of the International Conference on grounding and earthing*, Belo Horizonte, Brazil (pp. 85-89).

Longmire, C. L., & Longley, H. J. (1973). Time domain treatment of media with frequency parameters (Theoretical Notes 113). Defense Nuclear Agency, Santa Barbara, CA. Retrieved from https://www.ece.unm.edu/summa/notes/Theoretical.html

Messier, M. A. (1980). The propagation of an electromagnetic impulse through soil: Influence of frequency dependent parameters (Tech. Rep. MRC-N-415). Mission Res. Corp., Santa Barbara, CA.

Scott, J. H. (1966). Electrical and magnetic properties of rock and soil (Theoretical Note 18). U.S. Geological Survey. Retrieved from https://www.ece.unm.edu/summa/notes/Theoretical.html

Scott, J. H., Carroll, R. D., & Cunningham, D. R. (1964). *Dielectric Constant and Electrical Conductivity of Moist Rock from Laboratory Measurements, Sensor and Simulation Note 116*. NM: Kirtland AFB.

Smith, K.S., & Longmire, C.L. (1975). A universal impedance for soils (Topical Report for Period Jul). Defense Nuclear Agency, Alexandria, VA, USA.

Visacor, S. (2007). A Comprehensive Approach to the Grounding Response to Lightning Currents. *IEEE Transactions on Power Delivery*, *22*(1), 381–386. doi:10.1109/TPWRD.2006.876707

Alipio, R., & Visacro, S. (2012). How the frequency dependence of soil parameters affect the lightning response of grounding electrodes. In *Proceedings of the International Conference on Lightning Protection*, Vienna, Austria.

Chapter 1
Soil Resistivity Evaluation and Grounding System Resistance

ABSTRACT

This chapter contains the factors affecting the soil resistivity and grounding resistance such as the soil moisture content, soil mineral content and soil temperature. It discusses the methods of measuring of soil resistivity and grounding resistance using Wenner method. Method to obtain the required samples for obtaining accurate site resistivity is presented. Soil resistivity measurement procedure is given in this chapter. The chapter contains three electrode method or fall-of-Potential method, dead earth method, and ground resistance testing existing systems using 'Selective' Clamp-on-Measuring of high voltage transmission towers feet resistance. Methods of calculating the apparent soil resistivity of Multi-Layers, apparent soil resistivity of two layers and apparent soil resistivity of three layers are presented in this chapter.

INTRODUCTION

To design the most efficient and economical grounding systems for electrical networks and substations, it is necessary to obtain accurate values of the soil resistivity on the site. The soil at the most sites is non-uniform. The first part of this chapter includes different methods of soil resistivity measurements and methods used to compute the apparent resistivity of multi-layer soil structure and comparing the calculated values with the actual measurements.

DOI: 10.4018/978-1-5225-3853-0.ch001

Figure 1. Elements of simple grounding

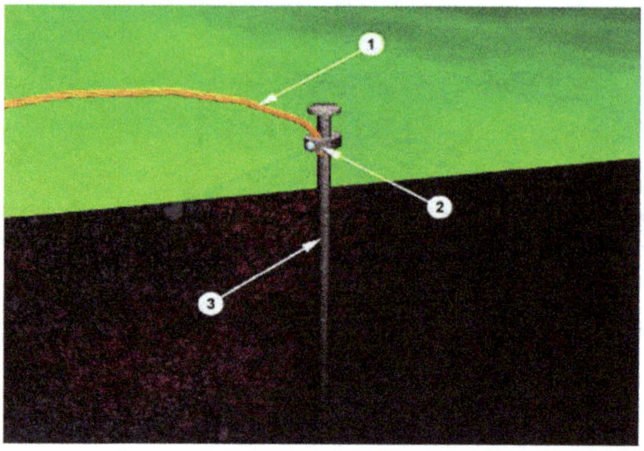

Also, the factors affecting the apparent soil resistivity of multilayer soil are investigated. Such these factors are:

1. The number of layers of soil structure (double and three layers are considered in this chapter) and their arrangement.
2. The thickness of each layer.
3. The reflection factor between each layer.

FACTORS AFFECTING THE SOIL RESISTIVITY AND GROUNDING RESISTANCE

As it is known Grounding and bonding are integral part of any modern electrical protection system design. An effective, low-impedance ground system is a key element of this system. It is crucial in ensuring personnel safety, as well as providing reliable protection for vital equipment and to minimize interruptions of service and costly downtime. As shown in (Figure 1), the grounding system consists of three basic components: 1-grounding conductor, 2-the connection/bonding of the conductor to the grounding electrode and 3- the electrode itself

The resistance of the grounding system has three basic components as follows:

1. The resistance of the grounding electrode itself and the connections to the electrode. This resistance can be ignored for its very low value in case of using electrodes and cable connections having high conductivity.

2. The contact resistance of the surrounding earth to the electrode, this resistance can also be neglected in case of making good contact between the grounding electrodes and the surrounding medium.
3. The resistance of the surrounding body of earth around the ground electrode. This resistance represents the actual value of grounding resistance and it is proportional with the soil resistivity and affected by grounding system configuration.

Many factors, both natural and human, affect the grounding resistance of electrical networks. Some of these factors are:

Soil Moisture

The samples of soil, when thoroughly dried, may in fact become very good insulators having a resistivity in excess of 10^9 ohm-centimeters. The resistivity of the soil sample is investigated by Engineering Standards Branch and Telecommunications Standards Division (ESBTSD). (1994). as given in Table 1 and (Figure 2) the soil resistivity decreases rapidly as the moisture content increases from very little moisture to approximately 30 percent.

Soil Mineral Content

In some locations, the resistivity of the earth is so high that low-resistance grounding can be obtained only at considerable expense and with an elaborated

Table 1. Effect of moisture content on soil resistivity

Moisture Content % by Weight	Resistivity Ω cm Top Soil	Resistivity Ω cm Sandy Loam
0	More than 10^9 ohm-centimeters	More than 10^9 ohm-centimeters
2.5	250000	150000
5	165000	43000
10	53000	185000
15	19000	10000
20	12000	6300
30	6400	4200
Morane gravel	3000	40 to 10000
Ridge gravel	15000	3000 to 30000
Solid granite	25000	10000 to 50000

Figure 2. Effect of moisture content on soil resistivity

Moisture Content (% by weight)	Resistivity (Ω-cm)	
0	Top Soil	Sandy Loam
	> 10^9	> 10^9
2.5	250,000	150,000
5	165,000	43,000
10	53,000	18,500
15	19,000	10,500
20	12,000	6,300
30	6,400	4,200

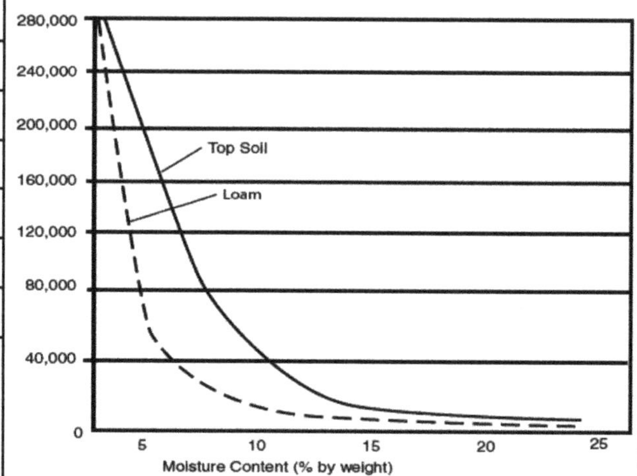

rounding system. In such situations, it may be economical to use a ground rod system of limited size and to reduce the ground resistivity by periodically increasing the soluble chemical content of the soil. Table 2 given by (ESBTSD) (1994) shows the substantial reduction in resistivity of sandy loam brought about by an increase in chemical salt content.

Table 2. The substantial reduction in resistivity of sandy loam brought about by an increase in chemical salt content, (sandy loam contains 15% moisture content by weight at 17°C temperature)

Added Salt % by Weight of Moisture	Resistivity ohm. Centimeters
0	10700
0.1	1800
1.0	400
5.0	190
10.0	130
20.0	100

Temperature

The resistivity of the soil is also influenced by temperature. Table 3, which its data are reported by J.H. Schon (1998) shows the variation of the resistivity of sandy loam, containing 15.2% moisture, with temperature changes from 20° to -15°C. In this temperature range the resistivity is seen to vary from 7200 to 330,000 ohm-centimeters. Since both temperature and moisture content become more stable at greater distances below the surface of the earth, it follows that a grounding system, to be most effective at all times, should be constructed with the grounding rod driven down a considerable distance below the surface of the earth. Best results are obtained if the grounding rod reaches the water table. Chemically treated soil is also subject to considerable variation of resistivity with temperature changes. If salt treatment is employed, it is necessary to use grounding rods which will resist chemical corrosion, (such as copper sulfate, sodium carbonate, and others. Salts must be EPA (Environmental Protection Agency) or local ordinance approved prior to use).

MEASURING OF SOIL RESISTIVITY AND GROUNDING RESISTANCE

Soil resistivity measurements have a three fold purpose. First, such data are used to make sub-surface geophysical surveys as an aid in identifying locations, depth to bed rock and other geological phenomena. Second, resistivity has a direct impact on the degree of corrosion in underground metal structures. A decrease in resistivity relates to an increase in corrosion activity and therefore dictates the protective treatment to be used. Third, soil resistivity

Table 3. The effect of temperature on the soil resistivity (sandy loam, 20% moisture content and 5% salt by weight)

Temperature °C	Resistivity ohm. Centimeters
20	110
10	142
0.0	190
-5	312
-13	1440

directly affects the design of a grounding system, and it is to that task that this discussion is directed.

When designing an extensive grounding system, it is advisable to locate the area of lowest soil resistivity in order to achieve the most economical grounding installation.

A good grounding system (also known as an earth electrode system) is important for the protection of an overall system facility. There are many factors that determine how well a grounding system performs. Two major parameters are its resistance to remote earth and the resistivity of the local soil. Each of these values can be measured to help in the design of the best grounding system. The resistance to remote earth of the grounding system needs to be at a minimum value in order to sustain its effectiveness. Different methods are used for the measurements of ground resistivity and grounding system resistance, such these methods are:

Wenner Method

The most commonly used technique is the Wenner four-pin method which is explained in details by IEEE 80 (2000). In brief, four probes are driven into the earth along a straight line at equal distances (s) and driven to a depth (a). The voltage between the two inner (potential) electrodes is then measured and it is divided by the current between the two outer (current) electrodes to give a value of mutual resistance R. The typical four-electrodes method of measuring soil resistivity is shown in (Figure 3).

To measure earth resistivity, it is natural to path current between the two outer (current) electrodes in the earth and measures the voltage between the two inner (potential) electrodes as indicated in (Figure 3).

$$\rho = \left[\frac{4\pi s}{1 + \frac{2s}{\sqrt{s^2 + 4a^2}} - \frac{s}{\sqrt{s^2 + a^2}}} \right] \cdot R_e \qquad (1)$$

where,

ρ = the resistivity of the soil, Ω.m

R_e = the resistance resulting from dividing the voltage between the potential probes by the current flowing between the current electrodes, Ω

Figure 3. Resistivity measurements by small sphere electrodes

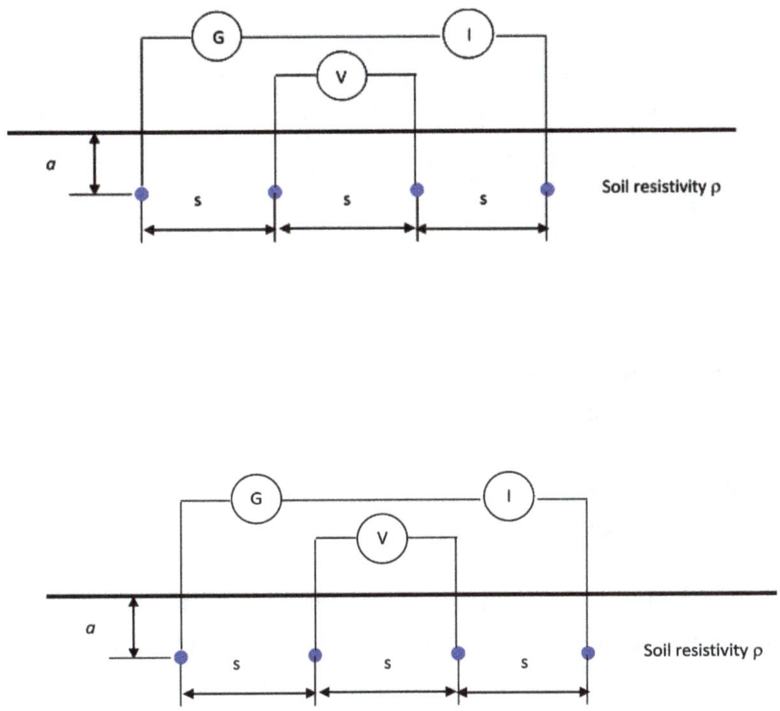

s = the distance between adjacent electrodes, m
a = the depth of the electrodes, m

If the depth of the electrodes is small compared to the distance between adjacent electrodes, as the case of penetrating ground probes at short distance, the above equation can be reduced to:

$$\rho = 2\ S\ \pi\ R_e \tag{2}$$

The current tends to flow near the surface for small probe spacing, whereas more of the current penetrates deeper soils for large spacing. Thus, it is customary to assume that the resistivity measured for a given probe spacing (a) represents the apparent resistivity of the soil to a depth of (a). Thus, equation 3 can be used to determine the apparent resistivity (ρ_a) at that depth.

For practical reasons, field-testing is generally done using short cylindrical rod electrodes rather than small spherical electrodes. The typical four-electrode method for measuring soil resistivity is shown in (Figure 3).

So that, the soil resistivity is

$$\rho = \left[\frac{2\pi l_r}{2\ln\left(\frac{2+\sqrt{4+\left(s/l_r\right)^2}}{1+\sqrt{1+\left(s/l_r\right)^2}}\right) + 2\sqrt{1+\left(s/l_r\right)^2} - \sqrt{4+\left(s/l_r\right)^2} - \left(s/l_r\right)} \right] \cdot R_e \qquad (3)$$

where,

ρ = the resistivity of the soil, Ω.m
R_e = the resistance resulting from dividing the voltage between the potential probes by the current flowing between the current electrodes, Ω
s = the distance between adjacent electrodes, m
l_r = the length of the electrodes, m

In most cases, the resistivity of a considerable volume of earth is required, so that the electrodes must be greater distance apart. The dimensions of the electrodes therefore will be small compared to the distance between them, so that the current distribution is practically the same as when the electrodes are considered as points.

The previous equations for soil measurements are based on the assumption that the soil resistivity is uniform. This requires that the soil resistivity is constant both laterally and with depth to infinity. Obviously, this is never the case. However, this assumption can be made without significant error if the soil is essentially uniform (both horizontally and vertically) to a distance measured from the edge of the grid or approximately 3-5 times the diagonal dimension of the grid. The uniform soil assumption can be used with less accuracy when the resistivity varies slightly with depth by using apparent resistivity (ρ_a). So, more exact theoretical approaches to situations where resistivity varies remarkably with depth of layers soil. It is anticipated that a two-layer model is reasonably valid for the actual soil conditions and the range of the resistivity variations found on the site.

The general formula given in equation 2 is a simplified version of some complex formulas derived by Professor H. B. Dwight (1936) of Massachusetts Institute of Technology and used to calculate the resistance to remote earth for a grounding system. The assumption in the general formula is that the resistivity of the soil is constant throughout the considered area, or averaged

Figure 4. Earth tester

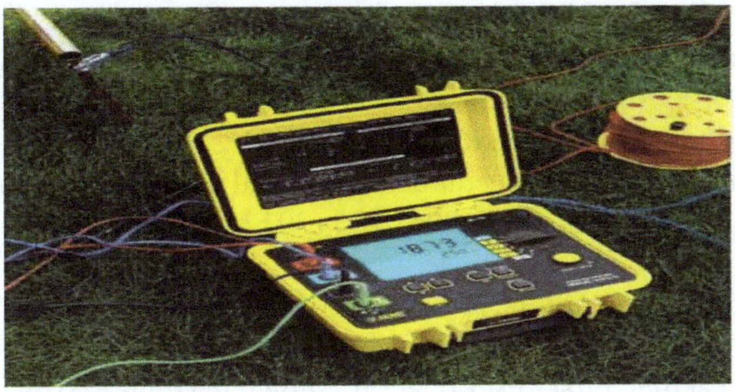

for the local soil. In the practical, soil resistivity is not constant, properties of electrodes and their connections vary, and complex equations just don't cut it. Therefore, an actual measuring technique is necessary. This technique is done with an earth resistance tester, often called a Megger. One example of the earth tester is seen in (Figure 4). This type of instrument can be used at various stages in the life of a grounding system, once during installation to see if it meets all specifications, and anytime thereafter to observe any possible changes.

Due to the effects of stray currents, buried water pipes, cable sheaths and other buried installations usually interfere and distort the readings; measurements should be taken along at least three directions. Figure 5 shows the recommended multiple sampling pattern to develop an accurate profile. Note that the more divergent the samples taken, the more accurate

Figure 5. Recommended multiple sampling patterns across site

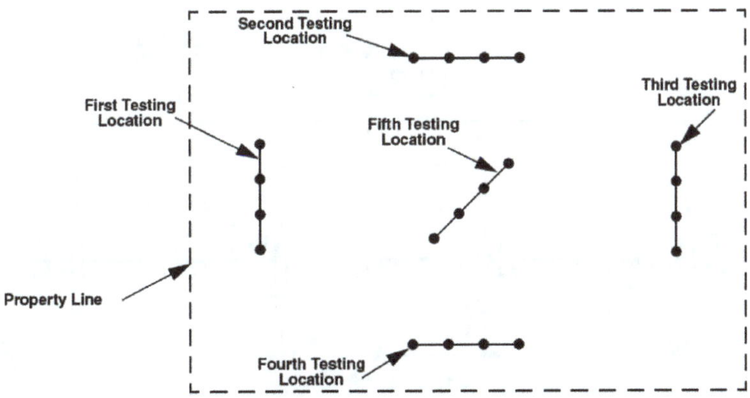

the measured soil resistivity will be as. This is reported by Appendix-B-Motorola-R56-09-01-05 (2012).

Follow the manufacturer's warning and caution information when using the ground or earth resistance tester. Follow furnished instructions when inserting and removing test rods into soil.

Perform the following procedure to obtain soil resistivity readings as given by Appendix-B-Motorola-R56-09-01-05 (2012).

1. On tester: verify that the jumper strap between the C1 and P1 terminals is disconnected (if applicable).
2. Starting at the "First Test Location" shown in (Figure 5), drive four test rods into the soil to a depth of 229 mm (9 in.), in a straight line, and spaced 1.52 m (5 ft.) apart.
3. Using test leads, connect the C1, P1, P2 and C2 terminals to their respective test rods, as shown in (Figure 6).
4. Turn the tester on: Press the test button and read the display. If the reading is not stable or displays an error indication, double-check the connection and the meter range setting. If the range is correct, try adjusting the test current. An effective way of decreasing the test rod resistance to ground is by pouring water around the rod. The addition of moisture is insignificant; it will only achieve a better electrical connection and will not influence the overall results.
5. Record the measurement obtained in the appropriate "Meter Readings" space on the photocopy of the "Soil Resistivity Worksheet" given in (Figure 7).

Figure 6. Four-terminal method of measuring earth resistivity

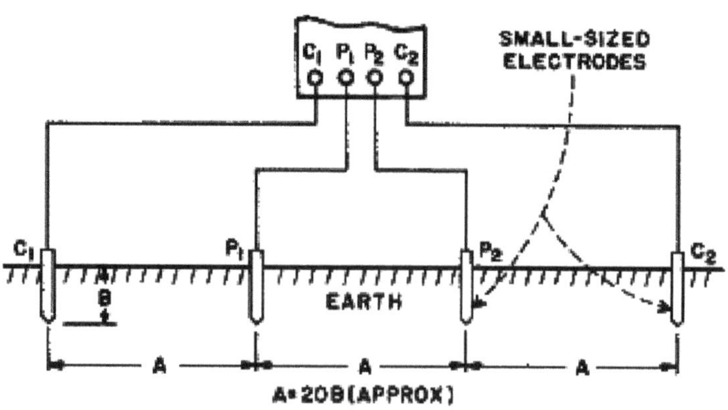

Figure 7. Soil resistivity worksheet as reported by Appendix-B-Motorola-R56-09-01-05 (2012)

Location	Spacing (Test Depth)				
	1.52 m(5 ft.)	3 m(10 ft.)	6.1 m(20 ft.)	9.1 m(30 ft.)	12.2 m(40 ft.)
Meter Readings (steps 2 through 5)					
1 of 5					
2 of 5					
3 of 5					
4 of 5					
5 of 5					
Soil Resistivity Calculations (step 10)					
$\rho = 191.5 \times A \times R$ ρ = soil resistivity in Ω-cm A = Distance between test rods (in feet) R = Resistance obtained from tester	OR		$\rho = 628 \times A \times R$ ρ = soil resistivity in Ω-cm A = Distance between test rods (in metres) R = Resistance obtained from tester		
1 of 5	ρ=	ρ=	ρ=	ρ=	ρ=
2 of 5	ρ=	ρ=	ρ=	ρ=	ρ=
3 of 5	ρ=	ρ=	ρ=	ρ=	ρ=
4 of 5	ρ=	ρ=	ρ=	ρ=	ρ=
5 of 5	ρ=	ρ=	ρ=	ρ=	ρ=
Test completed by:			Notes:		
Date:					
Client / Project:					
Site Location/ID:					
Ground Resistance Tester Model:________ S/N:________ Calibration date:________					
Soil Description:					
Ambient Conditions Temperature:________ Present conditions (dry, rain, snow):________ Date of last precipitation:________					

6. Remove the test rods from the soil.
7. In the same location on the site and along the same line as previous test, repeat steps 2 through 6 for remaining spacing listed on the Soil Resistivity Worksheet.

8. Prepare to take measurements for the next test location shown in (Figure7). Repeat steps 2 through 7 for this location.
9. Repeat steps 2 through 8 for all remaining test locations specified in (Figure 5)
10. On the Soil Resistivity Worksheet copy, calculate and record soil resistivity (ρ) for each of the 25 readings taken in the steps above.

Three Electrode Method or Fall-of-Potential Method

Assume that it is required to measure the earth resistance by using three rods as given in (Figure 8 a). First of all, it has to be driven into the earth some distance apart and a voltage applied, as shown in (Figure 8a). The current between rods 1 and 2 is measured by an ammeter; the potential difference

Figure 8. Principle of an earth resistance test

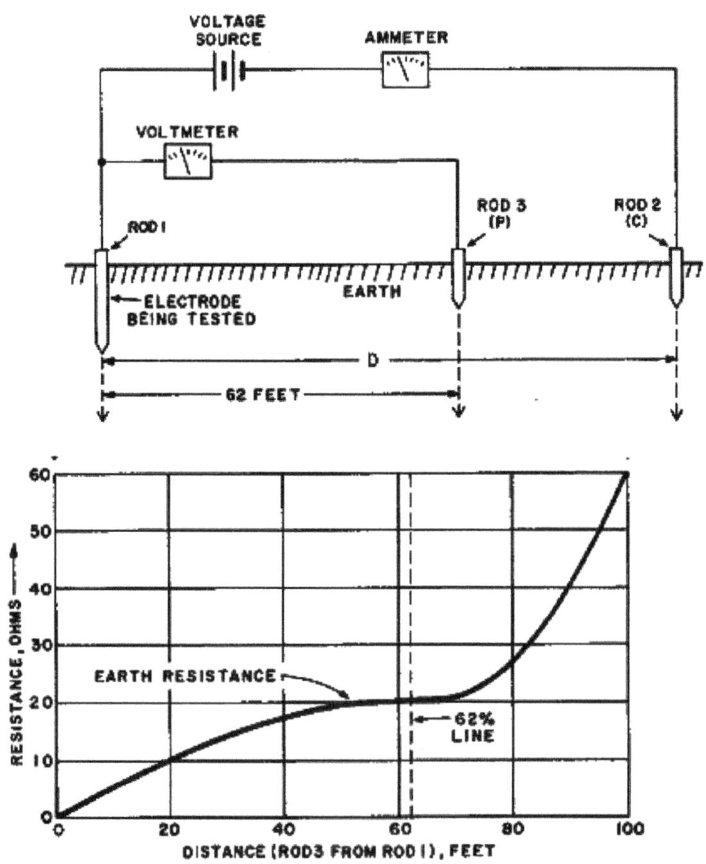

(voltage) between rods 1 and 3 is measured by a voltmeter. Rod 3 is usually located a straight line at various points between rods 1 and 2, a series of voltage readings can be obtained and by using Ohm's Law the earth resistance at any point measured can be determined.

The series of resistance values can be plotted against distance to obtain a curve as plotted in (Figure 8 b). Note that as rod 3 is moved away from rod 1, the resistance values increase, but the amount of increase gets less and less until a point is reached where the rate of increase becomes so small. The earth shells between the two rods (1 and 3) have so great a surface area that they add little to the total resistance. Beyond this point, as rod 3 approaches the earth shells of rod 2, resistance gradually picks up. Near rod 2, the values rise sharply.

Now, let's say that rod 1 is the earth electrode under test. From a typical earth-resistance curve, given in (Figure 8 b), the correct resistance is usually obtained if P (rod 3) is placed at a distance from the center of the earth electrode (rod 1) about 62 percent of the distance between the earth electrode and C (rod 2). At that point, the change in measured value with distance is approximately zero.

For example, in (Figure 8 a), the distance D from the earth electrode to C is 100 ft. Taking 62 percent of this distance, we get 62 ft. From (Figure 8 b), the resistance for this distance is 20 Ω. This is the measured resistance of the earth electrode. (Figure 9) and (Figure 10) show the using of the digital

Figure 9. Typical use of a Megger digital ground resistance tester to perform fall-of-potential testing

Figure 10. Checking the earth resistance of a ground system at a substation using a Megger ground tester

Megger in measuring the ground resistance by fall-of-Potential method. This method is used by Southey, Siahrang, Fortin et al. (2015).

The 62% Method

The fall of potential method can be adapted slightly for use with medium sized grounding systems. This adaptation is often referred to as the 62% Method, as it involves positioning the inner test stake at 62% of the earth electrode-to-outer stake separation (recall that in the Fall-of-Potential method, this figure was 50%).

All the other requirements of test stake location - that they be in a straight line and be positioned away from other structures - remain valid. When using this method, it is also advisable to repeat the measurements with the inner test stake moved ±10% of the earth electrode-inner test stake separation distance, as before.

The main disadvantage with this method is that the theory on which it is based relies on the assumption that the underlying soil is homogeneous, which in practice is rarely the case. Thus, care should be taken in its use and a soil resistivity survey should always be carried out. Alternatively, one of the other methods should be employed.

Other Test Methods

Many other methods exist for taking earth resistance measurements. Many of these methods have been designed in an attempt to alleviate the necessity for excessive electrode separations, when measuring large earth systems, or the requirement of having to know the electrical center of the system.

The Slope Method

This method is suitable for use with large grounding systems, such as substation earthing. It involves taking a number of resistance measurements at various earth systems to voltage electrode separations and then plotting a curve of the resistance variation between the earth and the current. From this graph, and from data obtained from tables, it is possible to calculate the theoretical optimum location for the voltage electrode and thus, from the resistance curve, calculate the true resistance.

The additional measurement and calculation effort tends to relegate this system to use with only very large or complex grounding systems.

The Star-Delta Method

This technique is well suited to use with large systems in built up areas or on rocky terrain, where it may be difficult to find suitable locations for the test electrodes, particularly over long distances in a straight line.

Three test electrodes set up at the corners of an equilateral triangle with the earth system in the middle, are used and measurements are carried out of the total resistance between adjacent electrodes, and also between each electrode and the grounding system.

Using these results, a number of calculations are performed and a result can be obtained for the resistance of the earth system.

Dead Earth Method

When using a four-terminal instrument, P1 and C1 terminals connect to the earth electrode under test; P2 and C2 terminals connect to an all-metallic water-pipe system. With a three-terminal instrument, connect X to the earth electrode, P and C to the pipe system as given in (Figure 11). If the water system is extensive, (covering a large area), its resistance shall be a fraction

Figure 11. Dead earth method or two-point resistance test earth

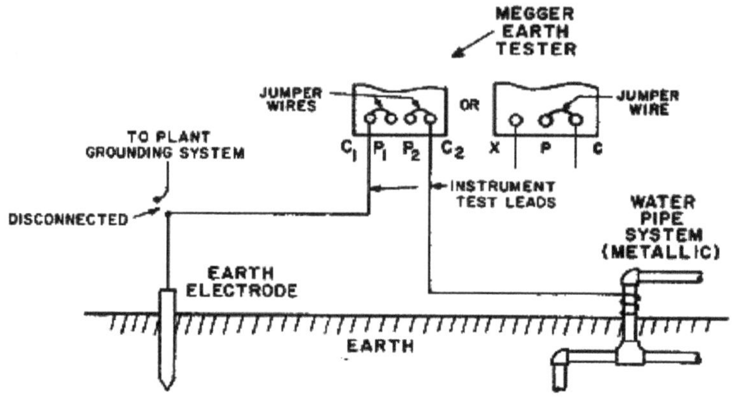

of ohms. The instrument reading is the resistance of the electrode under test. The dead earth method is the simplest way to make an earth resistance test. With this method, resistance of two electrodes in series is measured, the driven rod and the water system. But there are three important limitations:

1. The water pipe system must be extensive enough to have a negligible resistance.
2. The water pipe system must be metallic throughout, without any insulating couplings or flanges.
3. The earth electrode under test must be far enough away from the water-pipe system to be outside its sphere of influence.

In some locations, the earth electrode may be so close to the water-pipe system that you cannot separate the two by the required distance for measurement by the two-terminal method. Under these circumstances, if conditions 1 and 2 above are met, the connection to the water-pipe system cannot be carried out and obtain a suitable earth electrode. As a precaution against any possible future changes in the resistance of the water-pipe system, however, you should also install an earth electrode. Due to the many uncertainties associated with this method of testing, it should be considered using as a "last resort."

Ground Resistance Testing Existing Systems Using Clamp-On Ground Resistance Tester

The Ground Resistance Tester measures ground rod and small grid resistance through any season, without the use of auxiliary ground rods. Clamp-on ground resistance testers are used in multi-grounded systems without disconnecting the ground under test. The Clamp-on Ground Resistance Tester simply clamps around the ground conductor or rod and measures the resistance to ground. By performing measurements on intact ground systems, the user also verifies the quality of the grounding connections and bonds. Resistance and continuity of grounding loops around pads and buildings may also be measured. The tester includes a current measurement function. The probe's high sensitivity enables measurement of leakage current flowing to ground or circulating in ground loops down to 1 mA and neutral currents to 30Arms. This feature provides additional information which is becoming vital as distribution ground networks carry higher levels of noise and harmonics which affect power quality.

Battery life information is given at power-up and Auto-Off for power management. Additional features are also displayed on the large LCD to ensure precise measurements. The Buzzer and Auto-Off features may be disabled from the push-buttons at any time.

The tester body design permits one-handed operation. The guard provides additional strength, and prevents the hand from slipping or coming into contact with conductors under test.

The Electrical Specifications of such this instrument are as follows:

- Reference conditions: 23°C ± 3K, 50% RH ± 10%, battery at 8V ± 0.2V, external magnetic field < 40A/m, external electrical field < 1V/m, conductor centered, loop resistance non-inductive. Accuracy ± 2.5% of Reading, Frequency: 50/60Hz.
- Voltage Generated into the Loop: Approximately 60 mV (r. m. s)
- Resistance Overload: 1200 Ω
- Current Measurement: 1mA to 30 A (r. m. s)
- Current Overload: above 29.99 A (r. m. s)
- One type components of Clamp-on Ground Resistance Tester is given in (Figure 12).

The Clamp-On meter offers many advantages over the traditional method of testing. Much testing done with this method is invalid due to the operator not understanding the principles of operation and lack of training. When properly utilized, the following are a minimum of benefits:

- No need to disconnect the utility or isolate the grounding system. The meter can also be used with the equipment powered up.
- No probes to be driven or long conductors to be utilized.
- Access to large area is not required.
- Safe, non-contact measurement system reduces risk.
- Easy to use, accurate, one step measurement.

Figure 13 illustrates one good and two bad locations to test with the Clamp-On meter. Position 1 would not work because the path for the current flow caused by the meter would not be through the earth. The current would flow to the air duct and most likely the air duct would have electrical contact with building steel which is tied to the bus bar. It would essentially be a short circuit. Position 2 also would not work due to the current flowing to the equipment which should be isolated from the floor. It should read an open

Figure 12. Clamp-on ground resistance tester

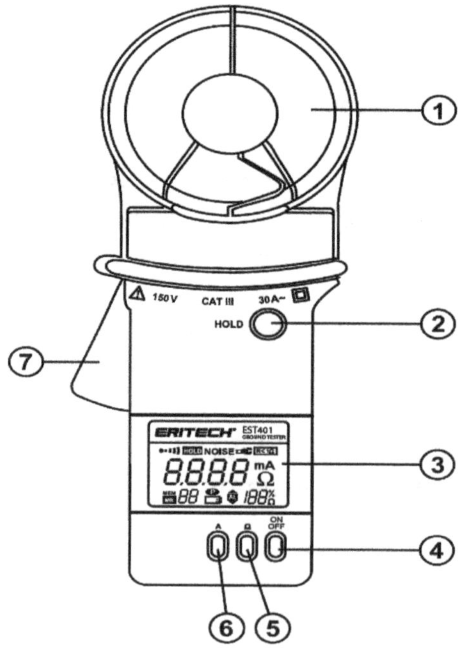

Figure 13. Locations to test with the clamp-on meter

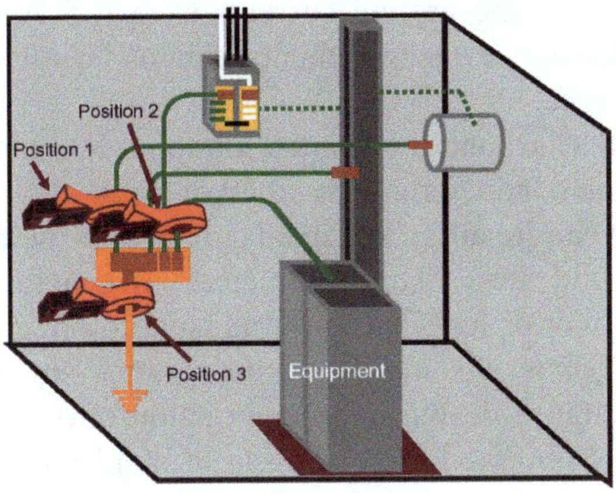

loop. The current in Position 3 would go through the ground and return via the utility neutral grounds to the facility and grounding conductor.

1. **Head Assembly:** Consists of two individually shielded magnetic cores.
2. **Hold:** Freezes last measured value on the display.
3. **Display:** 3000 count LCD, function indicators.
4. **ON/OFF:** Power ON or OFF, activates display self-test at power-up.
5. **Ω:** Resistance measurement.
6. **A:** Current measurement.
7. **Lever:** Opens or closes jaws.

Figure 13 illustrates one good and two bad locations to test with the Clamp-On meter.

Testing Procedures for Grounding Measurement Using Clamp-On Meter

- **Setting up the Meter for Measurement:**
    - Check that the surface of the instrument is free from contamination.
    - Check the calibration of the meter with the calibration loop (provided by the meter manufacturer), ensuring that the readings are within the tolerances indicated in the user manual.
- **Taking the Measurement:**
    - Rubber gloves are recommended for added safety.

   - Select the current option (A) on the meter and clamp around the conductor. If the ground currents exceed 5 A or noise exceeds 50 V, resistance measurements cannot be taken. Note this for maintenance.
   - Select the ground resistance option (Ω), clamp around the grounding conductor as described in (Figure 13) for the reading.
- **Interpreting the Results:** The display indicates the correct resistance of the system under test. Clamp-On meters will generally read accurately down to 0.1 ohm or less. It is very unusual to have a correct reading less than 1.0 ohm. Not many ground systems are below two ohms, certainly without a design and construction effort. If an unexpected low reading is observed the operating principles of the meter should be reviewed and the connection point should be reconsidered. Low resistance grounds do exist, they are just pretty rare.

If the reading on the meter is <1 ohm, there may be a continuity loop present and the test may have been performed at an incorrect location. Perform a visual inspection of the system under test to determine if any downstream neutral to ground bonds exist. If the reading on the meter is OL (Overload) the system under test does not have a complete conductive path, or loop for the current to flow in.

This unique exclusive method as given in (Figure 14), has been created to measure resistances of individual ground electrodes in all types of grounded systems including ground grids and wired meshes as are common in substations, high voltage pylons with ground cabling, and commercial settings with multiple grounds. By using a specialized clamp-on current transformer the effects of parallel resistances are eliminated from the measuring process and therefore do not influence the measuring results. A special rectification method is used to isolate or 'digitally filter' out other currents to significantly increase accuracy. As with the standard 3-Pole Fall of Potential testing the rules / guidelines for the setting of ground stakes apply for both simple and complex grounds. Connect the ground tester as shown in the picture above. Press START, and read out the resistance value. This is the actual value of the ground electrode under test. The "Ground under Test" does not have to be disconnected. (Figure 15) gives ground resistance testing existing systems using 'Selective' Clamp-on meter.

Figure 14. Measurements of resistances of individual ground electrodes

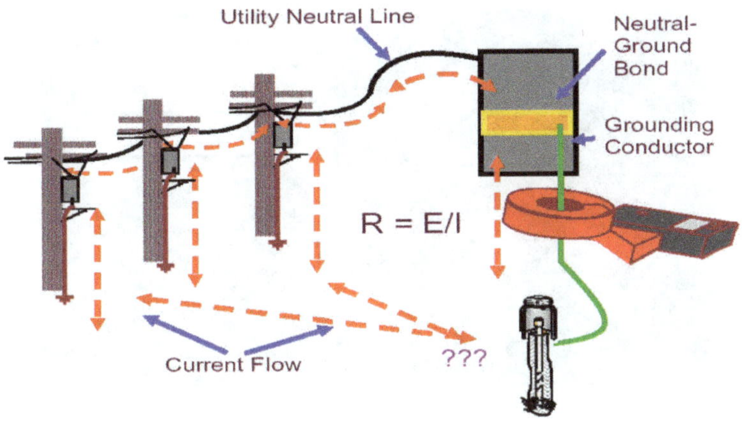

Figure 15. Ground resistance testing existing systems using 'selective' clamp-on

Measuring of High Voltage Transmission Towers Feet Resistance

Testing individual ground electrode resistances of high voltage transmission towers with overhead ground or static wire requires that these wires be disconnected. If a tower has more than one grounding electrodes at its base as given in (Figure 16), these must also be disconnected one by one and tested. The clamp-on current transformer can measure the individual resistances of each leg without disconnecting any ground leads or overhead static/ground wires. From (Figure 16) it is seen that the total resistance of an individual tower is the parallel sum of all grounds. The tower has four individual grounds have to be measured all, generating the individual resistance and then it can be calculated.

Figure 16. Method for measuring of high voltage transmission towers feet resistance

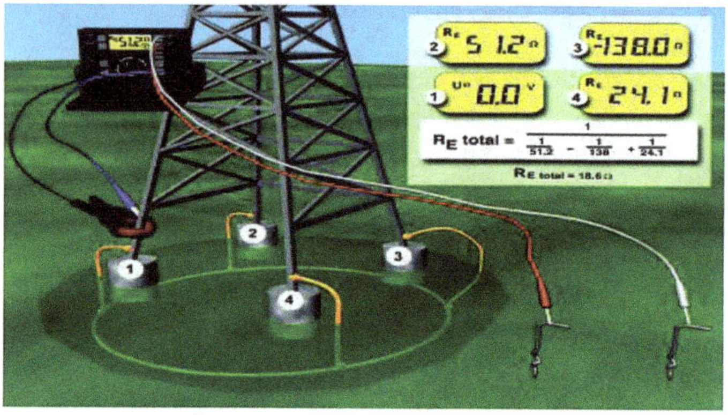

Once the resistance to remote earth for each electrode of tower legs in the entire grounding system is determined, one can calculate the resistance to remote earth for the entire grounding system in one of two ways. The first approach is in understanding that the electrodes are in parallel with each other (through the grounding grid and ground itself.) Because they are in parallel, the rule for parallel resistances can be used as given in equation 4

$$Rs = 1 / [(1/R1) + (1/R2) + (1/R3) + \ldots (1/Rn)] \quad (4)$$

where

Rs = resistance to remote earth for entire grounding system
R1, 2, 3…n = resistance to remote earth for each individual electrode

APPARENT SOIL RESISTIVITTY OF MULTI- LAYERS

Figure 17 shows the measured soil resistivity of multi-layer soil structure (three layers structures), first layer resistivity of ($\rho1 = 59\Omega.m$) with thickness ($d1= 1.1m$), the second layer has ($\rho2 = 104\Omega.m$) resistivity and thickness ($d2 = 4.6m$) finally the third layer has resistivity ($\rho3 = 35\Omega.m$) with infinite thickness. These measurements are done by Trevor Charlton (2000).

The following are the steps for calculating the apparent soil resistivity of multi-layer soil structure:

Figure 17. Wenner soil resistivity site result

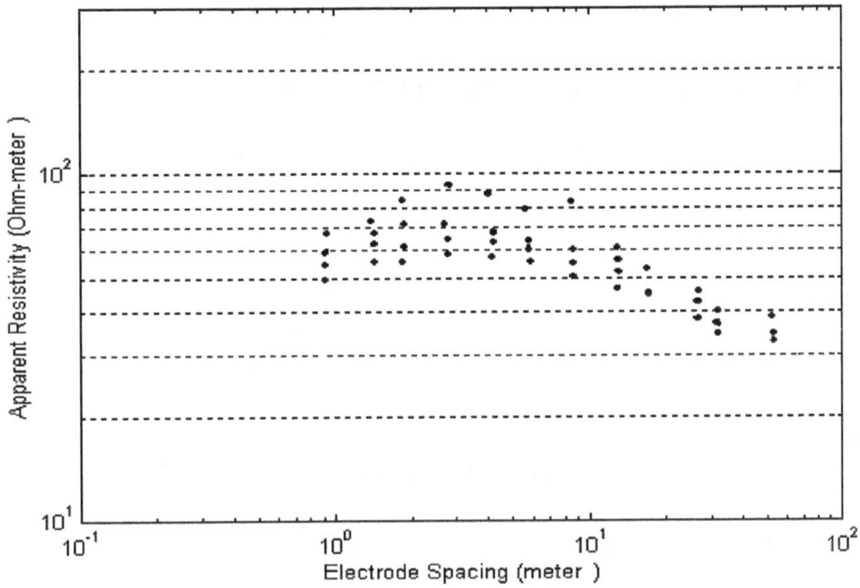

Apparent Soil Resistivity of Two Layers

The apparent soil resistivity of two layers soil can be evaluated using the following formulas as reported by IEEE 80 (2000)

$$\rho_{ao} = 2\left(\rho2 - \left(\rho2 - \rho1\right) \cdot e^{-js}\right) - \left(\rho2 - \left(\rho2 - \rho1\right) \cdot e^{-2js}\right) \tag{5}$$

where $J = \delta / 2(d1)$, $\delta = \dfrac{\ln(\rho1 / \rho2) - \ln(0.0176)}{3.5}$ and s is the spacing between probe electrodes of measuring soil resistivity.

However, this rule is not completely accurate in this application because of the extra resistance to remote earth through the grounding grid. The second approach used to calculate the resistance to remote earth for the entire grounding system assumes that the resistances to remote earth for each electrode are equal. Then, for the entire grounding system, the resistance to remote earth is 40% lower for a system with only two electrodes, 60% lower for a system with three electrodes, and 66% lower for a system with four electrodes (compared to the resistance to remote earth for one of the equal electrodes.) These values are slightly larger than the values given by the parallel resistance rule.

Apparent Soil Resistivity of Three Layers

The apparent soil resistivity of three layers soil can be evaluated using the following four methods [4].

First Method

The apparent soil resistivity of three layers is obtained in two steps as follows:

- **First Step:** Evaluating initial apparent soil resistivity between first and second layer represented by (ρ_{ao}) as indicate in (Figure 18) using equation (6)
- **Second Step:** Considering the initial apparent soil resistivity, which evaluated between first and second layers (ρ_{ao}) in the first step as a first layer and then, considering the third layer as a second one. Thus, the apparent soil resistivity of three layers will be as follows as reported by IEEE 80 (2000)

$$\rho_a = 2\left(\rho 3 - \left(\rho 3 - \rho_{ao}\right)\cdot e^{-qs}\right) - \left(\rho 3 - \left(\rho 3 - \rho_{ao}\right)\cdot e^{-2qs}\right) \tag{6}$$

where $q = \lambda / 2\left(d1 + d2\right)$, $\lambda = \dfrac{\ln\left(\rho_{ao} / \rho 3\right) - \ln\left(0.0176\right)}{3.5}$

Results of this method are plotted in (Figure 19) using the same parameters as actual measurements in (Figure 17).

In this method of evaluating apparent soil resistivity, it is found that the reduction of evaluating initial apparent soil resistivity occurs in first step whenever evaluating initial apparent soil resistivity between two resistivity of upper layers (first and second layers) in which their reflection factors are 0.9:1, this reduction is repeated again in second step.

Second Method

In this method, the apparent soil resistivity of three layers structure can be calculated also in two steps as follow:

Figure 18. Three-layer soil, first and second layer evaluated as ρ_{ao}

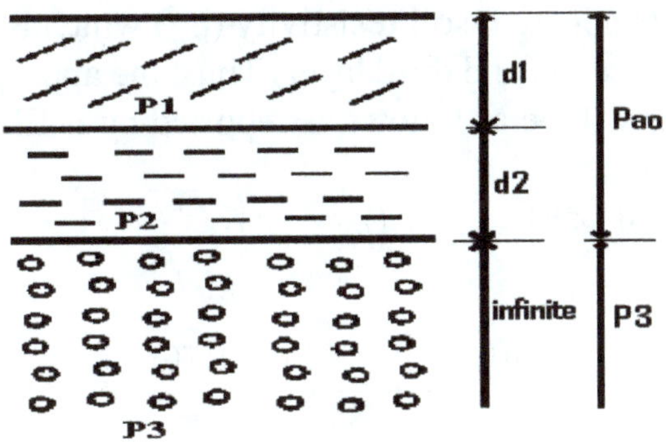

Figure 19. Resistivity of three-layer soil (method 1)

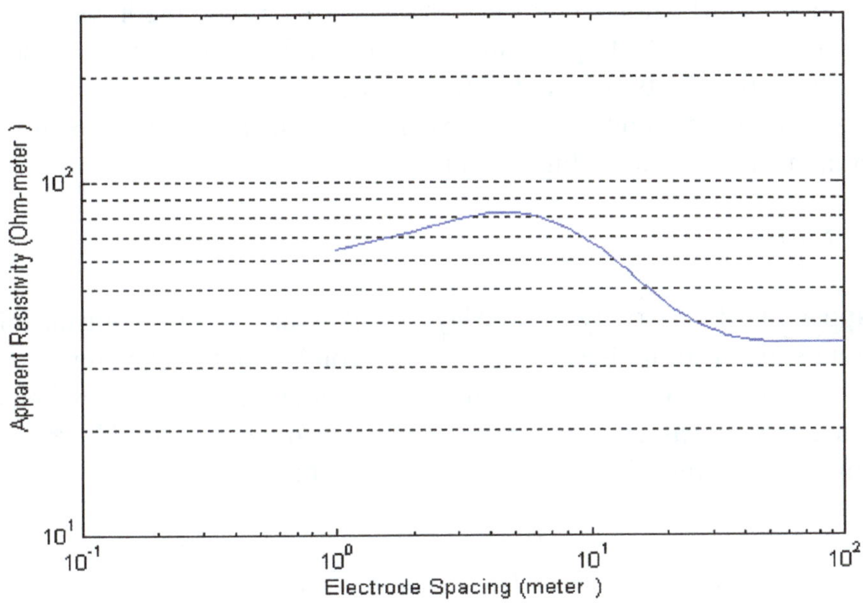

- **First Step:** Evaluating initial apparent soil resistivity between second and third layer represented by ρ_{ao} as indicate in (Figure 20)

$$\rho_{ao} = 2\left(\rho 3 - \left(\rho 3 - \rho 2\right) \cdot e^{-js}\right) - \left(\rho 3 - \left(\rho 3 - \rho 2\right) \cdot e^{-2js}\right) \tag{7}$$

where $J = \delta / 2\left(d1\right)$, $\delta = \dfrac{\ln(\rho 2 / \rho 3) - \ln(0.0176)}{3.5}$

- **Second Step:** Evaluating the apparent soil resistivity between first soil ($\rho 1$) and initial apparent soil resistivity (ρ_{ao}) which is evaluated in first step (between second and third layer) Thus, the apparent soil resistivity of three layers will be as follows as reported by IEEE 80 (2000).

$$\rho_a = 2\left(\rho_{ao} - \left(\rho_{ao} - \rho 1\right)e^{-qs}\right) - \left(\rho_{ao} - \left(\rho_{ao} - \rho 1\right)\cdot e^{-2qs}\right) \tag{8}$$

where $q = \lambda / 2\left(d1\right)$, $\lambda = \dfrac{\ln\left(\rho 1 / \rho_{ao}\right) - \ln(0.0176)}{3.5}$

Here, the reduction in evaluating initial apparent soil resistivity occurs also in two steps. First reduction occurs in first step whenever evaluating initial apparent soil resistivity (ρ_{ao}) between two resistivity of second and third layers in which their reflection factors are between 0.9:1 Second reduction occurs in second step in evaluating apparent soil resistivity (ρ_a) between resistivity of first layer resistivity ($\rho 1$) and initial apparent soil resistivity (ρ_{ao})

Results of this method are shown in (Figure 21) using the same parameters as actual measurement in (Figure 17).

Third Method

The apparent soil resistivity of three layers is evaluated by obtaining the initial apparent soil resistivity between first and second layer which is represented by (ρ_{ao}) and then, considering the initial apparent soil resistivity of first and second layers (ρ_{ao}) as a first layer and third layer ($\rho 3$) as a second layer. Thus, the apparent soil resistivity of three layers will be as follow:

Figure 20. Three layer soil, second and third layer evaluated as ρ_{ao}

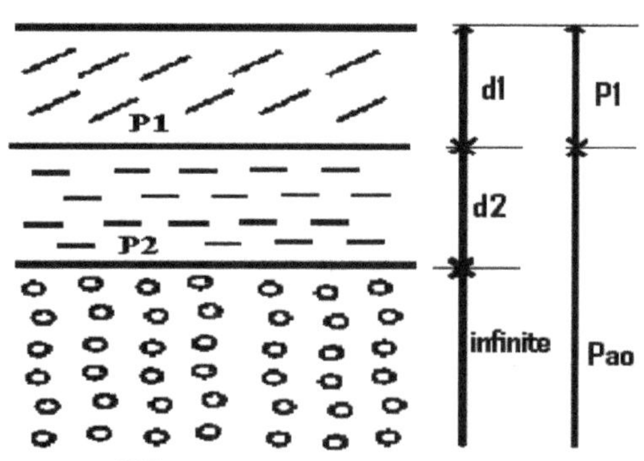

Figure 21. Resistivity of three layer soil (method 2)

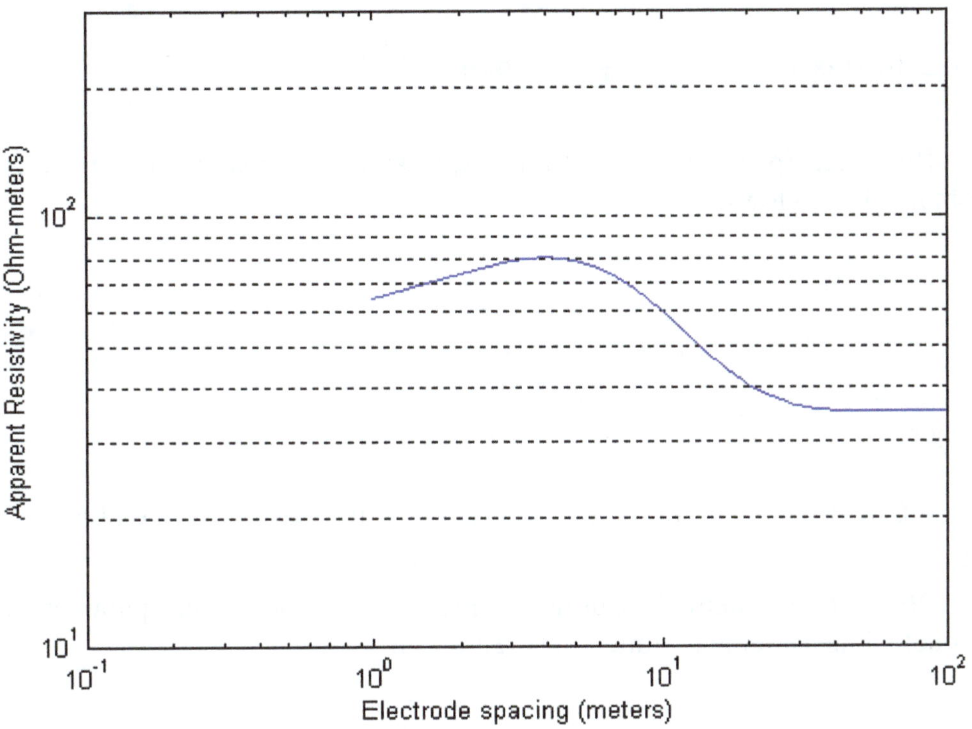

When (ρ3)> (ρ $_{ao}$)

$$\rho_a = \rho_{ao} + 4\rho_{ao} \cdot K_{ao3} \cdot s \cdot \left[\frac{1}{\sqrt{s^2 + 4(d1+d2)^2}} - \frac{1}{\sqrt{4s^2 + 4(d1+d2)^2}} \right] + 4 \cdot \pi \cdot V_b \cdot s \cdot$$

$$\left[\sqrt{\frac{c}{c + \left(s / \left(d1+d2\right)\right)^{\beta}}} - \sqrt{\frac{c}{c + \left(2s / \left(d1+d2\right)\right)^{\beta}}} \right] \tag{9}$$

where

$$K_{ao} = \frac{\left(\rho 2 - \rho 1\right)}{\left(\rho 2 + \rho 1\right)},\ K_{ao3} = \frac{\left(\rho 3 - \rho_{ao}\right)}{\left(\rho 3 + \rho_{ao}\right)},\ V_b = \frac{\rho_{ao}\left(-K_{ao} - \ln\left(1 - K_{ao3}\right)\right)}{2\pi(d1+d2)}$$

$$c = x_1 \cdot \left[\ln(\rho 3 / \rho_{ao})\right] \cdot x_3,\ \beta = 2 - x_2 \cdot \ln(\rho 3 / \rho_{ao})$$

$$x_1 = 16.4133, x_2 = 0.136074, x_3 = 0.393468$$

But when $(\rho 3) < (\rho_{ao})$ the finite expression for apparent resistivity is obtained as follows:

$$\rho_a = \rho 3 + \left(\rho_{ao} - \rho 3\right)\left(2e^{-u_s . s} - e^{-u_{2s} . 2s}\right) \tag{10}$$

where $u = u_m - \dfrac{(u_m - x_1).e^{-x_2\left(\frac{s}{d1+d2}\right)}}{d1 + d2}$, $u_m = x_3.x_4.\left(\rho 3 / \rho_{ao}\right).x_5$

$$x_1 = 0.673191, x_2 = 0.479513, x_3 = 1.33335, x_4 = 0.882645, x_5 = 0.697106$$

Results of this method are drawn in (Figure 22) using the same parameters as actual measurement plotted in (Figure 17).

Figure 22. Resistivity of three-layer soil (method 3)

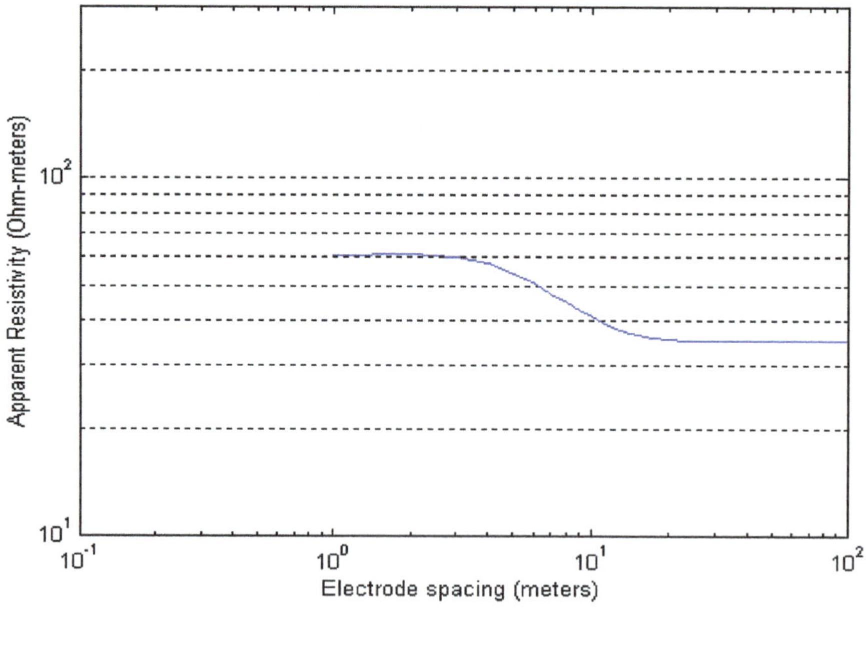

Fourth Method

The apparent soil resistivity of three layers is evaluated by obtaining the initial apparent soil resistivity between second and third layer that is represented by (ρ_{ao}) and then, taking the first layer ($\rho 1$) and second layers (ρ_{ao}) as a second layer. Thus, the apparent soil resistivity of three layers will be as follow:

When (ρ_{ao}) > ($\rho 3$)

$$\rho_a = \rho 1 + 4 \cdot \rho 1.K_{1ao}.s.\left[\frac{1}{\sqrt{s^2+4(d1)^2}} - \frac{1}{\sqrt{4s^2+4(d1)^2}}\right] + 4.\pi.V_b.s.$$

$$\left[\sqrt{\frac{c}{c+(s\,/\,d1)^{\beta}}} - \sqrt{\frac{c}{c+(2s\,/\,d1)^{\beta}}}\right] \tag{11}$$

where,

$$K_{1ao} = \frac{(\rho_{ao} - \rho 1)}{(\rho_{ao} + \rho 1)}, V_b = \frac{\rho_{ao}\left(-K_{1ao} - \ln\left(1 - K_{1ao}\right)\right)}{2\pi.(d1)}, \; c = x_1.\left[\ln\left(\rho_{ao}/\rho 1\right)\right].x_3$$

$$\beta = 2 - x_2.\ln\left(\rho_{ao}/\rho 1\right), \; x_1 = 16.4133, x_2 = 0.136074, x_3 = 0.393468$$

But when (ρ_{ao}) < ($\rho 1$) the finite expression for apparent resistivity is obtained as follows:

$$\rho_a = \rho_{ao} + \left(\rho 1 - \rho_{ao}\right)\left(2e^{-u_s.s} - e^{-u_{2s}.2s}\right) \tag{12}$$

where $u = u_m - \dfrac{\left(u_m - x_1\right).e^{-x_2\left(\frac{s}{d1}\right)}}{d1}$, $u_m = x_3.x_4.\left(\rho_{ao}/\rho 1\right).x_5$

$$x_1 = 0.673191, x_2 = 0.479513, x_3 = 1.33335, x_4 = 0.882645, x_5 = 0.697106$$

Results of this method are shown in (Figure 23) using the same parameters of the tested soils and its actual measurements are drawn in (Figure 17)

Figure 23. Resistivity of three-layer soil (method 4)

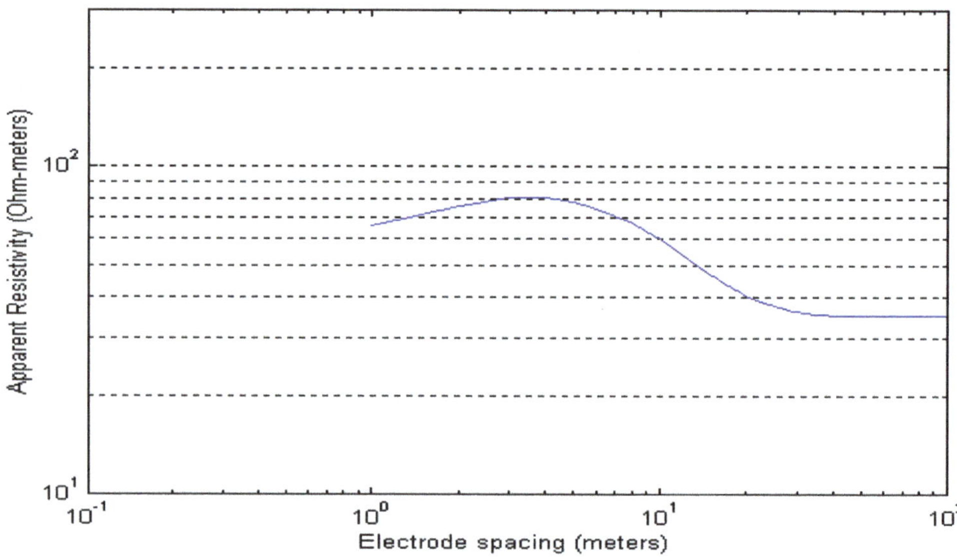

(Figure 24) shows the summary of the results that obtained by measurements of the site by Wenner technique and the calculated values using the described four methods, it is found that the measured and calculated results are much closer to each other. As shown in this figure the continuous lines represent the calculated soil resistivity and dots simulate the actual site measurements.

The accuracy of power system grounding analysis is critically dependent on the actual resistivity of local surrounding system. In this study, the factors affecting the apparent soil resistivity of multilayer soil are investigated. Such these factors are:

1. The number of layers of soil structure (double and three layers are considered in this study) and their arrangement.
2. The thickness of each layer.
3. The reflection factor between each layer.

FACTORS AFFECTING THE APPARENT SOIL RESISTIVITY OF MULTI-LAYER SOIL

The accuracy of power system grounding analysis is critically dependent on the actual resistivity of local surrounding system. In this study, the factors

Figure 24. Actual and calculates soil resistivity

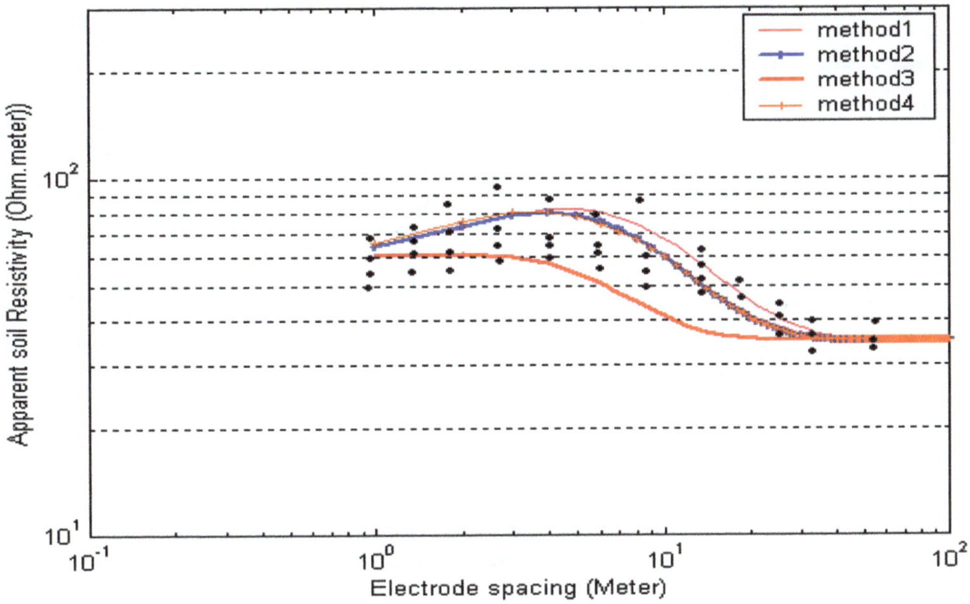

affecting the apparent soil resistivity of multilayer soil are investigated. Such these factors are:

1. The number of layers of soil structure (double and three layers are considered in this study) and their arrangement.
2. The thickness of each layer.
3. The reflection factor between each layer.

In this study, the factors affecting the apparent soil resistivity of multilayer soil are investigated. An example is presented for calculating the resistance of the grounding grid contains 50 m×50 m has a 300-meter total length with eight driven rods, each has10 meters length.

Double Layer Case

In this case two layers soil environment with upper layer thickness H in which ground rods penetrate the more conductive lower layer, in such case where $\rho_1 > \rho_2$ and the grid is buried in the upper layer ρ_1, but the ground rods are partly in ρ_2 and partly in ρ_1 the apparent soil resistivity ρ_a could be calculated by Schwarz's Formula mentioned in IEEE 80 (2000). According to IEEE 80

Figure 25. Relation between upper layer thickness and the apparent soil resistivity ($\rho 1 = 600, \rho 2 = 150\ \Omega.m$)

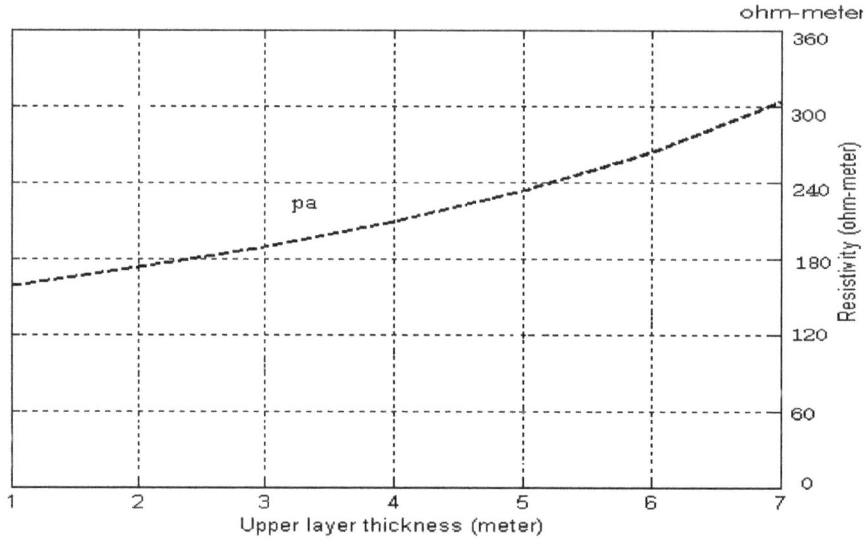

(2000). Nahman and Salamon (1984), the ground resistance of rods and the mutual resistance between the grid and rods are calculated with the use of an apparent soil resistivity. Figures (25 and 26) show the relation between the upper layer thickness and the apparent soil resistivity. From (Figure 25) it is noticed that increasing the upper layer thickness from 1 to 7 meter increases the apparent soil resistivity in case of $\rho 1 > \rho 2$ by 90%

In case of $\rho_1 < \rho_2$ as given in (Figure 26) increasing the upper layer thickness from 1 to 7 meter reduces the apparent soil resistivity in case of $\rho_1 > \rho_2$ by 60%.

$$K = \frac{\rho_2 - \rho_1}{\rho_2 + \rho_1} \tag{13}$$

(Figure 27) shows the relation between the reflection factor and the apparent soil resistivity, these values are calculated by Schwarz's Formula mentioned in IEEE 80 (2000).

K_1, K_2 are constants related to the geometry of the system and could be calculated from the relations where X is length to width ratio

$$K_1 = 1.41 - (0.04) \cdot X \tag{14}$$

Figure 26. Relation between upper layer thickness and the apparent soil resistivity ($\rho 1$ = 150,$\rho 2$ = 600 Ω.m)

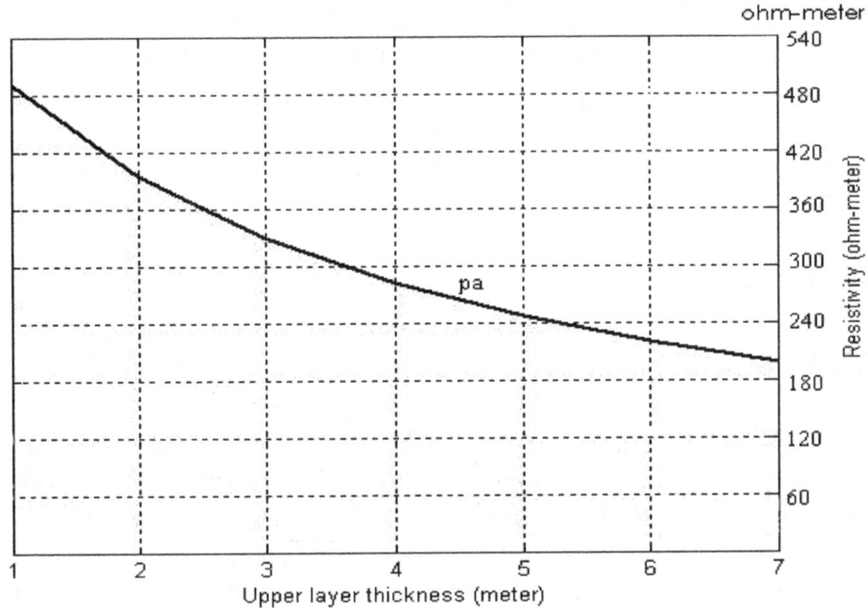

Figure 27. Relation between reflection factor and the apparent soil resistivity ($\rho 1$ = 50, $\rho 2$ = 10 to 250 Ω.m, H = 5m)

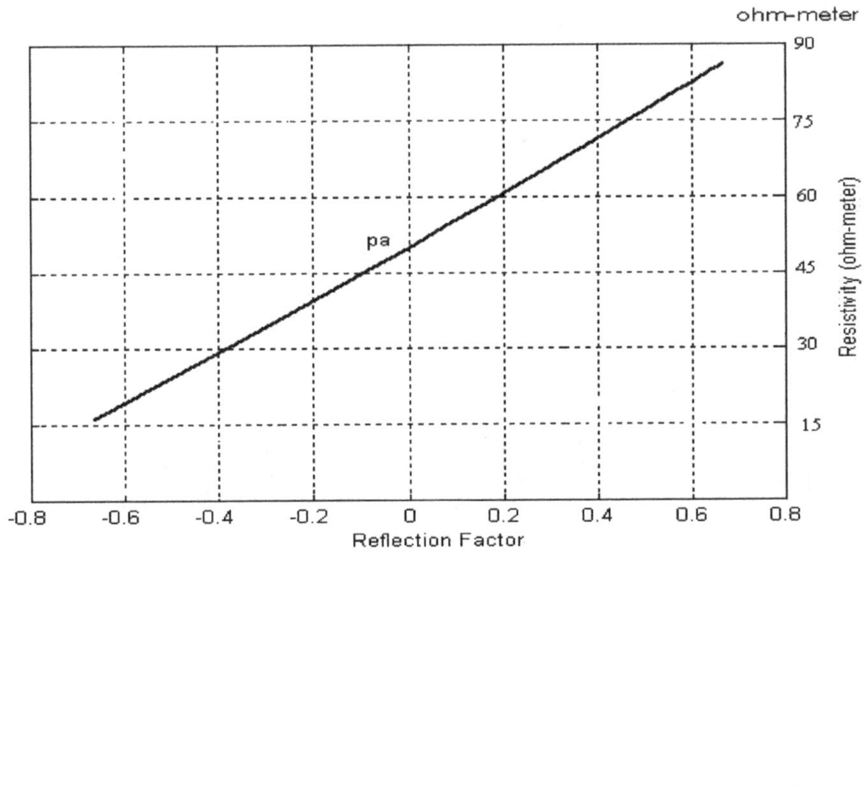

$$K_2 = 5.5 + (0.15) \cdot X \tag{15}$$

From (Figure 27) it is noticed that increasing the reflection factor from -0.7 to 0.7 increases the apparent soil resistivity by 500%. To study the effect of depth variation of laying grid on the apparent soil resistivity. The laying depth of the grid is changed between 0.1 and 1.5 m.

(Figure 28) shows the relation between the apparent soil resistivity and the laying depth of the grid.

From this figure, it is noticed that increasing the depth of laying grid slightly reduces the apparent soil resistivity seen by the grounding system by 7%.

Multi-Layer Case

In the analysis carried out in this section any number of layers could be reduced to two layers. As example in 3-layers, two layers will be reduced to one layer as follows: Evaluating initial apparent soil resistivity between first and second layer represented by ρ_{12} could be calculated by the relation given by Takahashi and Kawase (1991), Schon, (1998), Charlton, (2000), where ρ_1 and ρ_2 are the resistivity of first and second layer respectively, s is the space between probe electrodes of measuring the soil resistivity and is

Figure 28. Relation between depth of laying of grid and the apparent soil resistivity ($\rho_1 = 50$, $\rho_2 = 30\Omega.m$)

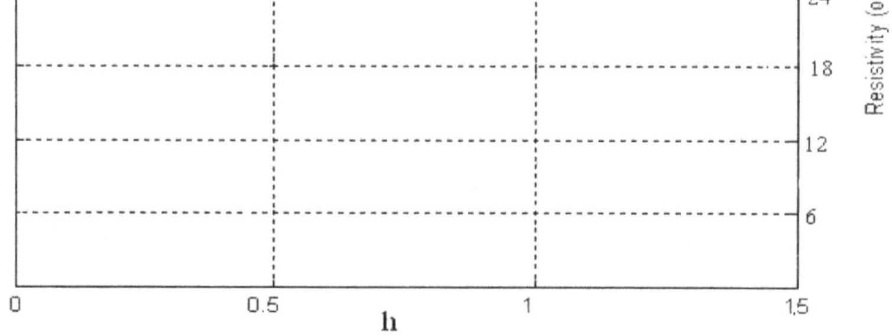

Figure 29. Relation between reflection factor and the apparent soil resistivity (ρ1 = 60,ρ2 = 50, ρ3 = 10 to 250 Ω.m, H = 2, H' = 4)

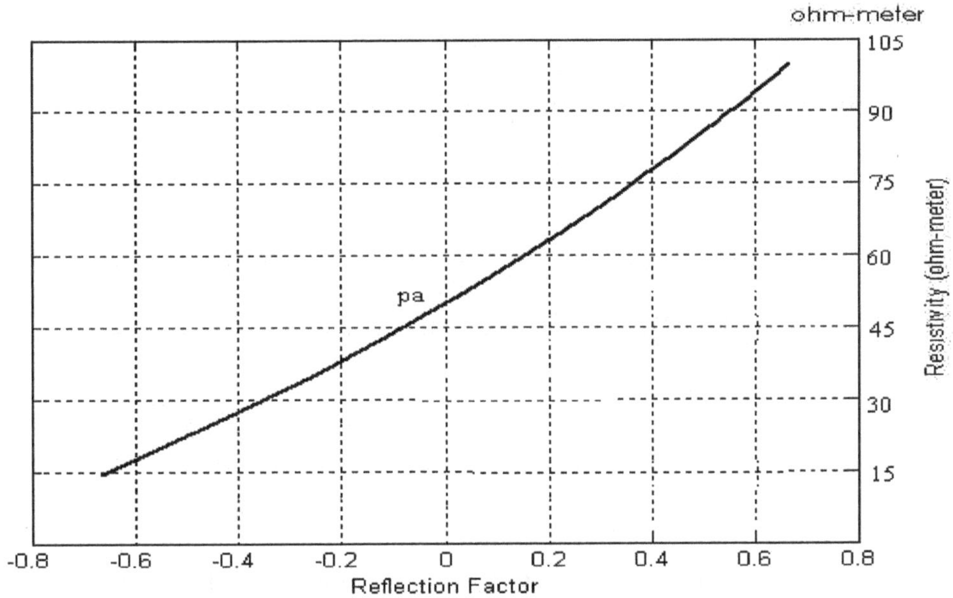

about 15m, H is thickness of first layer. To obtain the apparent soil resistivity of the three layers consider ρ_{12} is resistivity of one layer and then obtain the total apparent soil resistivity by:

$$\rho_a = l_2\left(\rho_{12}\cdot\rho_3\right)/\left(\rho_3\left(H'-h\right)+\rho_{12}\left(l_2+h-H'\right)\right) \tag{16}$$

where H' is the thickness of first and second layers. The relation between the reflection factors and apparent soil resistivity is given in (Figure28). The reflection factor is defined in this case as

$$K' = \frac{\rho_3 - \rho_{12}}{\rho_3 + \rho_{12}} \tag{17}$$

From (Figure 28) it is noticed that there is proportional relation between reflection factor and the apparent soil resistivity.

Changing the reflection factor from -0.7 to 0 increases the apparent soil resistivity by 300% and changing the reflection factor from 0 to 0.7 increases

Figure 30. Relation between depth of laying of grid and the apparent soil resistivity ($\rho 1 = 60$, $\rho 2 = 50$, $\rho 3 = 40$ Ω.m, $H = 2$, $H' = 4$)

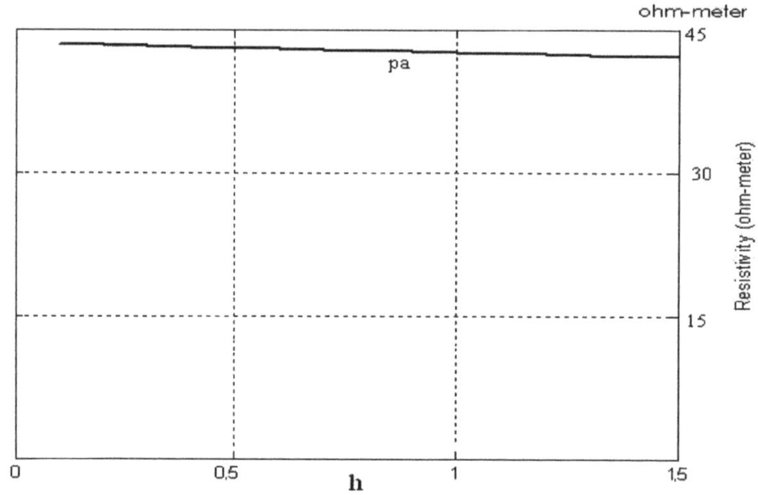

the apparent soil resistivity by 200% Similar relation is obtained between grid depth in first layer and apparent soil resistivity, (Figure 29) shows this relation. From this figure, it is noticed that increasing the depth of laying grid slightly reduces the apparent soil resistivity seen by the grounding system by 3%. Laboratory Investigations on Factors Affecting Soil Electrical Resistivity and Measurements are carried out by Zhou, Wang, Cai et al. (2015)

REFERENCES

Dwight, H. B. (1936). Calculation of resistance to ground. *Electrical Engineering*, *55*, 1319.

Engineering standards branch, telecommunications standards division (ESBTSD). (1994). Electrical Protection Grounding Fundamentals. United States department of agriculture, rural electrification administration. Retrieved from http://www.lmrsitestandard.com/wp-content/blogs.dir/2/files/2012/09/Appendix-B-Motorola-R56-09-01-05.pdf

IEEE. (2000). Guide for Safety in AC Substation Grounding.

Nahman, J. M., & Salamon, D. D. (1998). Analytical Expressions for the resistance of grounding grids in nonuniform soil. *IEEE Transactions on Power Apparatus and Systems*, *PAS-103*(4), 880–885. doi:10.1109/TPAS.1984.318367

Schon, J. H. (1998). Physical Properties of Rocks: Fundamentals and Principles of Petrophysics. In Handbook of Geophysical Exploration (Vol. 18). Pergamon.

Schon, J. H. (1998). Physical Properties of Rocks: Fundamentals and Principles of Petrophysics. In Handbook of Geophysical Exploration (Vol. 18). Pergamon.

Southey, R. D., Siahrang, M., Fortin, S., & Dawalibi, F. P. (2015). Using fall-of-potential measurements to improve deep soil resistivity estimates. *IEEE Transactions on Industry Applications*, *51*(6), 5023–5029. doi:10.1109/TIA.2015.2428679

Takahashi, T., & Kawase, T. (1991). Calculation of earth resistance for a deep driven rod in a multi-layer earth structure. *IEEE. Trans.*, *PWRD-6*(2), 608–614.

Trevor, C. (2000). Substation Earthing-shedding light on the black art. In *Proceedings of IEE Seminar*, Birmingham.

United States Environmental Protection Agency - Wikipedia. https://en.wikipedia.org/wiki/United_States_Environmental_Protection_Agency

Zhou, M., Wang, J., Cai, L., Fan, Y., & Zheng, Z. (2015). Laboratory Investigations on Factors Affecting Soil Electrical Resistivity and the Measurement. *IEEE Transactions on Industry Applications*, *51*(6), 5358–5365. doi:10.1109/TIA.2015.2465931

Chapter 2
Design of Grounding System

ABSTRACT

This chapter contains the types of grounding systems, single ground rod, single strip end connected, single strip center connected, radial grounding, single radials, equipotential mesh electrodes, radial grounding, multiple radials, and grid electrodes grid with ground rods, ring electrode, and grounding chain. It contains also the method of calculating single grounding electrode system resistance, length/depth of the ground electrode, diameter of the ground electrode and number of ground electrodes to obtain the required grounding system resistance. The chapter contains the grounding nomograph, example worksheet using nomograph, design of multiple grounding electrode in straight line and ring systems, examples layout and graph, multiple grounding electrode system are presented. It contains also calculating methods of multiple grounding electrode system resistance of different configurations. Color code technique to calculate the design parameters of the grounding rods is also presented.

TYPES OF GROUNDING SYSTEMS

The basic philosophy of any ground electrode installation should be an attempt to maximize the surface area contact with the surrounding soil. Not only does this help to lower the resistance of the grounding system, but it also greatly improves the surge impedance of the grounding system due to the large capacitive coupling which is achieved. The actual layout of the grounding

DOI: 10.4018/978-1-5225-3853-0.ch002

Figure 1. Single ground rod

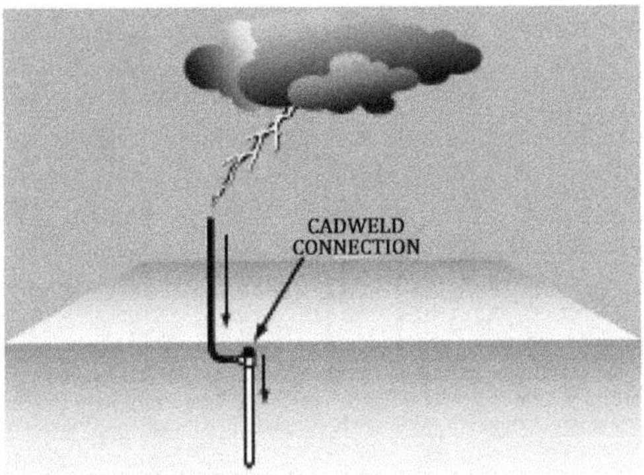

Figure 2. Single strip end connected

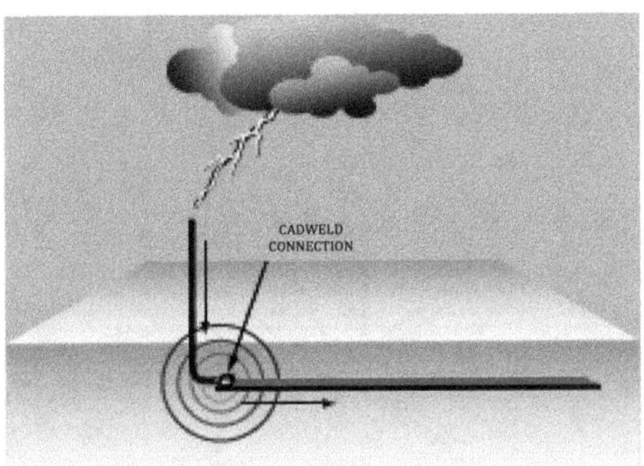

system, and the number, length and depth of ground rods will vary with the usages, type of soil and space availability. Some common designs include:

- **Single Ground Rod:** One single ground electrode may be sufficient for an electrical installation in a built-up area where the local supply authority utilizes a multiple or common multiple earth neutral system. However, it may not provide adequately low impedance for lightning current injection. See (Figure 1).
- **Single Strip End Connected:** This is a common option for installations where, because of rock, driving an electrode is impractical. It is not recommended for lightning protection systems as there is only one path. Very high ground voltages will be experienced at the injection point. Single strip end connected arrangement is given in (Figure 2).

Figure 3. Single strip center connected

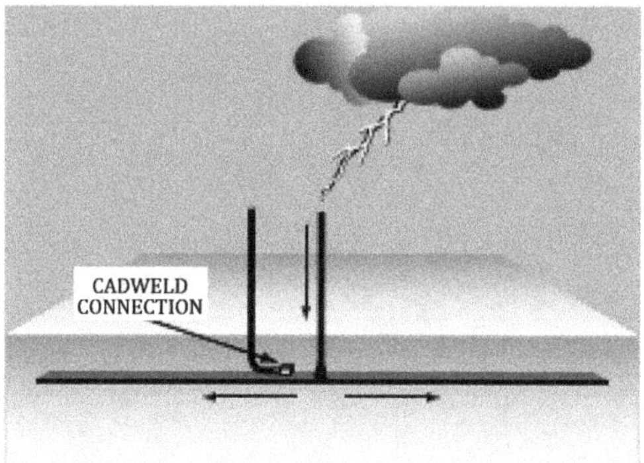

Figure 4. Radial grounding, single radials

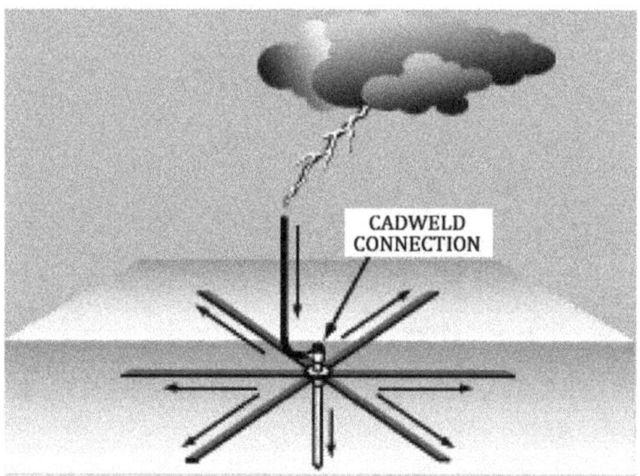

- **Single Strip Center Connected:** Since the connection to the strip is at the center, any fault/injection current travels in two directions. This layout, as given in (Figure 3) has lower impedance, but it is generally not adequate for lightning protection systems.
- **Radial Grounding, Single Radials:** A design that is well suited to lightning protection in areas of medium resistivity. The radials can run to 100 feet in length. See (Figure 4).
- **Equipotential Mesh Electrodes:** Minimize the risk of step and touch potential hazard by positioning a mat and bonding it to the structure or operating handle at locations where personnel may be required to operate switchgear or stand in the course of their duties. Low ground impedance can be obtained using mesh electrodes grounding system. as given in (Figure 5).

Figure 5. Equipotential mesh electrodes system

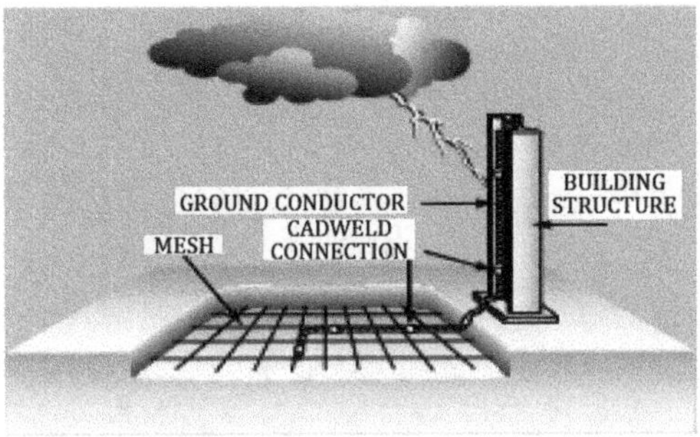

Figure 6. Radial grounding, multiple radials system

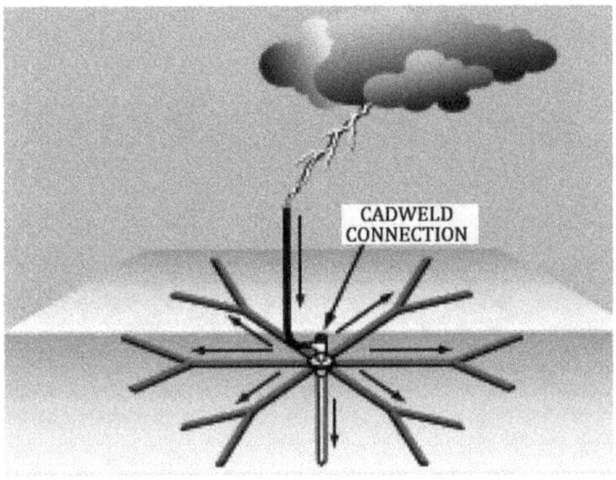

- **Radial Grounding, Multiple Radials:** Radial grounding has multiple radials design, as shown in (Figure 6), well suited to lightning as it allows energy to diverge as each conductor takes a share of the current, offering lower impedance. Voltage gradients leading away from the injection point will be lower, reducing danger from step potentials.
- **Grid Electrodes:** Grounding for installations where there is concentration of electrical equipment, such as electrical substations, is often designed to meet a specific value of resistance (typically 1 ohm). Under fault conditions, a grid can dissipate currents over a large area. (See Figure 7).
- **Grid With Ground Rods:** It may be advantageous to add ground rods to the grid. In doing so, it may be possible to access a low resistivity soil layer. Care must be taken to ensure each ground rod is spaced at least twice the installation depth. See (Figure 8).

Figure 7. Grid electrodes

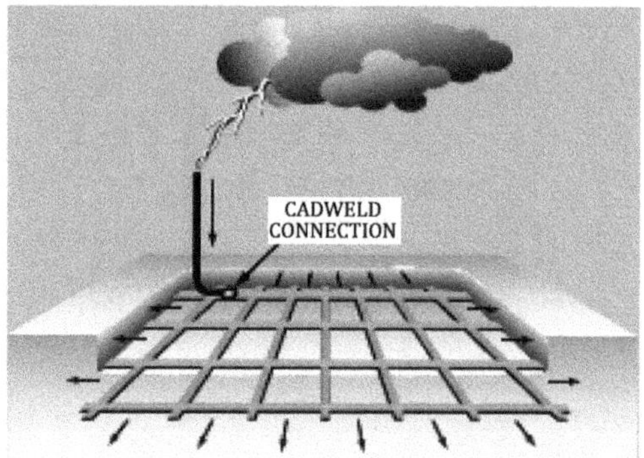

Figure 8. Grid with ground rods

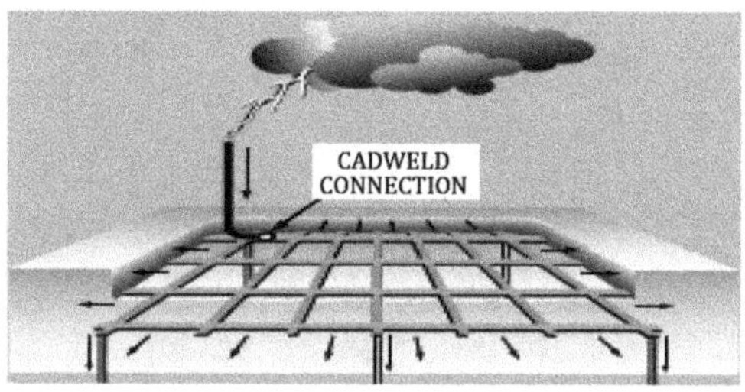

Figure 9. Ring electrode

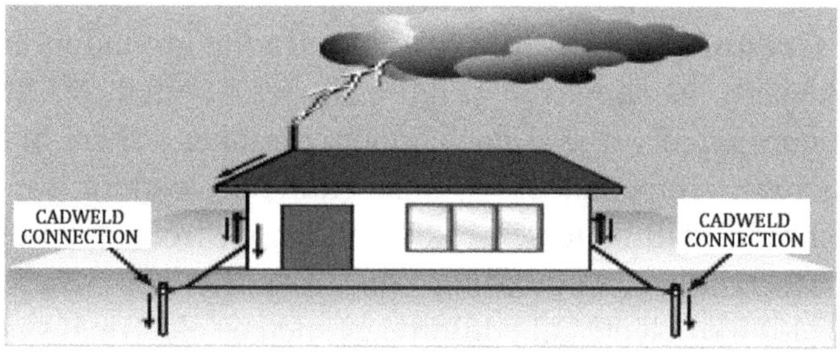

- **Ring Electrode:** Installations, including communications huts, pad mount transformers and fences surrounding high voltage installations, are generally surrounded by a ground ring. This practice also reduces the hazard of step and touch potential. See (Figure 9).

GROUNDING CHAIN

The performance of the grounding system is determined by the quality of the following five components all of which are of equal importance;

1. **The Grounding Electrode Conductor:** Typically made from copper or copper-clad steel, the grounding electrode conductor must be large enough to withstand the maximum available fault current over the maximum clearing time
2. **The Grounding Connections:** Often overlooked, the grounding connections are used to tie the elements of the electrode system together. Exothermically welded connections provide a molecular bond that will never loosen or corrode. Mechanical connectors, such as crimp, bolted, and wedge type, rely on physical point-to point surface contact to maintain the integrity of the electrical connection. IEEE Standard 837(2014) provides detailed information on the application and testing of permanent grounding connections.
3. **The Grounding Electrode:** The grounding electrode provides the physical connection to the earth and is the instrument used to dissipate current into it. There are two main types of electrodes. "Natural" electrodes are intrinsic to the facility and include metal underground water pipe, the metal frame of the building (if effectively grounded), and reinforcing bar in concrete foundations. "Made" electrodes are installed specifically to improve the performance of the ground system and include wire meshes, metallic plates, buried copper conductor and rods or pipes driven into the ground. The ground rod is the most widely used electrode.
4. **Electrode to Soil Resistance:** Amount of rod surface and rod replacement are the controlling factors. Doubling diameter reduces resistance by only 10% and is not cost effective. Doubling rod length, however, theoretically reduces resistance by 40%. The most common solution is proper placement of multiple rods that are driven to the required depths.

The Soil Characteristic

The soil resistivity, measured in ohm-centimeters or ohmmeters, plays a significant role in determining the overall performance of the grounding system and must be known before a proper grounding system can be engineered. Measuring soil resistivity allows the design engineer to locate an area with the most conductive soil and to determine the depth of the conductive soil so that electrodes can be placed accordingly.

Calculation of Single Grounding Electrode System Resistance

Single grounding electrode resistance can be calculated as reported by IEEE 80 (2000) standards in ohms using the relation

$$R = \frac{\rho}{2\pi l}\left[\ln\left(\frac{8l}{d}\right) - 1\right] \tag{1}$$

where ρ is the measured soil resistivity in ohm.m, l is the electrode length in meters and d is the electrode diameter in meters also. Some factors affecting the grounding by rods such as

Length/Depth of the Ground Electrode

One very effective way of lowering ground resistance is to drive ground electrodes deeper. Soil is not consistent in its resistivity and can be highly unpredictable. It is critical when installing the ground electrode, which it is below the frost line. This is done so that the resistance to ground will not be greatly influenced by the freezing of the surrounding soil. Generally, by doubling the length of the ground electrode you can reduce the resistance level by an additional 40%. There are occasions where it is physically impossible to drive ground rods deeper areas that are composed of rock, granite, etc. In these instances, alternative methods including grounding cement are viable.

Diameter of the Ground Electrode

Increasing the diameter of the ground electrode has very little effect in lowering the resistance. For example, you could double the diameter of a ground electrode and your resistance would only decreases by 10%.

Number of Ground Electrodes

Another way to lower ground resistance is to use multiple ground electrodes. In this design, more than one electrode is driven into the ground and connected in parallel to lower the resistance. For additional electrodes to be effective, the spacing of additional rods need to be at least equal to the depth of the driven rod. Without proper spacing of the ground electrodes, their spheres of influence will intersect and the resistance will not be lowered.

For a single grounding electrode system, the resistance can be easily calculated using a nomograph. An example calculation is shown in (Figure 10). If calculations show excessive resistance for a given electrode depth and diameter, recalculate substituting a larger diameter electrode and/or deeper electrode depth. In this manner, the proper size and depth of grounding electrode for a particular site can be determined. The steps of calculations are given in this figure

Figure 10. Single grounding electrode system calculation using a Nomo graph as reported by Appendix-B-Motorola-R56-09-01-05.pdf (2012)

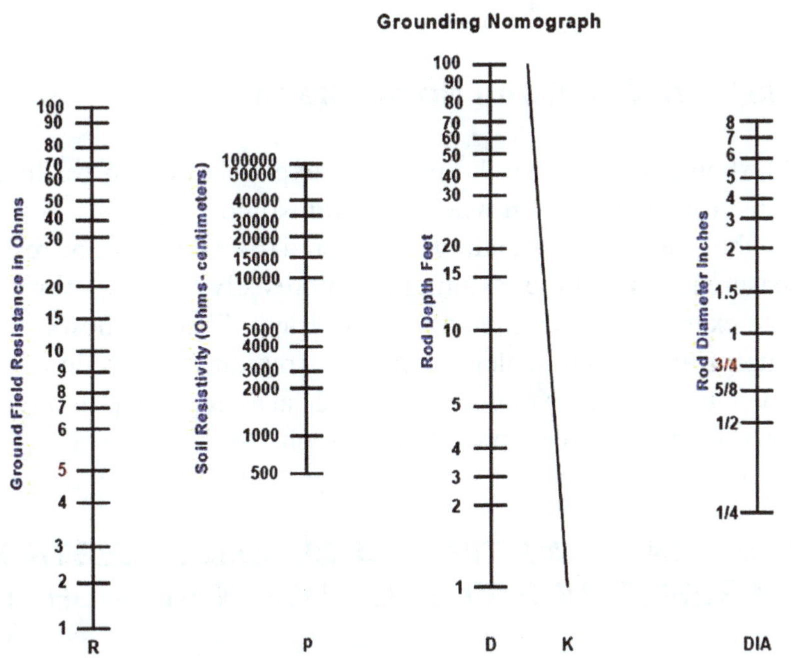

1. Select required resistance on "R" scale (2=Recommended and 5=Maximum)
2. Select measured or apparent resistivity on scale "P".
3. Lay a straightedge on "R" and "P" scale and intersect with "K" scale.
4. Lay straightedge on "K" scale intersect point and "DIA" scale 3/4 inch and intersect "D" scale.
5. Point on "D" scan will be rod depth required for resistance on "R" scale.

Figure 11. Example worksheet and Nomo graph (sheet 1 of 3) as reported by Appendix-B-Motorola-R56-09-01-05.pdf (2012)

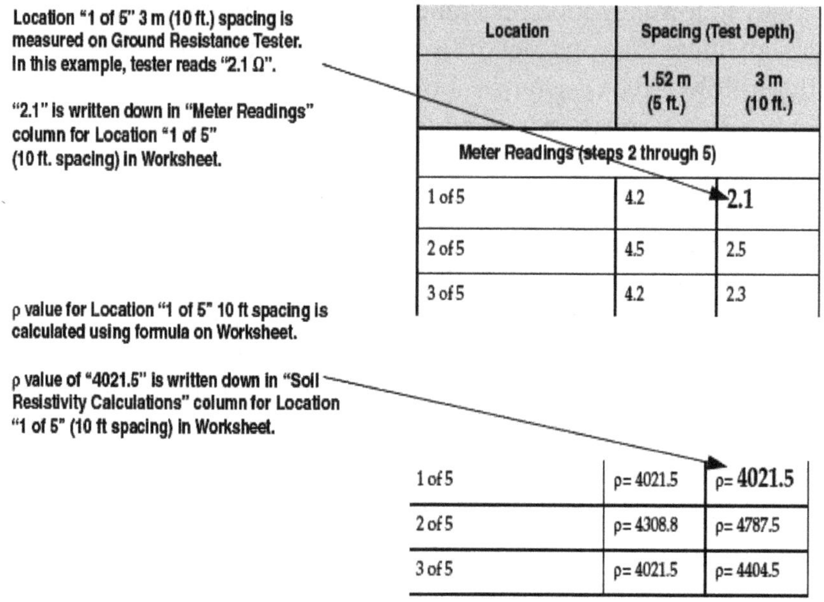

Example Work Sheet and Nomo Graph

Figure 11 shows sheets from 1 through 3 explain example readings and calculations from a completed worksheet and Nomo-graph.

Sheet 1 shows example of readings, as entered from field ground resistance tester measurements and the resulting of soil resistivity calculations. Sheet 2 shows an example of a completed Nomo-graph. Sheet 3 shows a second Nomo-graph filled-in with calculations for grounding electrode resistance improvement using a larger-diameter electrode at a deeper depth. (Figure 12) gives worksheet for soil resistivity Nomo graph.

CALCULATION OF MULTIPLE GROUNDING ELECTRODE SYSTEM RESISTANCE (ELECTRODES IN STRAIGHT LINE)

For a multiple grounding electrode system with multiple parallel electrodes in a straight line as shown in (Figure 15), the system resistance can be calculated as described in the following procedure:

Figure 12. Example worksheet and Nomo graph (sheet 2 of 3) as reported by Appendix-B-Motorola-R56-09-01-05.pdf (2012)

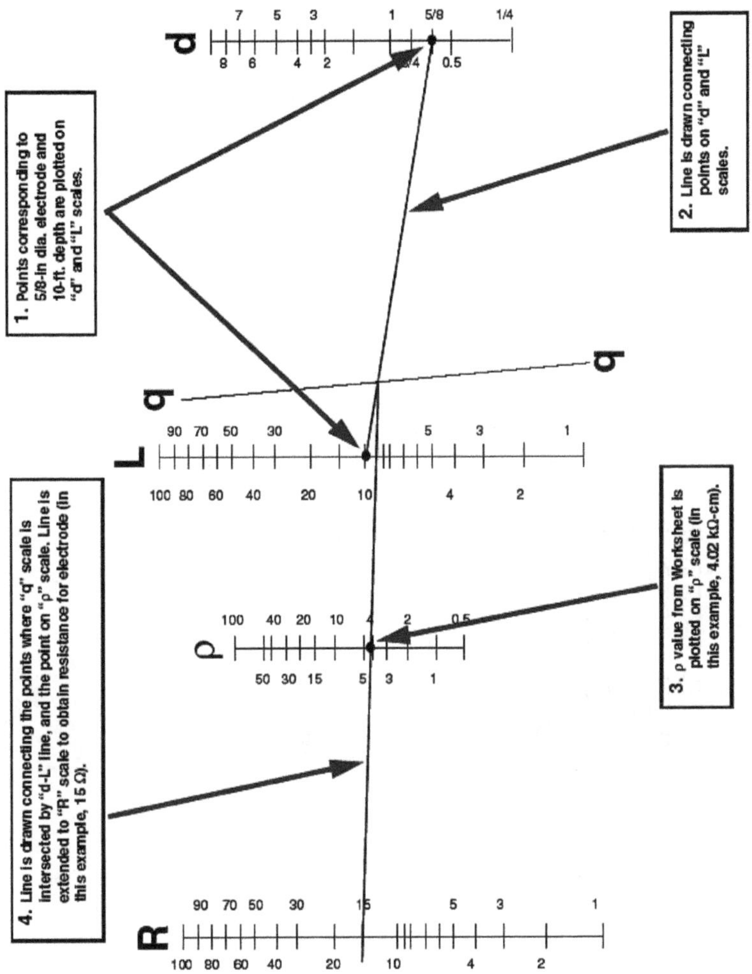

1. Perform soil resistivity test using winner method described before
2. Using the worst-case value obtained, calculate the resistance of one ground rod as explained before.
3. Sketch a proposed layout of the ground rod arrangement using equally spaced rods in a line.
4. Make a photocopy of (Figure 12)
5. Using the copy of Combined Resistance Graph (Ground Rods Arranged in Line or Ring), calculate the effective resistance of the proposed layout as follows:

Figure 13. Example worksheet and Nomo graph (sheet 3 of 3) as reported by Appendix-B-Motorola-R56-09-01-05.pdf (2012)

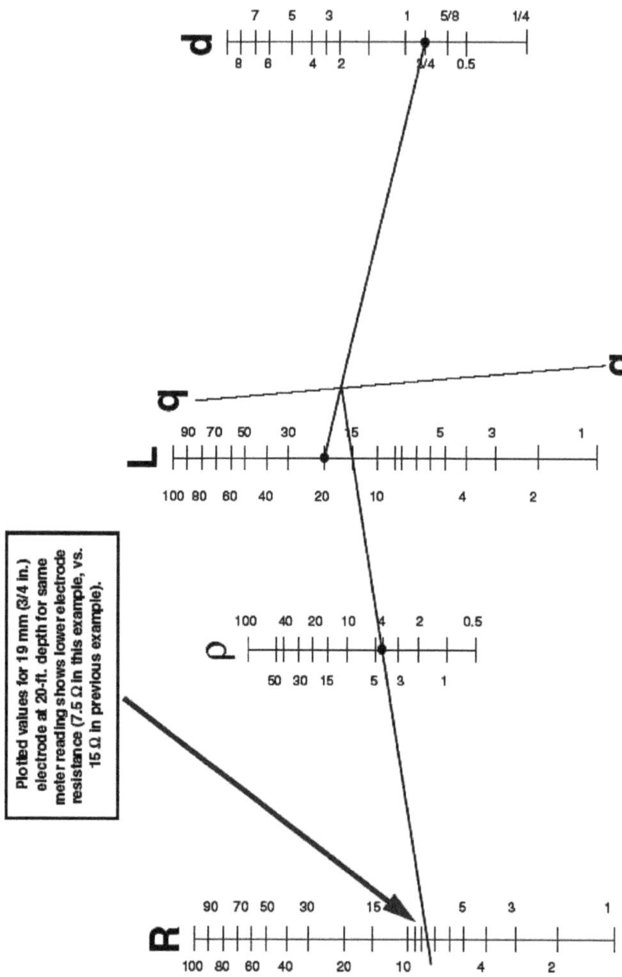

a. Determine the number of rods to be used, locate this number on the Number of Rods axis of the graph.
b. Note the spacing of the rods in the proposed layout in terms of spacing as related to length of rods. In graph, "s=L" is spacing equal to length of rod "s=2L" is spacing equal to twice the length of rod, and so on. Locate the spacing line on graph (s=L, s=2L, s=3L, s=4L) corresponding to proposed spacing. See (Figure 16).

Figure 14. Soil resistivity Nomo graph as reported by Appendix-B-Motorola-R56-09-01-05.pdf (2012)

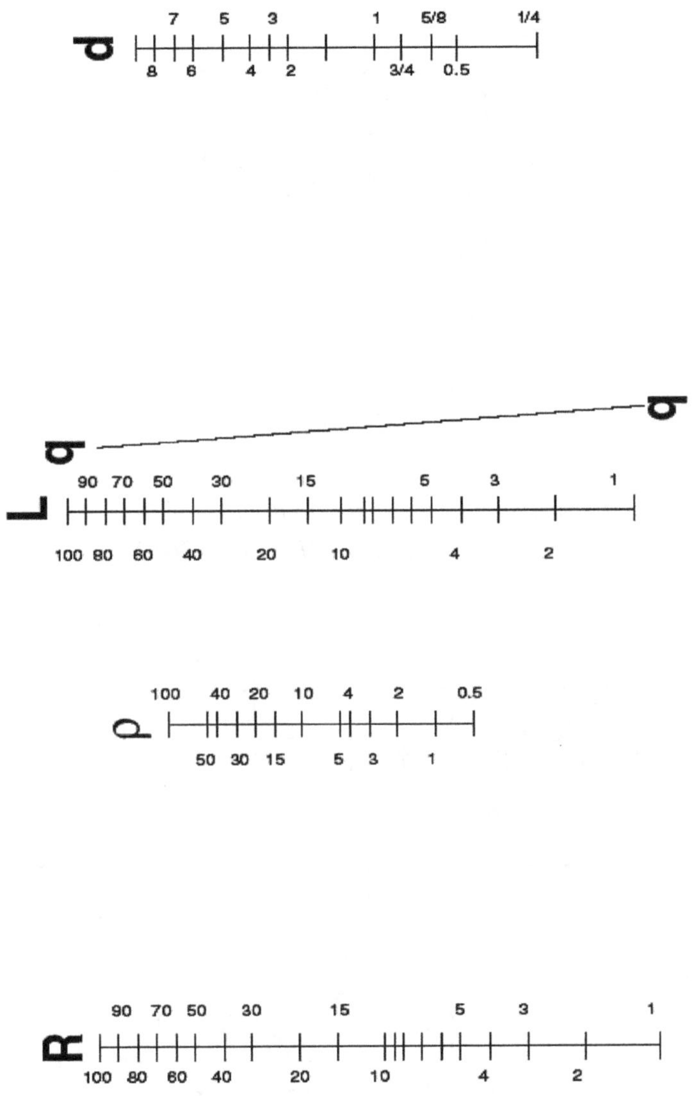

c. At the point on the graph where the Number of Rods line intersects the appropriate spacing line, note the Combined Resistance number at the left. See (Figure 16).
d. Multiply the Combined Resistance number by the resistance of a single ground rod noted in step 2 of this procedure. This is the worst-case resistance of the proposed grounding electrode system.

Figure 15. Example of multiple grounding electrodes in straight line

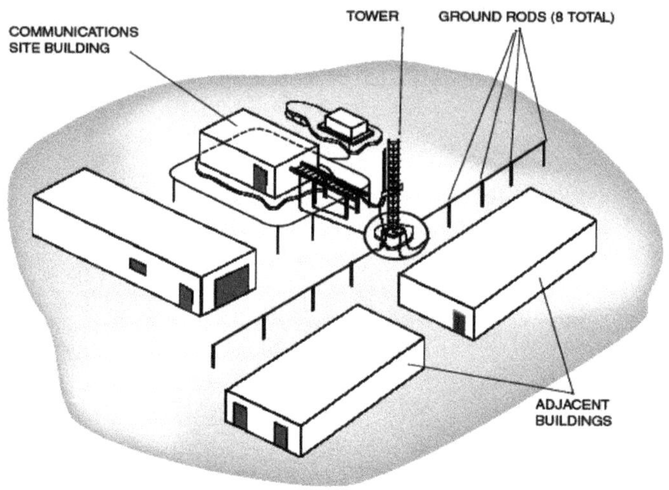

Figure 16. Combined resistance graph (ground rods arranged in line or ring) as reported by Appendix-B-Motorola-R56-09-01-05.pdf (2012)

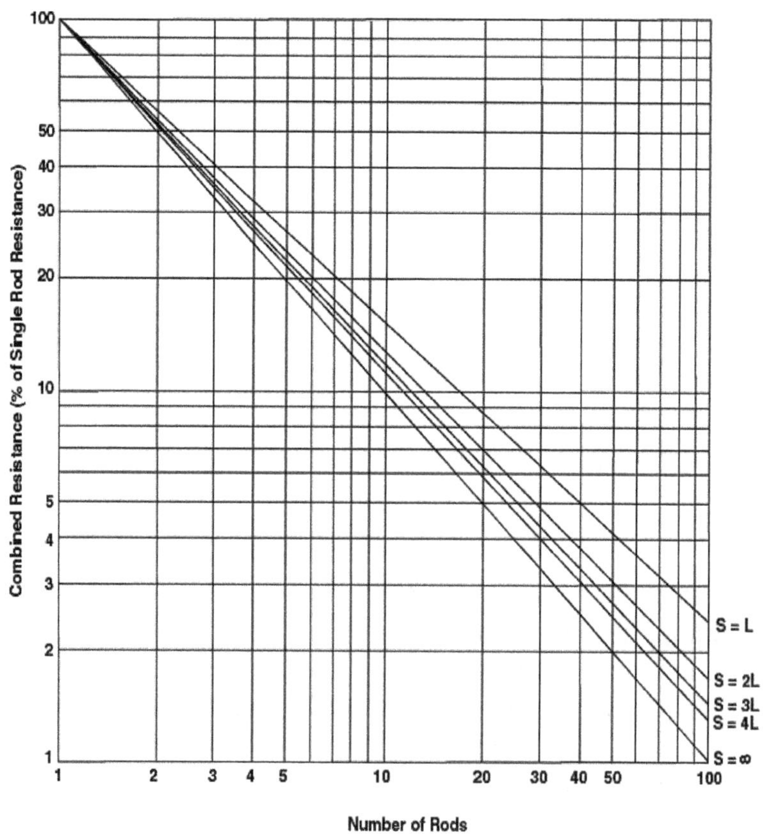

EXAMPLE LAYOUT AND GRAPH IN STRAIGHT LINE

Assuming a layout as shown in (Figure 16) with the following characteristics: Eight rods (each of 8-ft. length) are spaced at 4.9 m (16 ft.) points (or "2L" in terms of the graph) along a line. Worst-case soil resistivity measurement is 4021.5 Ω-cm and the resistance of single rod electrode that calculated using equation 1

$$R = \frac{\rho}{2\pi l}\left[\ln\left(\frac{8l}{d}\right) - 1\right] \tag{1}$$

is 15 Ω. System resistance is calculated as follows:

1. Using (Figure 14): because eight rods are required for the design, "8" line on Number of Rods in graph is selected.
2. Because rod spacing is 4.9 m (16 ft.), or "2L" of rod length, "s = 2L" line on graph is selected.
3. At the intersection of the "8" line and the "s = 2L" line, draw a horizontal line to the Combined Resistance axis at left. Note the point where the horizontal line crosses the Combined Resistance axis (in this case, at approximately "18" (or 18% of single rod resistance))

Single rod resistance of 15 Ω is then multiplied by 18% (0.18) to obtain the effective resistance of the system: 15Ω× 0.18= 2.7 Ω

In this example, effective overall resistance of the proposed system would be 2.7 Ω.

Complete explanation of example calculation of ground rods arranged in straight line is given in (Figure 15).

MULTIPLE GROUNDING ELECTRODE SYSTEM RESISTANCE CALCULATION (ELECTRODES IN RING CONFIGURATION)

For a multiple grounding electrode system with multiple electrodes installed in a ring configuration, as shown in (Figure 17), the system resistance is calculated in the same manner as electrodes placed in a straight line. When

Figure 17. Example calculations of ground rods arranged in straight line as reported by Appendix-B-Motorola-R56-09-01-05.pdf (2012)

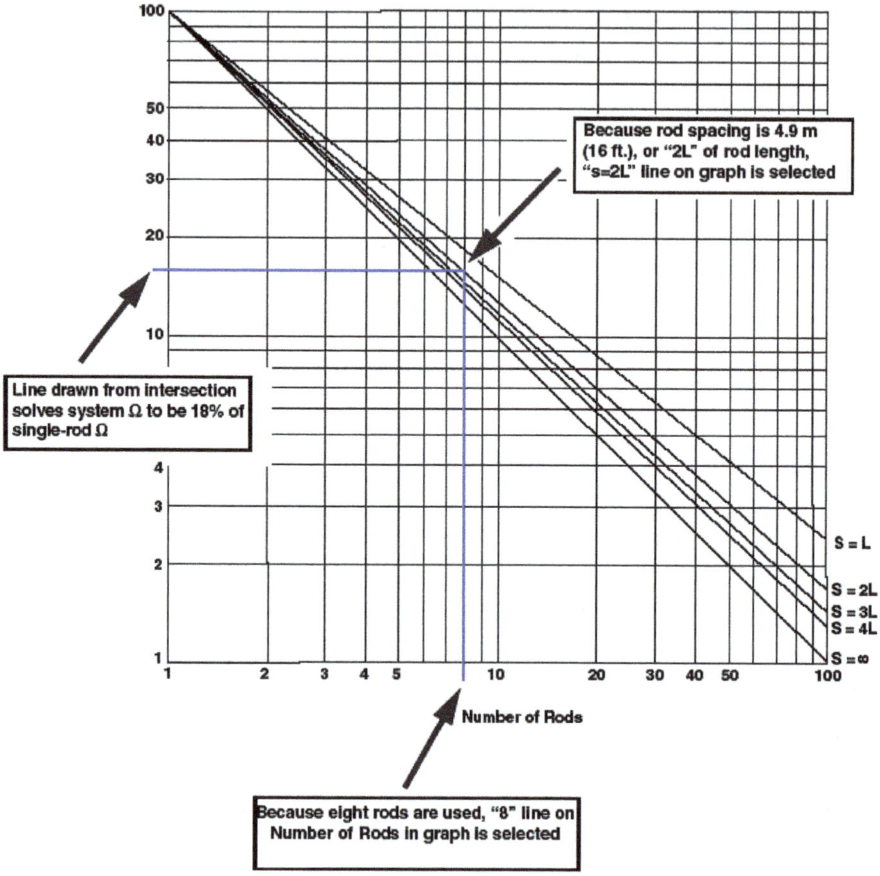

Figure 18. Ring configuration planning and resistance measurement considerations as reported by Appendix-B-Motorola-R56-09-01-05.pdf (2012)

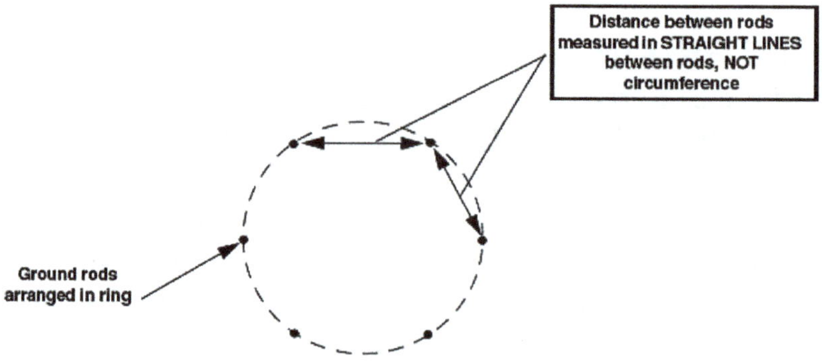

planning a ring configuration layout and performing calculations, note the following:

1. All rods in the system shall maintain equal or greater separation from adjacent rods.
2. The distance between rods shall be figured in a direct path to adjacent rods, not the circumference distance of the ring.

NOTE: The stipulations regarding rod spacing specified in "External Building and Tower Ground Ring".

MULTIPLE GROUNDING ELECTRODE SYSTEM RESISTANCE CALCULATION (GROUND ROD GRID CONFIGURATION)

For multiple grounding electrode system consisting of a ground rod grid configuration (as shown in (Figure 18), the system resistance can be calculated as described in the following procedure:

1. Perform soil resistivity test as described before by using Winner method
2. Using the worst-case value obtained, calculate the resistance of one ground rod as described before. Write down this number.
3. Sketch a proposed layout of the ground rod arrangement using equally spaced rods across the proposed area. Note the stipulations regarding rod spacing must be observed when planning rod layout.
4. Calculate the area of the proposed grid system in square feet. Note that this procedure requires grid measurements be entered in square feet. If metric measurements have been made, the measurements must be converted to feet.
5. Make a photocopy of (Figure 19).
6. Using the copy of Combined Resistance Graph (Ground Rods Arranged in Grid), calculate the effective resistance of the proposed layout as follows:
    a. Noting the number of rods to be used, locate this number on the Number of Rods axis of the graph.
    b. Note the square footage of the proposed rod layout. Locate the curve on the graph most closely corresponding to the proposed square footage.

Figure 19. Example of multiple grounding rod grid configurations

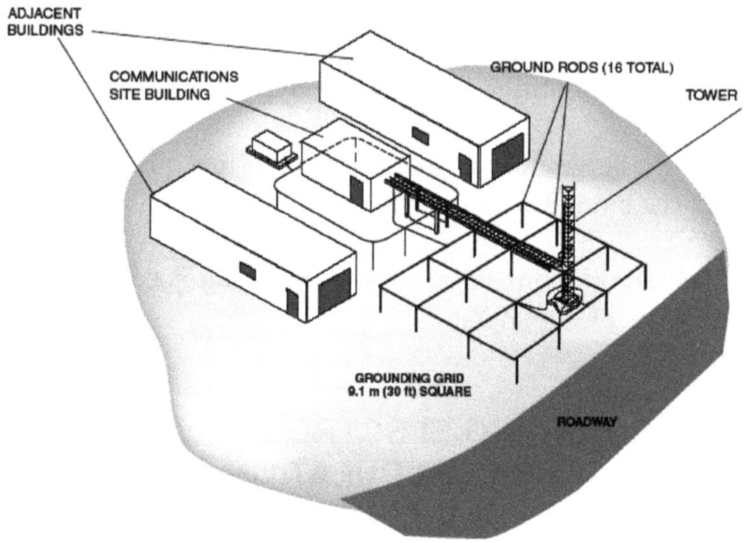

c. At the point on the graph where the Number of Rods line intersects the appropriate square footage curve, note the Resistance Ratio number at the left.
d. Multiply the Resistance Ratio number by the resistance of a single ground rod noted in step 2 of this procedure. This is the worst-case resistance of the proposed grounding electrode system.

EXAMPLE LAYOUT AND GRAPH

Assuming a layout as shown in (Figure 19) with the following characteristics:

- 16 rods equally spaced across a 30 × 30 ft. grid (900 sq. ft.). Worst-case soil resistivity measurement (step 1 above) is 4021.5 Ω-cm.
- Resistance of single rod tested (step 2 above) is 15 Ω. System resistance is calculated as follows:
    - See (Figure 19), because 16 rods are used, the point corresponding to “16” on Number of Rods in graph is selected. Draw a line vertically from the “16” point on the graph.
    - Because the grid is 900 sq. ft., a point just below the 1,000 sq. ft. curve on graph is plotted on the line drawn on the graph.

- At the point plotted in the previous step, (intersection of "900" sq. ft. and "16" rods), draw a horizontal line to the Resistance Ratio axis at left. Note the point where the drawn horizontal line crosses Resistance Ratio axis (in this case, at approximately ".17").

The single rod resistance of 15 W is then multiplied by 0.17 to obtain the effective resistance of the system: 15Ω× 0.17 = 2.55 Ω. In this example, the effective overall resistance of the proposed ground system would be 2.55 Ω.

Figure 20. Combined resistance graph (ground rods arranged in grid) as reported by Appendix-B-Motorola-R56-09-01-05.pdf (2012)

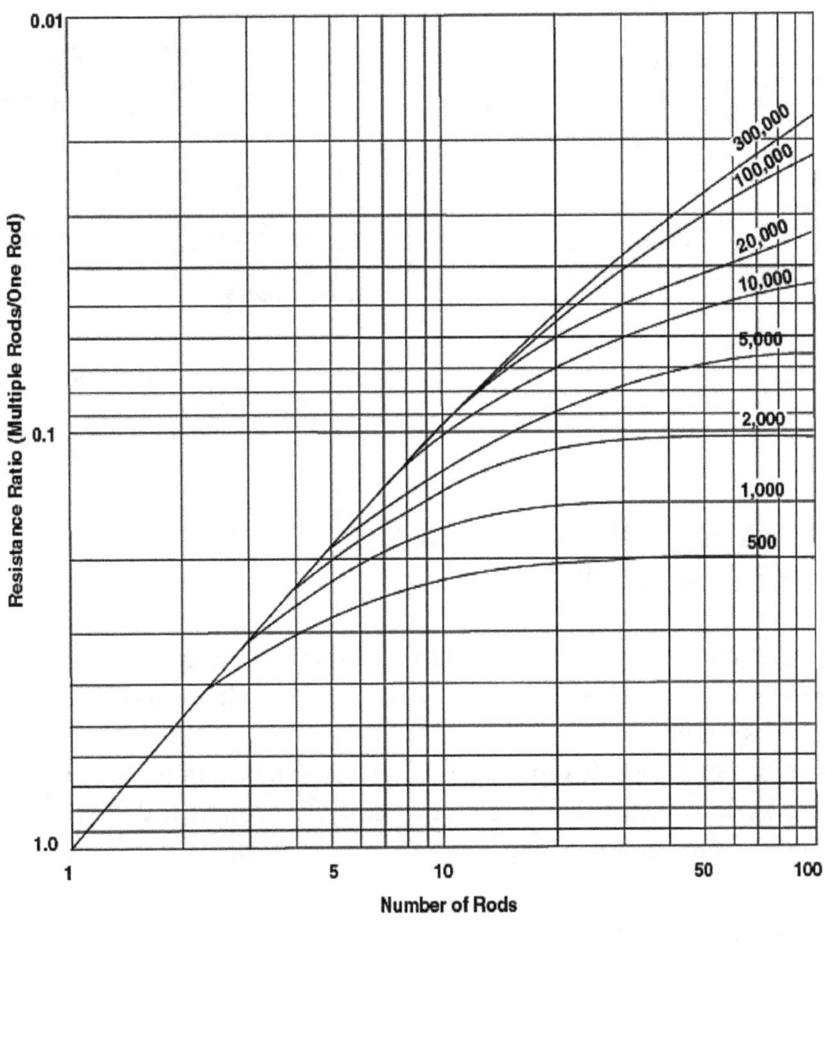

Figure 21. Example calculation of ground rod grid configuration as reported by Appendix-B-Motorola-R56-09-01-05.pdf (2012)

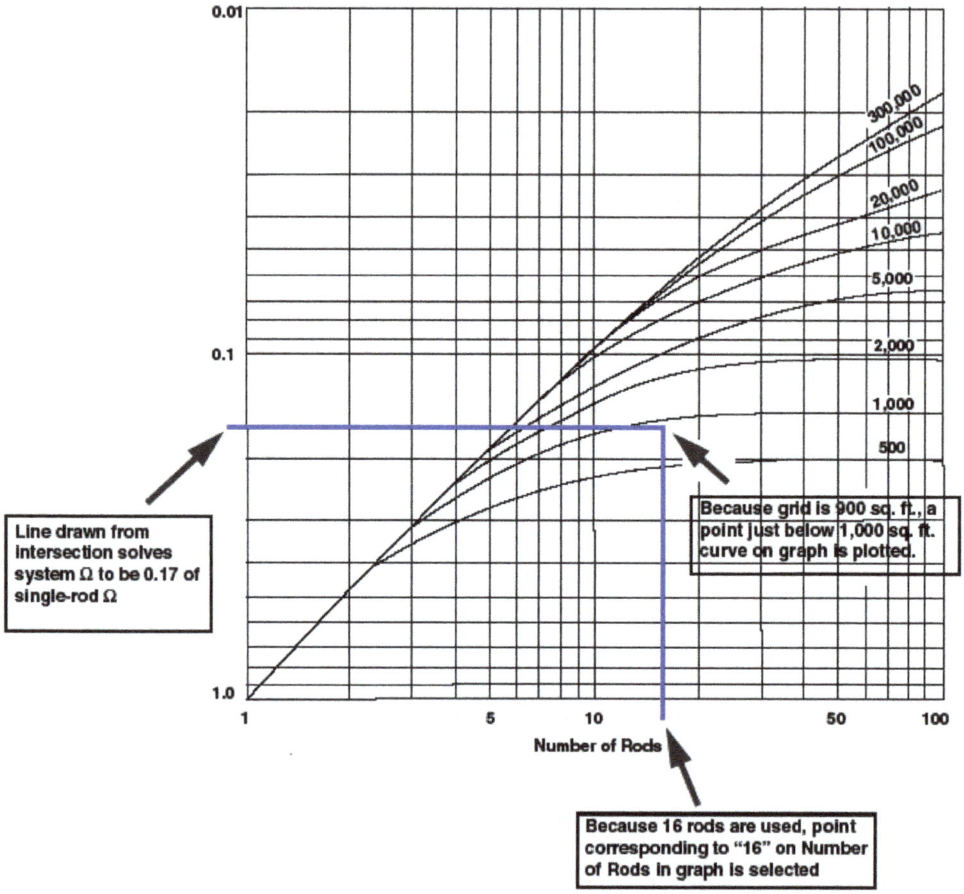

CALCULATION OF COMPLEX GROUNDING ROD SYSTEM RESISTANCE

Complex ground rod systems consist of multiple subsystems bonded together to form an overall site ground rod system. Figure 21 shows a typical complex ground rod system.

Resistance of a complex ground rod system can be calculated by breaking down the system into subsystems. Typically, a ground rod system can be broken down into the following individual subsystems:

1. Building ground ring

2. Tower ground ring
3. Tower radial grounding conductors

For a complex ground rod system consisting of the above subsystems or similar multiple subsystems, the overall system resistance can be approximated as described in the following procedure.

Adjacent subsystems should not be laid out closer than the ground rod spacing distance used within a particular subsystem. This is because as subsystems become closer than this distance, the adjacent subsystems begin to "act" as a single subsystem rather than two subsystems.

1. Perform soil resistivity test as described in "Soil Resistivity Measurement Procedure"
2. Using the worst-case value obtained, calculate the resistance of one ground rod as described in "Calculating Single Grounding Electrode System Resistance". Record this number; it is needed for following calculations.
3. Sketch a proposed layout of the ground rod arrangement using equally spaced rods across the proposed area.

The stipulations regarding rod spacing as specified in "Ground Rods" must be observed when planning rod layout.

4. Calculate the resistance of the building ground ring subsystem as described in "Multiple Grounding Electrode System Resistance Calculation (Electrodes in Ring Configuration)". Write down the result.
5. Calculate the resistance of the tower ground ring subsystem as described in "Multiple Grounding Electrode System Resistance Calculation (Electrodes in Ring Configuration)". Write down the result.
6. Calculate the resistance of the tower radial grounding conductor subsystem
7. Calculate and record the resistance of each individual tower radial grounding conductor as described in "Calculating Multiple Grounding Electrode System Resistance (Electrodes in Straight Line)".

If the radial grounding conductor does not contain ground rods, the resistance to earth of the radial grounding conductor can be calculated as follows:

$$R = \frac{\rho}{\pi L}(\ln 2L/(2aD)^{0.5}) - 1) \quad (2)$$

where

R = the resistance of electrode in ohm
ρ = the resistivity of soil in ohm meter, a is the wire radius in meters
L = the wire length in meter and D is the wire depth in meters

Using the results obtained in above step; calculate the combined (parallel) tower ground radials as follows:

1/Rtotal = 1/Rradial 1 + 1/Rradial 2 + ...1/Rradial n (3)

The ground rods associated with the tower ground ring are not included in the radial grounding conductors.

8. Noting the resistances determined for the subsystems, calculate the combined of all of the subsystems as follows:
9. subsystem 1 (building ground ring subsystem; step 4)
10. subsystem 2 (tower ground ring subsystem; step 5)
11. subsystem 3 (tower ground radial subsystem; step 6)

1/Rtotal = 1/Rsubsystem 1 + 1/Rsubsystem 2 + 1/Rsubsytem 3 (4)

Total resistance does not include incidental influence from site fencing, buried fuel tanks, or other objects not included in these calculations. More complex grounding systems, or highly accurate results where other objects exist, will require the assistance of an appropriate engineering consultant. $(\Sigma X)^2$

EXAMPLE OF COMPLEX SYSTEM CALCULATIONS

Assuming a layout as shown in (Figure 20) with the following characteristics:

1. Worst-case soil resistivity measurement (step 1 above) is 4021.5 Ω-cm.
2. Building ground ring (step 4) using four rods, each with a resistance of 15 Ω. Building ground ring subsystem calculates to approximately 4.35 Ω.

3. Tower ground ring (step 5 above) using three rods, each with a resistance of 15 Ω. Tower ground ring subsystem calculates to approximately 5.55 Ω.
4. Tower radial grounding conductor subsystem (step 6 above) as shown in(Figure 20). Total resistance of this subsystem is as follows:
   a. Radial "A" has three ground rods. Resistance of this radial calculates to approximately 5.55 Ω.
   b. Radial "B" has two ground rods. Resistance of this radial calculates to approximately 8.1 Ω.
   c. 4. c. Radial "C" has two ground rods. Resistance of this radial calculates to approximately 8.1 Ω. Tower ground radial calculates to 2.34 Ω, as shown below using the formula provided in step 6:

$$1 / R\text{tower radial} = 2.34\ \Omega = 1/\ (5.55\Omega) + 1/\ (8.1\Omega) + 1/\ (8.1\Omega)$$

Overall system resistance is calculated as follows:
The individual resistances of the three subsystems are noted:

Building ground ring = 4.35 Ω
Tower ground ring subsystem = 5.55 Ω
Tower ground radial = 2.34 Ω

The combined (parallel) resistance of all of the subsystems is now calculated as follows:

$$1/\ R\text{total} = 1.19\ \Omega = 1/\ (4.35\Omega) + 1/\ (5.55\Omega) + 1/\ (2.34\Omega)$$

In this example, the calculated effective overall resistance of the proposed system would be 1.19 Ω.

CHARACTERISTICS OF VERTICAL ELECTRODES

Electrode Material

The material composition of a grounding electrode is based primarily on corrosion and strength (durability) concerns. As far as resistance-to-earth is concerned an electrode could be made of most any metal because the majority of the total resistance occurs in the earth immediately adjacent to

Figure 22. Complex grounding system
A: *Grounding Radials*
B: *Tower Ground Bus Bar and Down Conductor*
C: *Generator Grounding Conductor*
D: *Buried Fuel Tank Grounding Conductor*
E: *External Ground Bus Bar*
F: *Shelter Ground Ring*
G: *Fence Grounding Conductor*
H: *Ground Ring Bonding Conductors (2 minimum)*
I: *Tower Ground Ring*
J: *Grounding Electrodes (Ground Rods)*

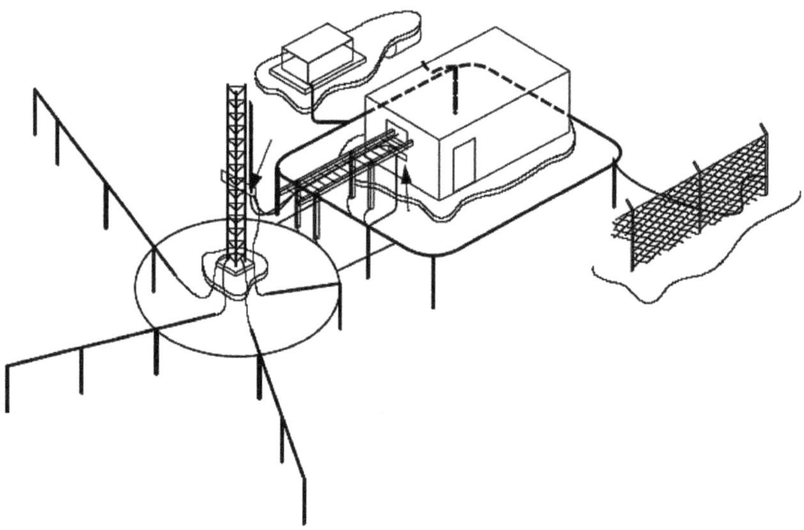

the electrode surface. Corrosion studies have shown that copper performs well in most soils. Unfortunately, solid copper electrodes would be too soft for driving in most soils. As a result, copper is usually used as a cladding material over steel and mates strength with good corrosion performance.

Steel has also proven to be an effective grounding electrode. However, since steel corrodes more readily than copper, steel electrodes are almost always galvanized (coated) with a special formulation of zinc.

Multiple Vertical Electrodes Buried in Earth

There are locations where the objective resistance to earth is less than the value attainable with a single electrode. Among such locations might be the sites of carrier system repeaters, central office switching systems, remote switching terminals, concentrators, etc. The installation of two or more electrodes connected in parallel provides a means of reducing the grounding system resistance. Multiple vertical electrode systems can be installed in a straight-line, circular, square, or rectangular configuration. The electrodes may be interconnected with either insulated or bare buried conductors. The specific grounding system design is usually based on the space available for installation and the desired resistance-to-ground objective.

For multiple vertical electrodes buried in a straight line, grounding systems of this type are typically used at installations have to be completed along a road or within the limits of a narrow right-of-way. Spacing of vertically installed electrodes is important. Close spacing will result in considerable mutual resistance which adds to the overall resultant parallel resistance of a multiple rod installation. Placing vertical ground electrodes with a separation of twice the electrode length will minimize the unwanted effect from the mutual resistance between electrodes on the grounding system as reported by Sunde (1949) and Tagg (1964).

CHARATS FOR GROUNDING SYSTEM DESIGN USING DRIVEN RODS

The aim of such calculation and measurements is to obtain the number of driven rods and weight of the copper used in grounding system.

In practice, it is desirable to drive ground rods deep into the ground to reach more conductive soil. The following equation could be used to calculate the equivalent resistance for n electrode according to IEEE 80 (2000)

$$R_n = \left(\frac{\rho_a}{2n\pi l_2}\right)\cdot\left(\ln\left(\frac{8l_2}{d_2}\right) - 1 + 2k_1\left(\frac{l_2}{\sqrt{A}}\right)\cdot\left(\sqrt{n}-1\right)^2\right) \tag{4}$$

where ρa is the apparent soil resistivity as seen by the ground rod, in case of uniform soil equal ρ1 Ωm, n = number of ground rods placed in parallel area A, l2 and d2 are the length and diameter of the driven rods in meter respectively, K_1 = constants related to the geometry of the system could be obtain from the equation

$$K_1 = 1.41 - (0.04)\cdot X \tag{5}$$

X = length to width ratio

As indication to the cost, the weight of the grounding system copper is calculated based on the length and diameter according to the relation.

$$w = \left[\left(n \cdot \frac{\pi}{4} \cdot d^2 \cdot L\right) + \left(m \cdot L_x \cdot a\right) + \left(z \cdot L_y \cdot a\right)\right] \cdot \rho_c \tag{6}$$

where a is the cross section area of conductors forming the grid, w is the copper weight, d is the conductor diameter, L is electrode length, n=number of electrodes, ρc is the copper density, L_x, L_y are length and width of the grid respectively and m, z are the number of conductor in each direction.

(Figure 21) gives charts for grounding system design by using rods connected together. In case of non-uniform soil, the apparent soil resistivity can be calculated for double layer soil by the relation given in equation 7 according to IEEE 80 (2000) which is in agreement with the experimental measurements.

$$\rho_a = 2 \cdot \left(\rho 2 - \left(\rho 2 - \rho 1\right) \cdot e^{-js}\right) - \left(\rho 2 - \left(\rho 2 - \rho 1\right) \cdot e^{-2js}\right) \tag{7}$$

where

$$J = \delta / 2(d1), \quad \delta = \frac{\ln(\rho 1 / \rho 2) - \ln(0.0176)}{3.5}$$

d1 = the depth of the first layer in meter and s is the distance between two adjacent rods

The flowcharts can be modified to be used in non-uniform case as given in (Figure 21) by calculating the apparent soil resistivity using equation (7).

To obtain driven rods grounding resistance of 3Ω as example in non-uniform soil with ρ1 = 150 Ω.m and ρ2 = 90 Ω.m then ρa = 120 Ω.m this charts could be used to obtain 12 driven rods of 6.5 m with total weight 1044 kg or 20 driven rods of 4.2 m with total weight 1060 kg.

Using computer program to compute R_g with the variation of n, L and ρ and store the result in 3D matrix as shown in (Figure 24).

Figure 23. Charts for grounding system design by using rods connected together

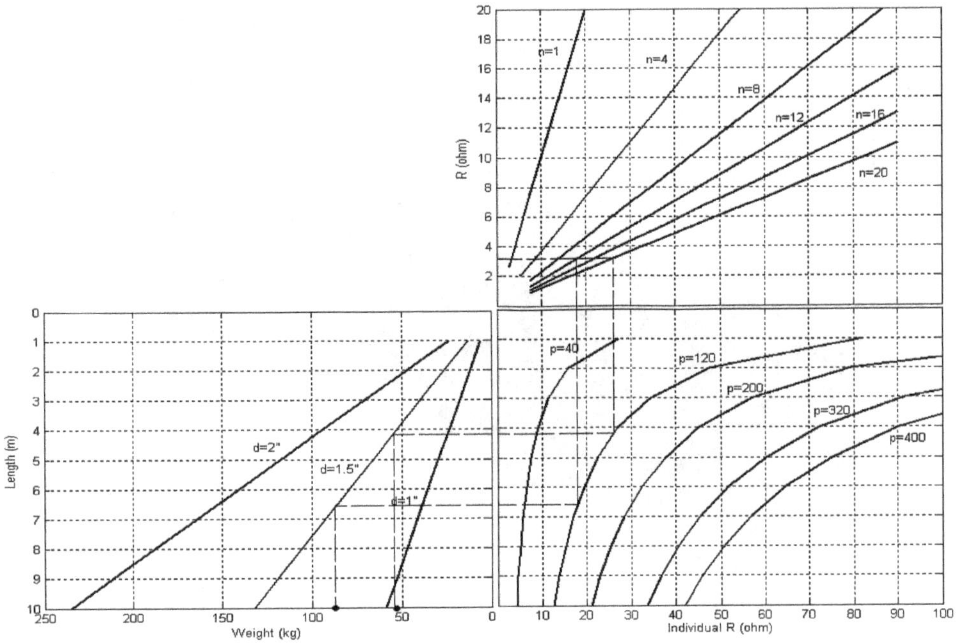

Figure 24. 3D matrix to store Rn with variations of n, L and ρ

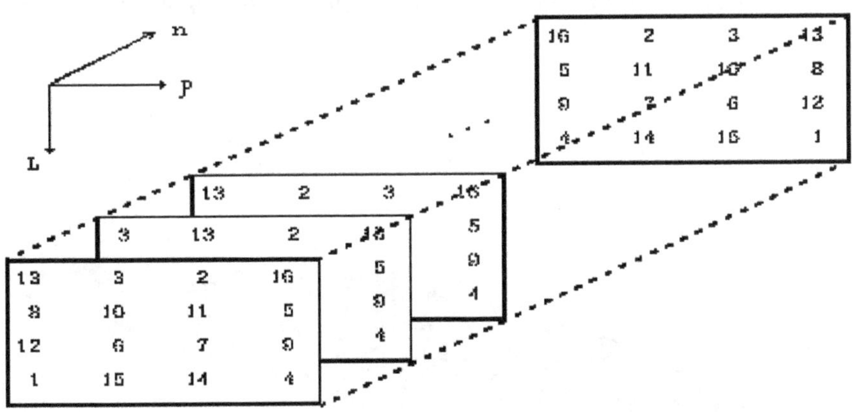

(Figure 25) and (Figure 26) show 3D Charts for grounding system design by using driven rods for different soil resistivity and different numbers of driven rods.

Figure 25. 3D charts for grounding system design by using driven rods (n = 1, 3 and 7)

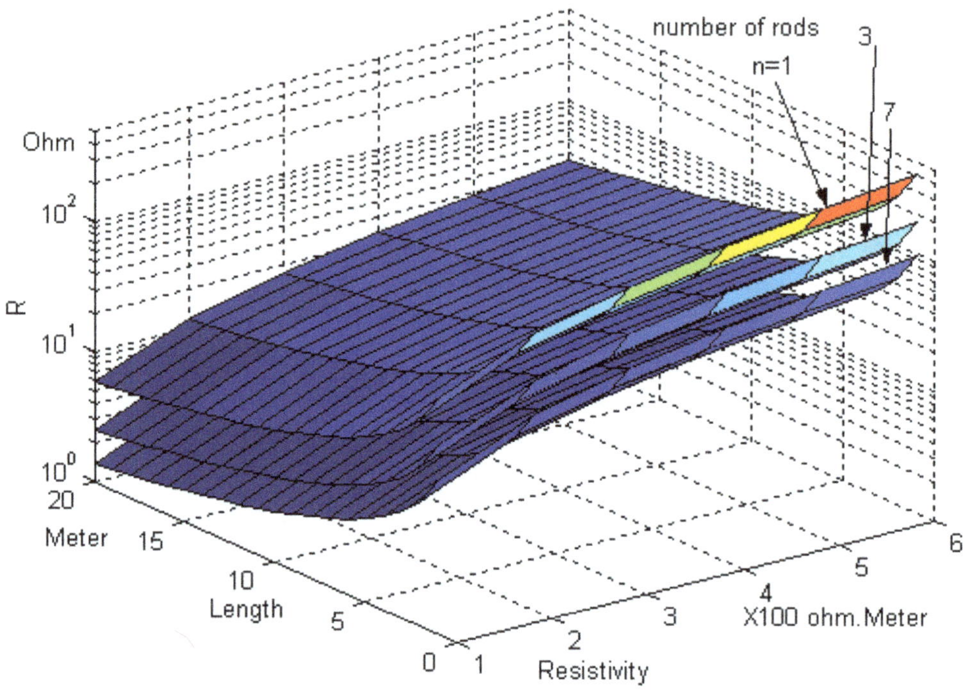

Figure 26. 3D charts for grounding system design by using driven rods (n = 8, 12 and 16)

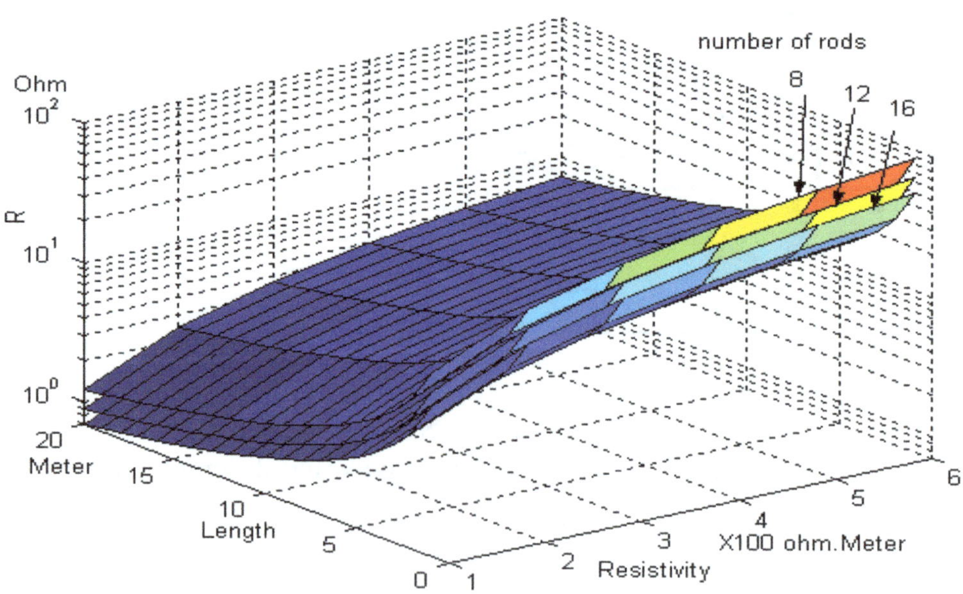

COLOR CODE

Color code technique has been developed to calculate the design parameters of the grounding rods in uniform and non-uniform soil and does not require the use of complex mathematics and time consuming computations and it gives accurate results.

Compute Rn with all variations of L and ρ and present the resistance in pseudo color graph as shown in (Figures 25 and 26).

The values of the resistance R specify the color in each cell of the plot (4-bit color), the smallest and largest elements of matrix are assigned the first and last colors given in the color table; colors for the remainder of the elements in matrix are determined by table-lookup within the remainder of the color table.

The X-axis of (Figures 25 and 26) represents the soil resistivity values (scale 5) in Ω.m and the Y-axis represents the rod length in meter.

(Figure 25) presents 4 rods system and (Figure 26) presents the 7 rods system, color bar located in the right side of each figure to indicate each color value (table-lookup) for example in (Figure 24) white color = 2 Ω, gray color = 4 Ω, red color = 6 Ω, … etc.).

In (Figure 25) white color = 1.3 Ω, gray color = 2.6 Ω, red color = 3.9 Ω, … etc.).

To obtain grounding system with resistance of 4 Ω using 4 driven rods, (Figure 25) should be used, from the color bar in the right side gray color = 4 Ω so all gray color area in the graph gives the resistance needed i.e. for soil resistivity = 50 Ω.m, 4 rods with 4 m length are required and for soil resistivity = 100 Ω.m, 4 rods with 9 m length are required. Chart employs seven rods grounding system is represented in (Figure 25).

For example; to obtain grounding system with resistance close to 3 Ω, from the color bar in the right side red color = 2.6-3.9 Ω so all red color area in the graph gives the resistance needed i.e. for soil resistivity = 75 Ω.m, 7 rods with 3.5 m length are required.

(Figure 27) presents the colored parametric surface defined by four matrix arguments (X, Y, Z and color). The range of the matrix determines the color scaling.

Figure 27. Pseudo color graph for 4 rods grounding system

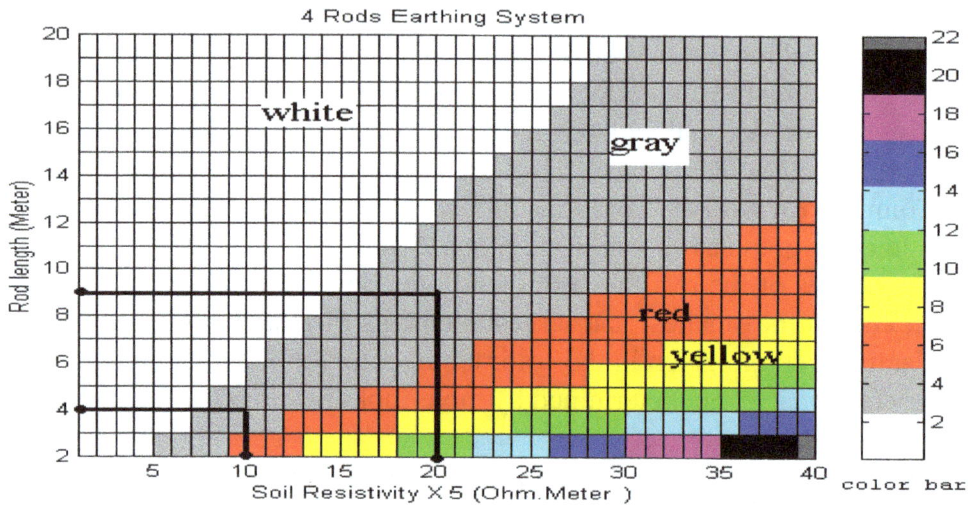

Figure 28. Pseudo color graph for 7 rods grounding system

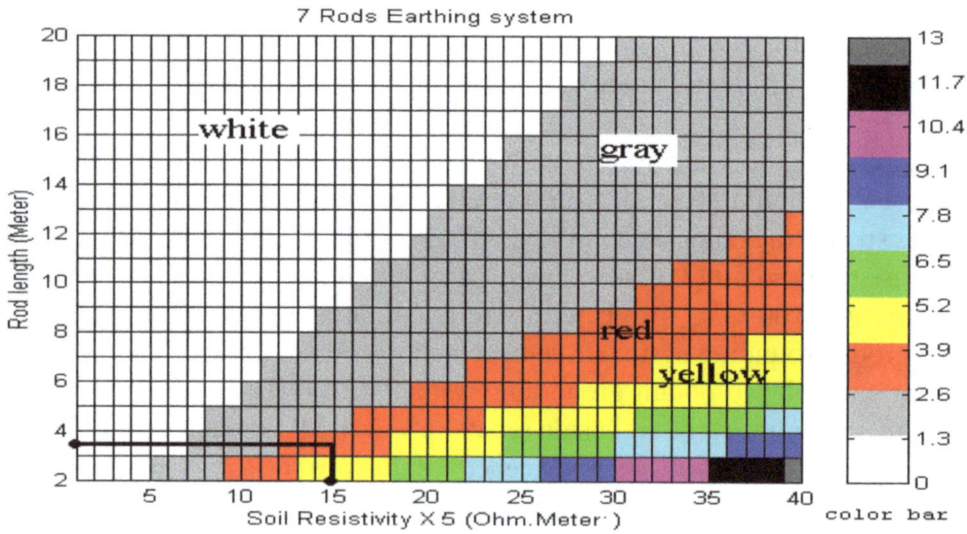

DETERMINATION OF ELECTRODE LOCATION

The low-resistance earth electrode depends upon the soil resistivity. There are two approaches to picking the grounding location as reported by Mugger (2014):

Figure 29. 3D surface color graph for 7 rods grounding system

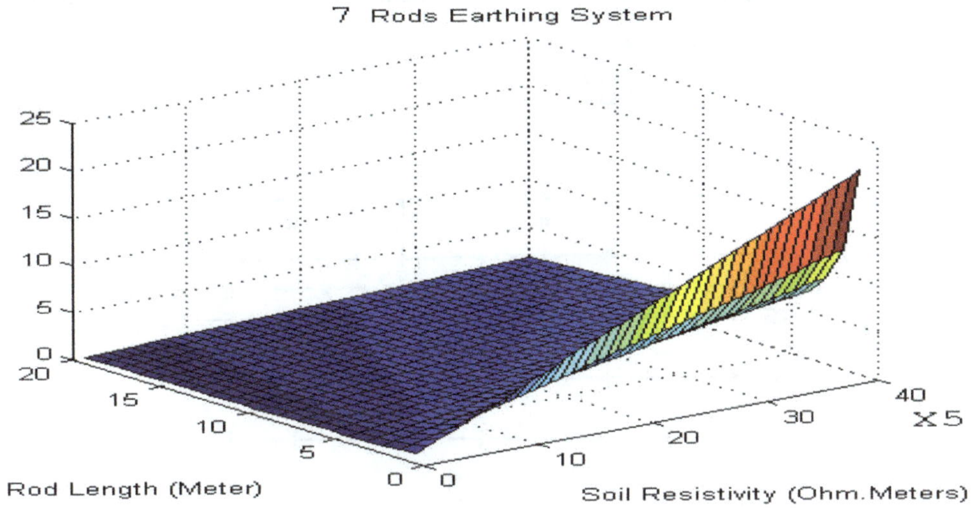

1. Drive rods in various locations to such depths as may be required and test their resistances while they are being driven.
2. Measure the earth resistivity before driving ground rods. Then calculate the number and length of rods required.

To get a low-resistance electrode in an unfavorable location, lay out straight lines 10 ft apart, covering the area. Drive four stakes 10 ft apart, but not more than 6 in. deep, along a line a-b-d-c, as shown in (Figure 28). Measure the grounding resistance between stakes b and c, using the method described for earth resistivity. Then, shift the stakes along the line in question to points b-c-d-e, c-d-e-f, etc. and test until the entire line has been covered. Next, move to the next line and repeat the process until the whole chosen area has been covered. The location giving the lowest value for grounding resistance has the lowest specific resistance for the soil to the chosen depth of 10 ft. The spot is likely to give you the best earth electrode.

If you want results affected by the average earth resistivity to a depth of 20 ft, repeat the survey on lines 20 ft apart and with stakes spaced 20 ft apart. Such surveys do not require much time and can pay off in ensuring a good grounding system. Another way is to drive rods or pipes in various locations to such depths as may prove practicable, testing their resistance while they are being driven. In this manner, you can usually tell at once when moisture or other good conducting earth is reached. However, the work involved is apt to be much more than with the first method.

Figure 30. Method of prospecting for best earth electrode location to a depth a; location giving lowest reading on the Megger ground tester is the most desirable.

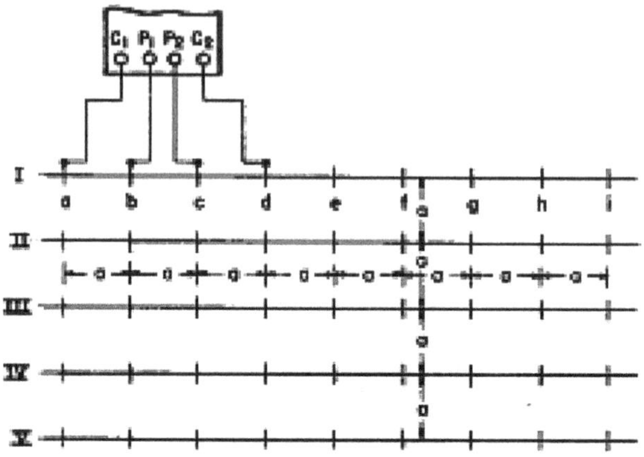

DESIGN OF GROUNDING SYSTEM IN UNIFORM AND NON-UNIFORM SOILS

Design of Grounding System

There are two main design goals to be achieved by any substation ground system under normal as well as fault conditions. These goals are

1. To provide means to dissipate electric currents into the earth without exceeding any operating and equipment limits.
2. To assure that a person in the vicinity of grounded facilities is not exposed to the danger of critical electric shock.

The design procedures are aimed at achieving safety from dangerous step and touch voltages within a substation. It is possible for transferred potentials to exceed the GPR of the substation during fault conditions. Thus, the design procedure described here is based on assuring safety from dangerous step and touch voltages within, and immediately outside, the substation fenced area. Because the mesh voltage is usually the worst possible touch voltage inside the substation (excluding transferred potentials), the mesh voltage will be used as the basis of this design procedure.

Step voltages are inherently less dangerous than mesh voltages. If, however, safety within the grounded area is achieved with the assistance of a high resistivity surface layer (surface material), which does not extend outside the fence, then step voltages may be dangerous. In any event, the computed step voltages should be compared with the permissible step voltage after a grid has been designed that satisfies the touch voltage criterion.

Traditional Steps to Design the Ground Grid

1. The property map and general location plan of the substation should provide good estimates of the area to be grounded. A soil resistivity test, will determine the soil resistivity profile and the soil model needed (that is, uniform or two-layer model).
2. The conductor size is determined. The fault current $3I_0$ should be the maximum expected future fault current that will be conducted by any conductor in the grounding system, and the time t_c, should reflect the maximum possible clearing time (including backup).
3. The tolerable touch and step voltages should be calculated. The choice of time, t_s, is based on the judgment of the design engineer.
4. The preliminary design should include a conductor loop surrounding the entire grounded area, plus adequate cross conductors to provide convenient access for equipment grounds, etc. The initial estimates of conductor spacing and ground rod locations should be based on the current I_G and the area being grounded.
5. Estimates of the preliminary resistance of the grounding system in uniform soil should be calculated. For the final design, more accurate estimates of the resistance may be desired. Computer analysis based on modeling the components of the grounding system in detail can compute the resistance with a high degree of accuracy, assuming the soil model is chosen correctly.
6. The current I_G should be calculated. To prevent over design of the grounding system, only that portion of the total fault current, $3I_0$, that flows through the grid to remote earth should be used in designing the grid. The current I_G should, however, reflect the worst fault type and location, the decrement factor, and any future system expansion.
7. If the GPR of the preliminary design is below the tolerable touch voltage, no further analysis is necessary. Only additional conductor required to provide access to equipment grounds is necessary.

8. The calculation of the mesh and step voltages for the grid as designed can be done by the approximate analysis techniques for uniform soil, or by the more accurate computer analysis techniques.
9. If the computed mesh voltage is below the tolerable touch voltage, the design may be complete. If the computed mesh voltage is greater than the tolerable touch voltage, the preliminary design should be revised.
10. If both the computed touch and step voltages are below the tolerable voltages, the design needs only the refinements required to provide access to equipment grounds. If not, the preliminary design must be revised.
11. If either the step or touch tolerable limits are exceeded, revision of the grid design is required. These revisions may include smaller conductor spacing, additional ground rods, etc.
12. After satisfying the step and touch voltage requirements, additional grid and ground rods may be required. The additional grid conductors may be required if the grid design does not include conductors near equipment to be grounded. Additional ground rods may be required at the base of surge arresters, transformer neutrals, etc. The final design should also be reviewed to eliminate hazards due to transferred potential and hazards associated with special areas of concern.

Refinement of Preliminary Design

If calculations based on the preliminary design indicate that dangerous potential differences can exist within the substation, the following possible remedies should be studied and applied where appropriate:

1. **Decrease Total Grid Resistance:** A decrease in total grid resistance will decrease the maximum GPR and, hence, the maximum transferred voltage. The most effective way to decrease ground grid resistance is by increasing the area occupied by the grid. Deep driven rods or wells may be used if the available area is limited and the rods penetrate lower resistivity layers. A decrease in substation resistance may or may not decrease appreciably the local gradients, depending on the method used.
2. **Closer Grid Spacing:** By employing closer spacing of grid conductors, the condition of the continuous plate can be approached more closely. Dangerous potentials within the substation can thus be eliminated at a cost. The problem at the perimeter may be more difficult, especially at a small substation when resistivity is high. However, it is usually possible, by burying the grid ground conductor outside the fence line, to ensure

that the steeper gradients immediately outside this grid perimeter do not contribute to the more dangerous touch contacts. Another effective and economical way to control gradients is to increase the density of ground rods at the perimeter. This density may be decreased toward the center of the grid. Another approach to controlling perimeter gradients and step potentials is to bury two or more parallel conductors around the perimeter at successively greater depth as distance from the substation is increased. Another approach is to vary the grid conductor spacing with closer conductors near the perimeter of the grid. This is investigated by AIEE Working Group on Substation Grounding Practices (1954) and Sverak (1976).

3. **Diverting a Greater Part of the Fault Current to Other Paths:** By connecting overhead ground wires of transmission lines or by decreasing the tower footing resistances in the vicinity of the substation, part of the fault current will be diverted from the grid. In connection with the latter, however, the effect on fault gradients near tower footings should be weighed.
4. **Limiting Total Fault Current:** If feasible, limiting the total fault current will decrease the GPR and all gradients in proportion. Other factors, however, will usually make this impractical. Moreover, if accomplished at the expense of greater fault clearing time, the danger may be increased rather than diminished.
5. **Barring Access to Limited Areas:** Barring access to certain areas, where practical, will reduce the probability of hazards to personnel.

Use of Computer Analysis in Grid Design

Computer algorithm has been developed based on the previous steps, figure (A.5) in appendix A shows flowchart for the program. The accuracy of the computer algorithm is dependent on how well the soil model and physical layout reflect actual field conditions. There are several reasons that justify the use of more accurate computer algorithms in designing the grounding system. These reasons include

1. Parameters exceed the limitations of the equations.
2. A two-layer or multilayer soil model is preferred due to significant variations in soil resistivity.

3. Uneven grid conductor or ground rod spacing cannot be analyzed using the approximate methods.
4. More flexibility in determining local danger points may be desired.
5. Presence of buried metallic structures or conductor not connected to the grounding system, which introduces complexity to the system. (Figure 28) contains a summary of the design parameters used in the design procedure.

Figure 31. Design procedure program flowchart (IEEE 80)

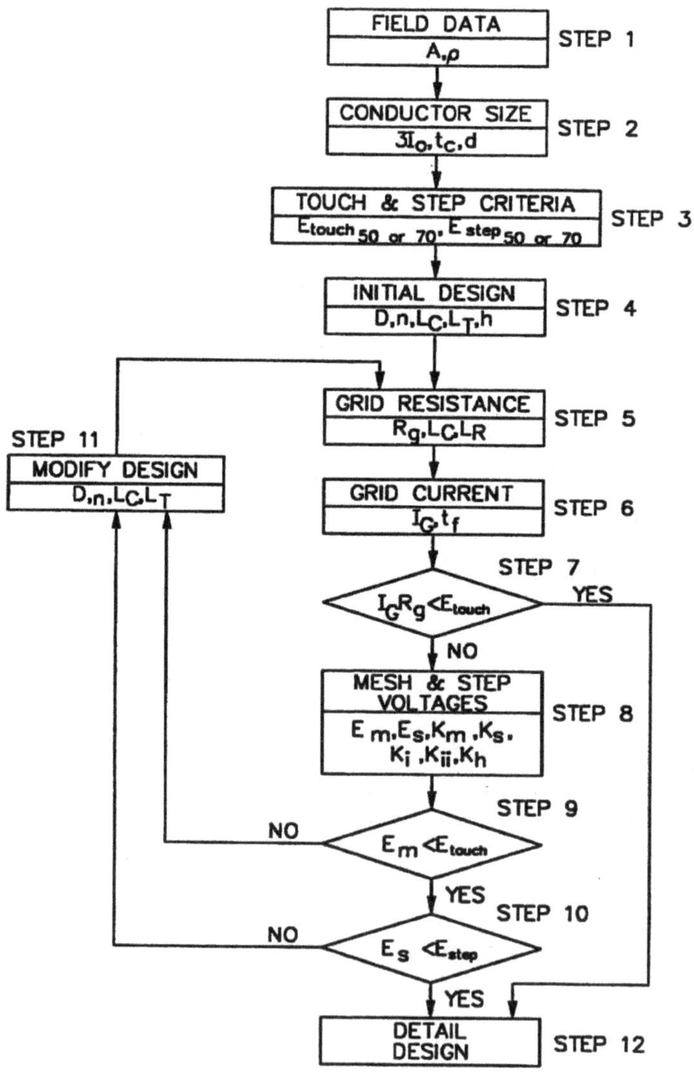

Table 1. Test cases parameters

Parameters	Case 1	Case 2	Case 3	Case 4
Fault duration *tf*	0.5	0.5	0.5	0.5
Zero sequence equivalent system impedance Z0	10 + 40i	10 + 40i	10 + 40i	10 + 40i
Positive sequence equivalent system impedance Z1	4 + 10i	4 + 10i	4 + 10i	4 + 10i
Negative sequence equivalent system impedance Z1	4 + 10i	4 + 10i	4 + 10i	4 + 10i
Available width of the grounding area	63	70	120	42
Available length of the grounding area	84	70	120	90
Current division factor *Sf*	0.6	0.6	0.6	0.6
Decrement factor	1	1	1	1
Line-to-line voltage at worst-fault location (kV)	115	115	66	66
Soil resistivity ρ	400	40	600	4.71
Crushed rock resistivity (wet) ρ_s	2500	3000	2000	8000
Thickness of crushed rock surfacing *hs*	0.1	0.1	0.1	0.1
Depth of grid burial *h*	0.5	0.5	0.5	0.5

Test and Verify the Computer Program

Different cases have been tested on the program; four of them are mentioned below, case number one matched the case of (IEEE STD 80-2000), that is reported in annex B, case number four used actual field data reported by Egyptian Ministry of electrical power report, (2005), cases number two and three are test cases. Table 1 shows the parameters used to test the computer program.

Design Results for Case One

Tables 2 and 3 show the program results for case number one.

Design number 5 is acceptable designs, because the mesh voltage (796.09V) is below the maximum safe touch voltage limit (837.59V)

Table 2. Design case number one

Estep70 (V)	2684.3
Etouch70 (V)	837.59

Table 3. Design case number one

Item	Design Iteration Number				
	1	2	3	4	5
Number of parallel conductor in a direction	4	7	10	13	16
Number of parallel conductor in b direction	5	9	13	17	21
distance between conductor in a or b direction	21	10.5	7	5.25	4.2
Total length of grounding system conductors	651	1155	1659	2163	2667
Grid resistance	3.0368	2.7687	2.6635	2.6073	2.5724
The mesh voltage	2347.4	1518.9	1158.9	944.02	796.09
Step voltage	191.31	201.83	222.79	247.02	272.46
GPR	5794	5282.4	5081.7	4974.5	4907.8
situation	not Safe	not Safe	not Safe	not Safe	Safe

Design Results for Case Two

Table 4 and 5 show the program results for case number two.

The designs are acceptable, because the mesh voltages are below the maximum safe touch voltage limit, and this is due to the lower soil resistivity and the wide design areas

Table 4. Design case number two

Estep70 (V)	2994.8
Etouch70 (V)	915.23

Table 5. Design case number two

Item	Design Iteration Number
Number of parallel conductor in a direction	4
Number of parallel conductor in b direction	4
Distance between conductor in a or b direction	23.333
Total length of grounding system conductors	560
Grid resistance.	0.323
The mesh voltage	251.26
Step voltage	18.896
GPR	616.3
Situation	Safe

Design Results for Case Three

Table 6 and 7 show the program results for case number three.

Designs number 7 is acceptable design, because the mesh voltage is below the maximum safe touch voltage limit.

Design Results for Case Four

Tables 8 and 9 show the program results for case number four.

Preliminary design is acceptable designs, because the mesh voltages is below the maximum safe touch voltage limit, and this is due to the lower soil resistivity as the station closer to the sea and the wide design areas.

Table 6. Design case number three

Estep70 (V)	2307.6
Etouch70 (V)	743.42

Table 7. Design case number three

Item	**Design Iteration Number**						
	1	**2**	**3**	**4**	**5**	**6**	**7**
Number of parallel conductor in a direction	4	5	6	7	8	9	10
Number of parallel conductor in b direction	4	5	6	7	8	9	10
distance between conductor in a or b direction	40	30	24	20	17.143	15	13.333
Total length of grounding system conductors	960	1200	1440	1680	1920	2160	2400
grid resistance.	2.8406	2.7156	2.6323	2.5728	2.5281	2.4934	2.4656
The mesh voltage	1416.8	1188.6	1035.6	924.63	839.68	772	716.43
Step voltage	85.912	82.416	80.867	80.392	80.584	81.227	82.192
GPR	3110.4	2973.5	2882.3	2817.1	2768.2	2730.2	2699.8
situation	Not Safe	Not Safe	Not Safe	Not Safe	Not Safe	Not Safe	Safe

Table 8. Design case number four

Estep70 (V)	7574
Etouch70 (V)	2060

Table 9. Design case number four

Item	Design Iteration Number
	1
Number of parallel conductor in a direction	8
Number of parallel conductor in b direction	16
distance between conductor in a or b direction	6
Total length of grounding system conductors	1392
Grid resistance.	0.037043
The mesh voltage	10.036
Step voltage	2.2929
GPR	40.561
Situation	Safe

UNEQUALLY SPACED DESIGN TO MINIZE THE COST

In this case the aim of design optimization is to reduce the cost of the overall grounding system, this could be achieved by reorganizing the grid conductor in unequally spaced manner to decrease the total length of grid conductors without increasing the mesh voltage and this could save about 34% of grounding grid material. It also costs less to install the grid; this is discussed by Hung, Chen and Yan (1995). A set of formulae for determining the conductor spacing is derived.

In (Figure 30), L_1 and L_2 are the side lengths of the grid: n_1 is the number of conductors in parallel with the x axis, and n_2 is the number of conductors in parallel with the y axis; k_1 and k_2 are the total number of segments of a conductor which is divided by cross conductors In (Figure 30), k_1 is the number of segments of the conductor in parallel with the x axis, so $k_1 = n_2$-l; k_2 is the number of segments of the conductor in another direction. So $k_2 = n_1-1$. i is the segment sequence; L_{ik} is the length of the ith segment. It has been found that unequal grid conductor spacing depends on the segment number (k_1 or k_2), the segment sequence (i), and the grid side length (L_1 or L_2). Their relationships are found as follows. If a conductor is divided by k segments

Figure 32. Unequally spaced grounding grid (K1 = n2-1; K2 = n1-1)

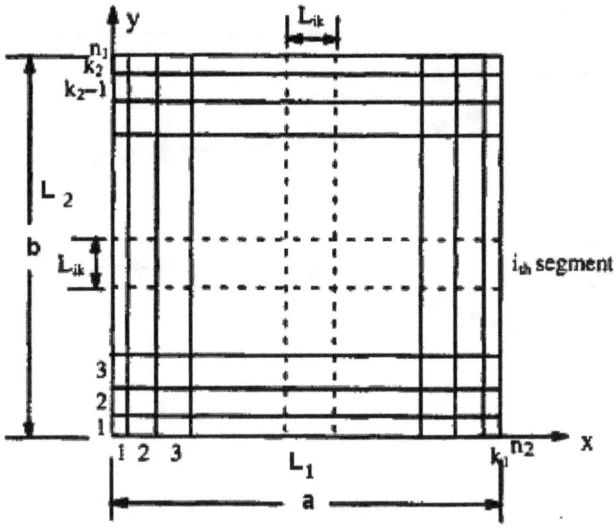

the length of the ith segment, L_{ik}, is expressed in percent of the total length of the conductor, as reported by Hung, Chen and Yan (1995):

$$S_{ik} = \frac{L_{ik}}{L} \cdot 100\% \tag{8}$$

Taking into account the symmetry of the grids, we have

$$i \leq \begin{pmatrix} \dfrac{k}{2} & for - k - even \\ \dfrac{k+1}{2} & for - k - odd \end{pmatrix}$$

where L = L_1 or L_2; k = k_1 or k_2 depending on whether the conductor is parallel to the x or y axis.

S_{ik} calculated grid conductor spacing suitably adjusted so that the mesh potentials above the grid remain the same. Relationship can be expressed as

$$S_{ik} = b_1 \cdot e^{-b_2 i} + b3 \tag{9}$$

The coefficients b1, b2 and b3 are dependent on k as shown in (Figure 30).

Figure 33. Relations between b and k

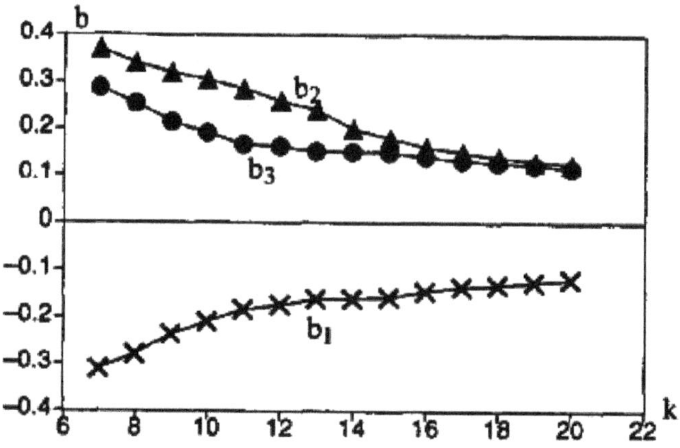

Procedure to Design an Unequally Spaced Grid

A simple design procedure can be described as follows as reported by Hung, Chen and Yan (1995):

1. Calculate the maximum allowable touch voltage Etouch70
2. Calculate the total conductor length, L, of the equally spaced grid:

$$L = \frac{\eta.\rho.I_s}{E_{touch70}} \tag{10}$$

where the value of the empirical coefficient, η, is equal to 1.7.

3. Determine the total number, n, of conductors of the equally spaced grid
   a. Square grounding grid:

$$n = \frac{L}{\sqrt{A}} \tag{11}$$

where A, is the grid area

b. Rectangular grounding grid:

$$n_1 = \frac{L + L_1 - L_2}{2L_1}, \ n_2 = \frac{L + L_2 - L_1}{2L_2}, \ n = n_1 + n_2 \tag{12}$$

where L_1 and L_2 are the grid side lengths (m). n_1 and n_2 are the number of parallel conductors in each direction of the equally spaced grid.

4. Calculate the percentage, λ, of the saved grounding grid material

Keeping the safe parameters, E_{touch} and E_{step} below but close to those of the equally spaced grid, unequal spacing can save grid material (%) as reported by Hung, Chen and Yan (1995):

$$\lambda = \frac{\alpha.\frac{\beta}{100}}{1 + \alpha.\frac{\beta}{100}} 100\% \tag{13}$$

The values of empirical coefficients α, β change with the total number, n, of conductors:

$$\alpha = \begin{pmatrix} 121.2.e^{-0.4n} + 1.4864 & 8 \leq n \leq 20 \\ 121.2.e^{-0.3n} + 1.2664 & 20 \leq n \leq 30 \\ 121.2.e^{-0.4n} + 1.14 & n \succ 30 \end{pmatrix} \tag{14}$$

$$\beta = -92.6.e^{-0.07n} + 66.337 \tag{15}$$

5. Determine the total number, n', of conductors of the unequally spaced grid

a. Square grounding grid:

$$n' = n(1 - \lambda) \tag{16}$$

b. Rectangular grounding grid:

$$n_1' = n_1(1-\lambda)$$

$$n_2' = n_2(1-\lambda) \quad (17)$$

$$n' = n_1' + n_2'$$

where n_1' and n_2' are the number of the parallel conductors in each direction of the unequally spaced grid; and the values of n, n_1 and n_2 are obtained from equations. 11 and 12.

6. Determine unequal grid conductor spacing using equations. 8 and 9.

Based on the previous steps computer program (Figure 32) shows the flowchart of the program. Using case number one and three from Table 1 to test the program

Figure 34. Improved design procedure program flowchart

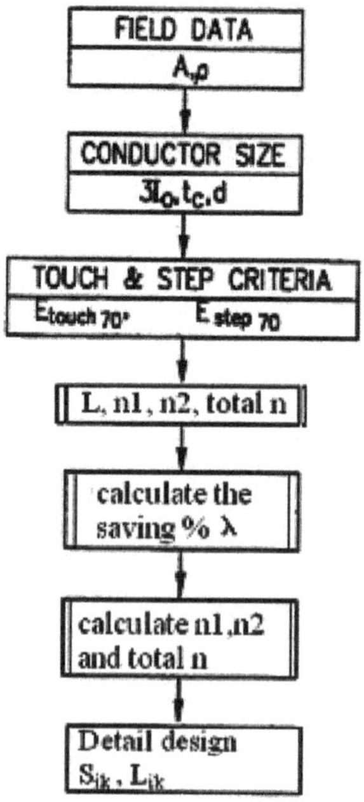

Examples of Design an Unequally Spaced Grid

Design Results for Case One

Table 10 and 11 show the program results for case number one.

Design Results for Case Three

Table 13, 14 and 15 show the program results for case number three.

Table 10. Design case number one

Item	Design
Number of parallel conductor in a direction	6
Number of parallel conductor in b direction	8
Total length of grounding system conductors	1050
Situation	Safe

Table 11. Design case number one

L_{1k1}	L_{2k1}	L_{3k1}	L_{4k1}	L_{5k1}
7 m	12.9 m	16.9 m	12.9 m	7 m

Table 12. Design case number one

L_{1k2}	L_{2k2}	L_{3k2}	L_{4k2}	L_{5k2}	L_{6k2}	L7k2
6 m	11.6 m	15.5 m	18.2 m	15.5 m	11.6 m	6 m

Table 13. Design case number three

Item	Design
Number of parallel conductor in a direction	4
Number of parallel conductor in b direction	4
Total length of grounding system conductors	960
Situation	Safe

Table 14. Design case number three

L_{1k1}	L_{2k1}	L_{3k1}
32 m	56 m	32 m

Table 15. Design case number three

L_{1k2}	L_{2k2}	L_{3k2}
32 m	56 m	32 m

Ackerman, Sen and Oertli (2013), reported how to Designe safe and reliable grounding grid in AC substations with poor soil resistivity and Gross and Wallman (1961), explained the grounding grid resistances of small and large substations.

REFERENCES

Ackerman, A., Sen, P. K., & Oertli, C. (2013). Designing safe and reliable grounding in AC substations with poor soil resistivity: An interpretation of IEEE Std. 80. *IEEE Transactions on Industry Applications*, *49*(4), 1883–1889. doi:10.1109/TIA.2013.2256092

AIEE Working Group on Substation Grounding Practices. (1954). Application guide on methods of substation grounding. *AIEE Transactions on Power Apparatus and Systems*, *11*, 271–278.

Appendix-B-Motorola-R56-09-01-05.pdf. (n. d.). Retrieved from http://www.lmrsitestandard.com/wp-content/blogs.dir/2/files/2012/09/

Egyptian Ministry of electrical power report. (2005). Electrical safety report.

United States department of agriculture, rural electrification administration. (n. d.). Electrical Protection Grounding Fundamentals. Retrieved from www.iesalc.org/?wpfb_dl=251

Gross, E. T. B., & Wallman, E. J. (1961). Grounding grid resistance of small and large substations. *Electrical Engineering*, *80*(3), 198–199. doi:10.1109/EE.1961.6433685

Hung, L., Chen, X., & Yan, H. (1995). Study of Unequally Spaced Grounding Grid. *IEEE Transactions on Power Delivery*, *10*(2).

IEEE. (2000). 80-2000 Guide for Safety in AC Substation Grounding.

IEEE. (2014). 837-2014 Standard for Qualifying Permanent Connections Used in Substation Grounding. Retrieved from https://standards.ieee.org/findstds/standard/837-2014.html

Megger. (n. d.). A practical guide to earth resistance testing. Retrieved from https://portalvhds963slh4m3fqg2.blob.core.windows.net/megger.../GettingDownToEa

Sunde, E. D. (1949). *Earth Conduction Effects in Transmission System, book. English*. New York, Toronto, London: D. Van Nostrand Company.

Sverak, J. G. (1976). Optimized grounding grid design using variable spacing technique. *IEEE Transactions on Power Apparatus and Systems*, *95*(1), 362–374. doi:10.1109/T-PAS.1976.32113

Tagg, G. F. (1964). *Earth Resistances*. London: George Newness Limited.

Chapter 3 Grounding Connections

ABSTRACT

This chapter sheds light on the following: TT system (earthed neutral), automatic disconnection for TT system, TN system (exposed conductive parts connected to the neutral), TN-C system, TN-S system- TN-C-S system, IT system (isolated or impedance-earthed neutral), IT system (impedance-earthed neutral), grounding transformer connection and design, grounding of industrial and commercial generators, Zigzag grounding transformer earthed systems (solid grounding, resistance grounding, reactance grounding, resonant grounding (ground-fault neutralizer), location of system grounding points selection and grounding locations specified by the NEC and multiple power sources. The chapter contains also the different methods used in determining the levels of short circuit current. It contains also details about the grounding through resistances for transformers and generators and reactances for overhead transmission lines.

GROUNDING CONNECTIONS ACCORDING TO STANDARDS

International standards (IEC 60364) clearly define the various elements of grounding connections. The different grounding schemes (often referred to as the type of power system or system grounding arrangements) described characterize the method of grounding the installation downstream of the secondary winding of a MV/LV transformer and the means used for grounding

DOI: 10.4018/978-1-5225-3853-0.ch003

the exposed conductive-parts of the LV installation supplied from it. The following terms are definitions of standardized grounding schemes.

TT System (Earthed Neutral)

One point at the supply source is connected directly to earth. All exposed- and extraneous-conductive-parts are connected to a separate earth electrode at the installation. This electrode may or may not be electrically independent of the source electrode. The two zones of influence overlap without affecting the operation of protective devices.

TT system is technique for the protection of persons. The exposed conductive parts are earthed and residual current devices (RCDs) are used. It is used also in the interruption for the first insulation fault.

The TT system is the simplest solution to design used in installations supplied directly by the public LV distribution network. It does not require continuous monitoring during operation (a periodic check on the RCDs may be necessary). The protection is ensured by special devices, the residual current devices (RCD), which also prevent the risk of fire when they are set to ≤ 500 mA. Each insulation fault results in an interruption in the supply of power, however the outage is limited to the faulty circuit by installing the RCDs in series (selective RCDs) or in parallel (circuit selection). Loads or parts of the installation which, during normal operation, cause high leakage currents, require special measures to avoid nuisance tripping, i.e. supply the loads with a separation transformer or use specific RCDs. (Figures 1 and 2) give the characteristics of TT system as reported by Wiki Electrical Installation and Schneider (2016) and Guide to the wiring regulations, 17th Edition IEE wiring regulations (2008).

Figure 1. Characteristics of TT system

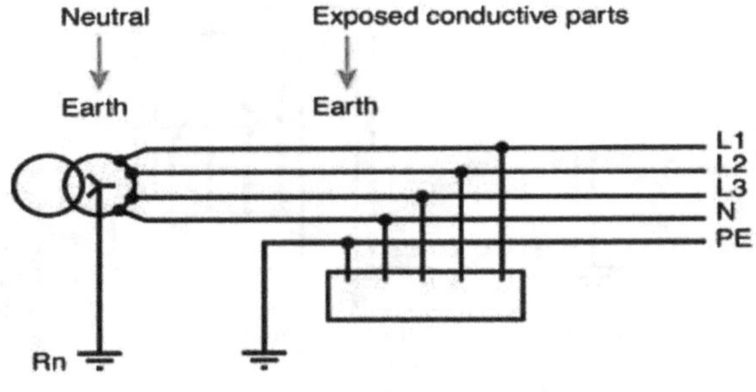

Figure 2. Main characteristics of TT system

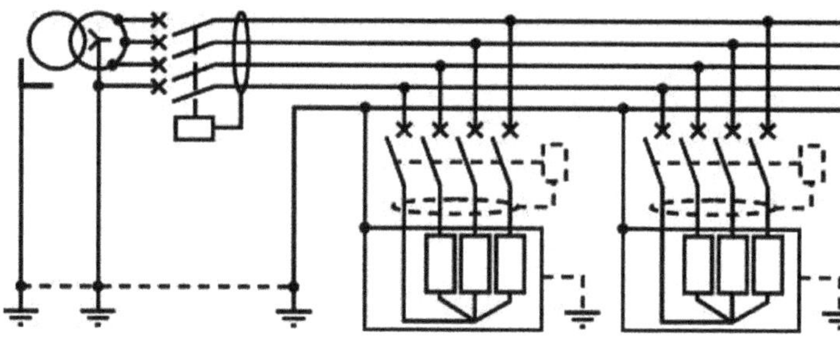

Automatic disconnection for TT system is achieved by RCD having a sensitivity of $I_{\Delta n}$ is less than or equal ($50/R_{A}$), where R_A is the resistance of the earth electrode for the installation, $I_{\Delta n}$ is the rated residual operating current of the RCD. For temporary supplies (to work sites,) and agricultural and horticultural premises, the value of 50 V is replaced by 25 V. An example of TT system is given in (Figure 3), in this figure, the resistance of the earth electrode of substation neutral R_n is 10 Ω. The resistance of the earth electrode of the installation R_A is 20 Ω. The earth-fault loop current $I_d = 7.7$ A and the fault voltage $U_f = I_d \times R_A = 154$ V and therefore dangerous, but $I_{\Delta n} = 50/20 = 2.5$ A so that a standard 300 mA RCD will operate in about 30 ms without intentional time delay and will clear the fault where a fault voltage exceeding appears on an exposed-conductive-part. The upper limit of resistance for an installation grounding electrode which must not be exceeded, for given sensitivity levels of RCDs at U_L voltage limits of 50 V and 25 V. Table 1 gives the maximum resistance of the earth electrode. (Figure 3) gives an example of TT system as reported by Wiki Electrical Installation and Schneider (2016).

Figure 3. Example of TT system

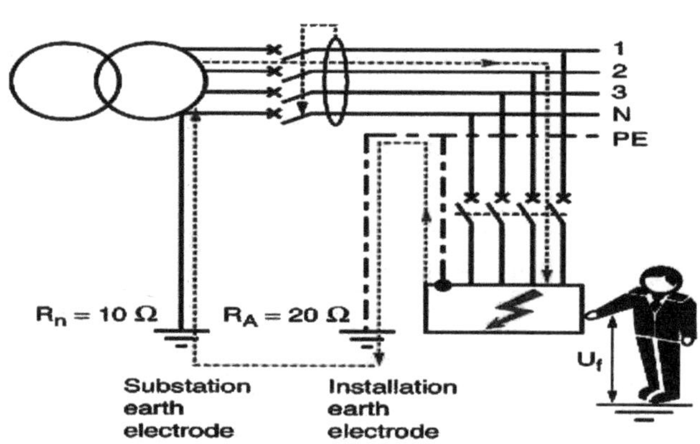

Table 1. Maximum resistance of the earth electrode

$I_{\Delta n}$	(50 V)	(25 V)
3 A	16 Ω	8 Ω
1 A	50 Ω	25 Ω
500 mA	100 Ω	50 Ω
300 mA	166 Ω	83 Ω
30 mA	1666 Ω	833 Ω

TN System (Exposed Conductive Parts Connected to the Neutral)

The several versions of TN systems are shown below:

TN-C System

The neutral conductor is also used as a protective conductor and is referred to as a PEN (Protective Earth and Neutral) conductor. This system is not permitted for conductors of less than 10 mm^2 or for portable equipment. The TN-C system requires an effective equi-potential environment within the installation with dispersed earth electrodes spaced as regularly as possible since the PEN conductor is both the neutral conductor and at the same time carries phase unbalance currents as well as 3rd order harmonic currents (and their multiples).

The PEN conductor must therefore be connected to a number of earth electrodes in the installation. Caution: In the TN-C system, the “protective conductor” function has priority over the “neutral function”. In particular, a PEN conductor must always be connected to the grounding terminal of a

Figure 4. Example of TN-C system

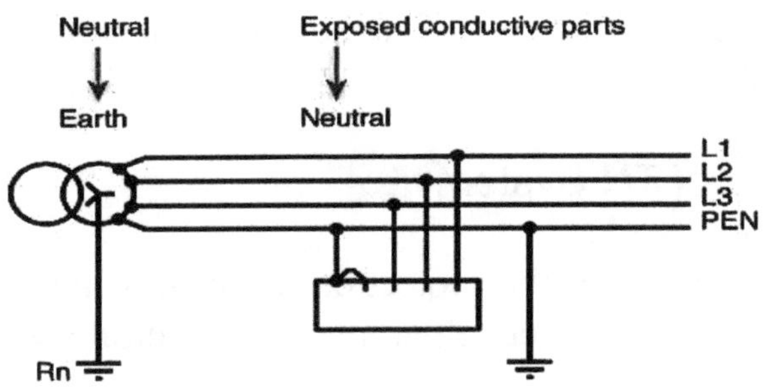

Figure 5. Example of TN-S system

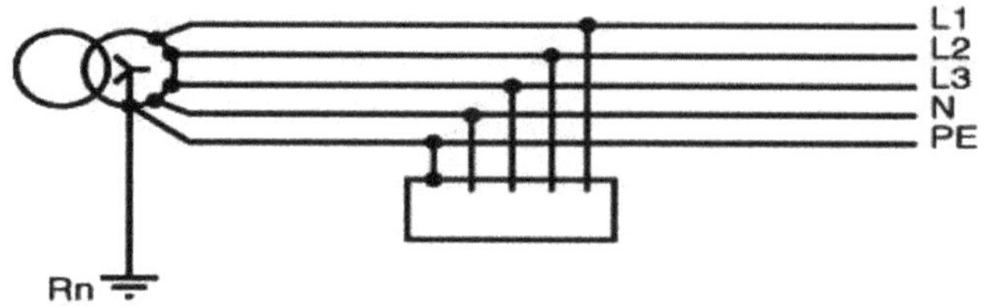

load and a jumper is used to connect this terminal to the neutral terminal. Example of TN-C system is given in (Figure 4) as reported by Wiki Electrical Installation and Schneider (2016)

TN-S System

The TN-S system (5 wires) is obligatory for circuits with cross-sectional areas less than 10 mm^2 for portable equipment. The protective conductor and the neutral conductor are separate. On underground cable systems where lead-sheathed cables exist, the protective conductor is generally the lead sheath. The use of separate PE and N conductors (5 wires) is obligatory for circuits with cross-sectional areas less than 10 mm^2 for portable equipment. Example of TN-S system is drawn in (Figure 5) as reported by Wiki Electrical Installation and Schneider (2016)

TN-C-S System

The TN-C and TN-S systems can be used in the same installation. In the TN-C-S system, the TN-C (4 wires) system must never be used downstream of the TN-S (5 wires) system, since any accidental interruption in the neutral on the upstream part would lead to an interruption in the protective conductor in the downstream part and therefore a danger. Example of TN-C-S system is given in (Figure 6) and connection of the PEN conductor in the TN-C system is shown in (Figure 7) as reported by Wiki Electrical Installation and Schneider (2016)

Characteristics of TN System

The TN system is used for the protection of persons, interconnection and grounding of exposed conductive parts and the neutral are mandatory and

Figure 6. Example of TN-C-S system

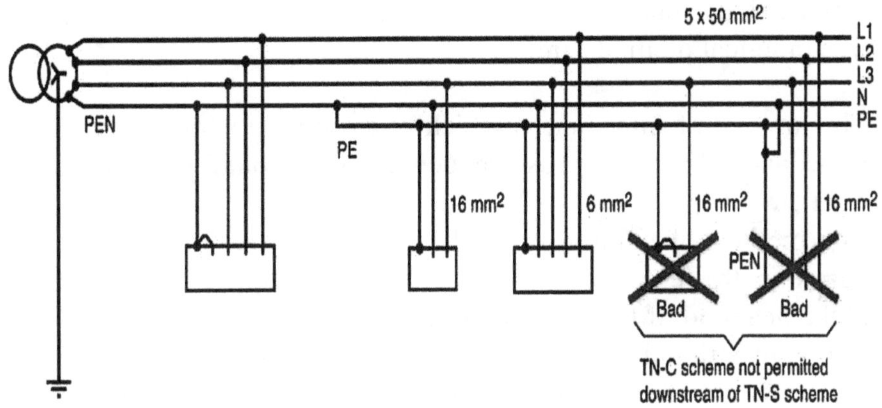

Figure 7. Connection of the PEN conductor in the TN-C system

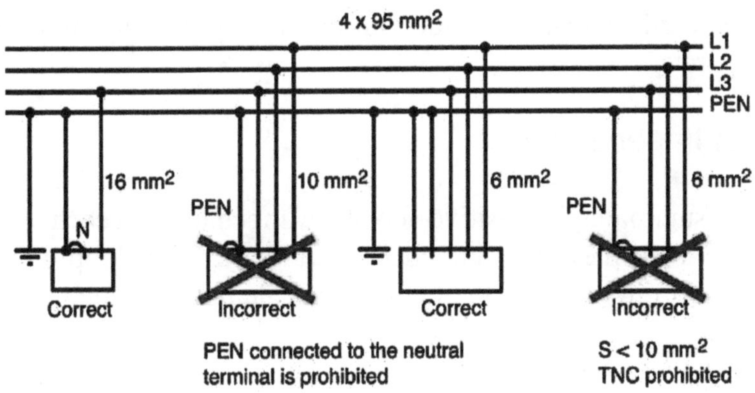

interruption for the first fault using over current protection (circuit- breakers or fuses). The operating technique is for the interruption for the first insulation fault.

Methods of Determining Levels of Short-Circuit Current

In TN-earthed systems, short-circuit to earth will, in principle; always provide sufficient current to operate an over current device. The source and supply mains impedances are much lower than those of the installation circuits, so that any restrictions in the magnitude of earth-fault currents will be mainly caused by the installation conductors, i.e., long wire loops lead to appliances greatly increase the "fault-loop" impedance, with a corresponding reduction of short-circuit current.

Three practical methods are:

1. The method of impedances which based on the summation of all the impedances (positive-phase-sequence only) around the fault loop, for each circuit
2. The method of composition which is an estimation of short-circuit current at the remote end of a loop, when the short-circuit current level at the near end of the loop is known
3. The conventional method that is based on calculating the minimum levels of earth-fault currents, together with the use of tables of values for obtaining rapid results

These methods are only reliable for the case in which the cables that make up the earth-fault-current loop are in close proximity (to each other) and not separated

Method of Impedances

This method summates the positive-sequence impedances of each item (cable, conductor, transformer, etc.) included in the earth-fault loop circuit from which the short-circuit earth-fault current is calculated, using the formula:

$$I = Uo / ((\Sigma R)^{2} + (\Sigma X)^{2})^{0.5} \qquad (1)$$

where

$(\Sigma R)^{2}$ = (the sum of all resistances in the loop)^2 at the design stage of a project.
$(\Sigma X)^{2}$ = (the sum of all inductive reactance in the loop)^2
Uo = nominal system phase-to-neutral voltage.

The application of the method is not always easy, because it supposes knowledge of all parameter values and characteristics of the elements in the loop. In many cases, a national guide can supply typical values for estimation purposes.

Method of Composition

This method permits the determination of the short-circuit current at the end of a loop from the known value of short-circuit at the sending end, by means of the approximate formula:

$$Isc = I\,(Uo/\,(U + Zs\,Isc)) \quad (2)$$

where

Isc = upstream short-circuit current, I = end-of-loop short-circuits current
Uo = nominal system phase voltage and Zs = impedance of loop

Note: In this method, the individual impedances are added arithmetically as opposed to the previous "method of impedances" procedure.

In this method, the resultant calculated current is less than that it would actually flow. If the over current settings are based on this calculated value, then operation of the relay, or fuse, is assured

Conventional Method

The maximum length of any circuit of a TN-earthed installation is:

$$Lmax = (0.8\,Uo\,Sph)\,/\,(\rho\,(1+ m)\,Ia) \quad (3)$$

where:

m = Sph / SPE
SPE = cross-sectional area of the protective conductor concerned in mm^2.
Lmax = maximum length in meters,
Sph = cross-sectional area of the phase conductors of the circuit concerned in mm^2
Uo = phase volts = 230 V for a 230/400 V system,
ρ = resistivity at normal working temperature in ohm-mm^2/meter = 22.5×10^{-3} for copper; = 36×10^{-3} for aluminum.
Ia = trip current setting for the instantaneous operation of a circuit-breaker, or
Ia = the current which assures operation of the protective fuse concerned, in the specified time.

Figure 8. Calculation of L max for TN-earthed system, using the conventional method

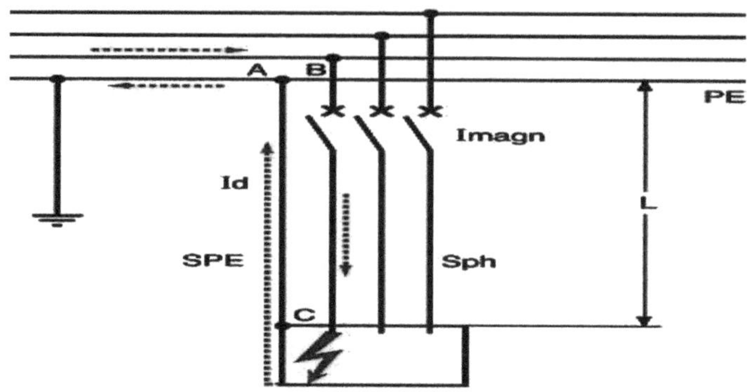

This method is generally considered to be sufficiently accurate to fix the upper limit of cable lengths. The principle of this method based on the assumption that the voltage at the origin of the circuit concerned (i.e. at the point at which the circuit protective device is located) remains at 80% or more of the nominal phase to neutral voltage. The 80% value is used, together with the circuit loop impedance, to compute the short-circuit current. This coefficient takes account of all voltage drops upstream of the point considered. In LV cables, when all conductors of a 3-phase 4-wire circuit are in close proximity (which is the normal case), the inductive reactance internal to and between conductors is negligibly small compared to the cable resistance. This approximation is considered to be valid for cable sizes up to 120 mm^2. (Figure 8) gives how to calculate L max for a TN-earthed system, using the conventional method.

IT System (Isolated or Impedance-Earthed Neutral)

In IT systems, there is no connection between the neutral point of the supply source and safety grounding.

In IT systems, it is recommended to connect the conductive-parts of the installation with an earth electrode. In practice, all circuits have leakage impedance to earth, since no insulation is perfect. In parallel with this (distributed) resistive leakage path, there is the distributed capacitive current path, the two paths together constituting the normal leakage impedance to earth. (Figure 9) gives an example of IT system (isolated neutral) as reported by Wiki Electrical Installation and Schneider (2016)

Figure 9. Example of IT system (isolated neutral)

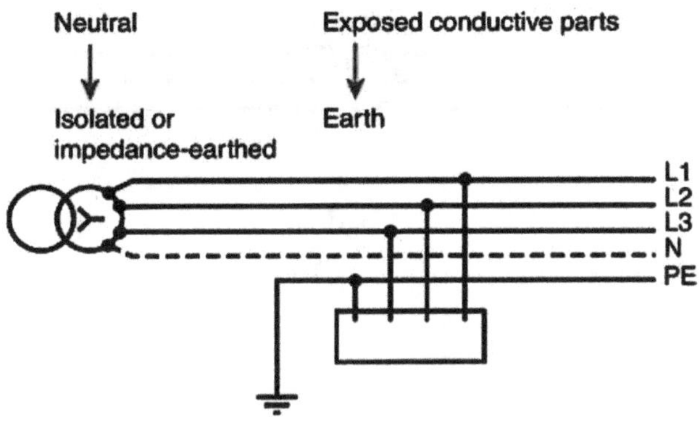

Figure 10. IT system (isolated neutral)

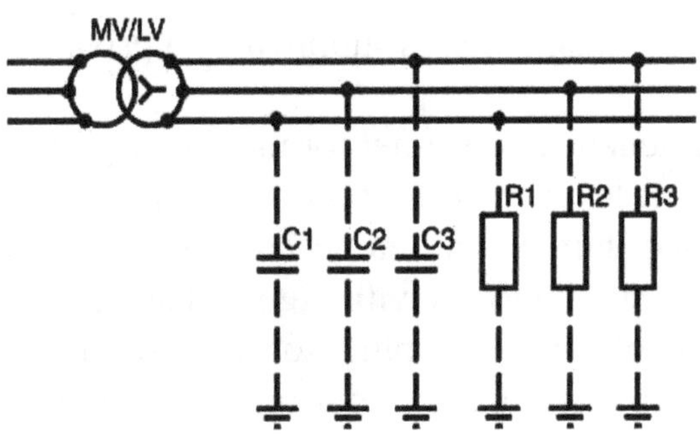

In a LV 3-phase 3-wire system, 1 km of cable will have leakage impedance due to C1, C2, C3 and R1, R2 and R3 equivalent to neutral earth impedance Zct of 3,000 to 4,000 Ω, without counting the filtering capacitances of electronic devices. See (Figure 10).

The Protection technique of IT system is as follows:

- Interconnection and grounding of exposed conductive parts
- Indication of the first fault by an insulation monitoring device (IMD)
- Interruption for the second fault using overcurrent protection (circuit-breakers or fuses). The operating technique of IT system is as follows:
    - Monitoring of the first insulation fault.
    - Mandatory location and clearing of the fault.
    - Interruption for two simultaneous insulation faults.
- Solution offering the best continuity of service during operation

Figure 11. Impedance equivalent to leakage impedances in an IT system

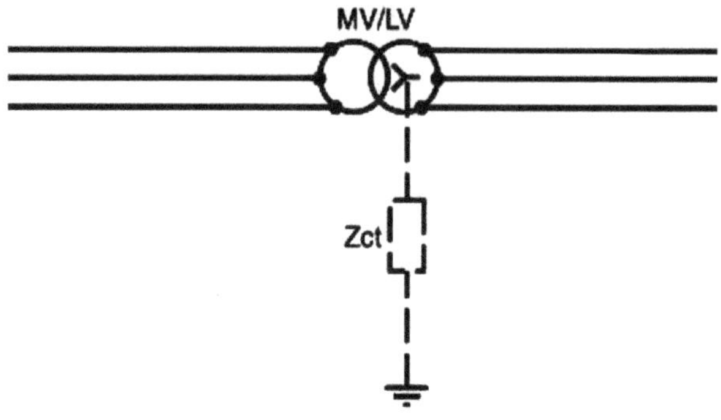

- Indication of the first insulation fault, followed by mandatory location and clearing, ensures systematic prevention of supply outages
- Generally used in installations supplied by a private MV/LV or LV/LV transformer
- Requires maintenance personnel for monitoring and operation
- Requires a high level of insulation in the network (implies breaking up the network if it is very large and the use of circuit-separation transformers to supply loads with high leakage currents)
- The check on effective tripping for two simultaneous faults must be carried out by calculations during the design stage, followed by mandatory measurements during commissioning on each group of interconnected exposed conductive parts.

(Figure 12) gives the IT system main characteristics and Table 2 gives the IT system - Preliminary conditions.

Figure 13 shows the position of each protection device of IT system.

Figure 12. IT system main characteristics

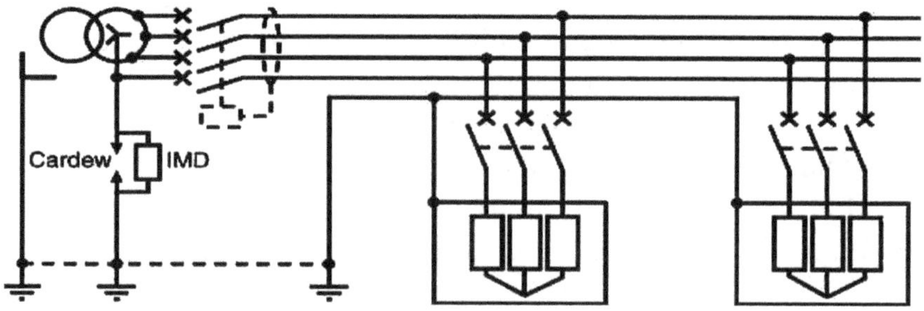

Table 2. IT system: preliminary conditions

Minimum Functions Required	Components and Devices
Protection against over voltages at power frequency	(1) Voltage limiter
Neutral grounding resistor (for impedance grounding variation)	(2) Resistor
Overall earth-fault monitor with alarm for first fault condition	(3) Permanent insulation monitor PIM with alarm feature
Automatic fault clearance on second fault and protection of the neutral conductor against overcurrent	(4) Four-pole circuit-breakers (if the neutral is distributed) all 4 poles trip
Location of first fault	(5) With device for fault-location on live system, or by successive opening of circuits

Figure 13. Positions of essential protection devices

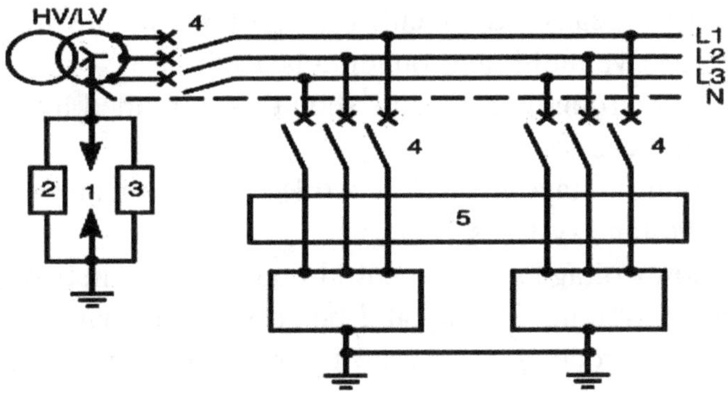

LOW VOLTAGE GROUNDING

The most dangerous circumstance is when the break occurs on the neutral service wire entering consumer's premises, for if the appliance is switched on, the full phase voltage will appear on the metalwork, although the appliance will not operate.

The danger can be minimized by connecting the framework of the appliance to the earth impedance on the consumer's premises as well, the resistance of which should be as low as possible, as the lower it is, so the lower will be the voltage appearing on the framework should a break occur on the neutral.

The following points should be taken into consideration during the design of a grounding system:

1. The size of the neutral distribution should not be less than that of the phase wire.
2. The neutral should be connected to earth at consumer's terminals and the earth electrode consisting of single copper or mild- steel rod or pipe with 1.8 meter long
3. Water mains should not be used for grounding
4. An earth conductor should be used throughout each installation and should be connected to the neutral conductor and grounding lead at the meter board.
5. No switch or fuse should be used in the neutral wire throughout the installation
6. The resistance of the earth conductor from an appliance to the meter board should not exceed 0.5 ohm and this should be verified in each installation by means of a current of at least 10 A flowing for five minutes before protective multiple grounding is used.
7. Two separate grounding systems should be provided for grounding in the following cases:
    a. The high voltage neutral network and any metal work of low or high voltage systems
    b. The low voltage system neutral point and the network of the low voltage system. This should be so situated that the low voltage earth electrode is not within the resistance area of high voltage earth electrode. If this precaution is not taken then a breakdown on the high voltage system causing current to flow through the high voltage electrode would result in the potential rise of the high voltage earth electrode to be communicated to the neutral of the low voltage and all apparatus frames connected to the network.
8. Earth continuity conductor through the factory, connected to each item of plant to be protected. This conductor should then be continued, preferably underground to the substation, and connected directly to the neutral terminal of the low voltage winding of the substation transformer. This means the fault current return path is provided independent of the neutral wire, which will still be required to carry the out-of-balance current of the supply network. The earth return wire will carry no load current under normal circumstances but will carry all fault currents and provide short circuit conditions in the case of earth faults, thus the operation of protective equipment is insured even when a very large capacity plant is used.

METHODS OF SYSTEM NEUTRAL GROUNDING

Most grounded systems employ some method of grounding the system neutral at one or more points. These methods can be divided into two general categories: solid grounding and impedance grounding. Impedance grounding may be further divided into several subcategories: reactance grounding, resistance grounding, and ground-fault neutralizer grounding. Each method, as named, refers to the nature of the external circuit from system neutral to ground rather than to the degree of grounding. In each case the impedance of the generator or transformer whose neutral is grounded is in series with the external circuit. Thus, a solidly grounded generator or transformer may or may not furnish effective grounding to the system, depending on the system source impedance.

Isolated System

Many standards have discussed grounding of an electrical distribution system. The electrical distribution system may or may not include a reference to ground. In those instances, allowed by the National Electrical Code, some designers view the opportunity to use an ungrounded system as a way of providing higher reliability than would a grounded system. The thinking is that since a ground would not create a fault, an ungrounded system would therefore provide higher system reliability.

That must be realized, however, is that even if no current carrying conductor is intentionally grounded, current still flows from any or all phase conductors to ground. This is done through phase to ground capacitance. While not enough current flows to cause damaging fault currents, it does provide enough current to create a unique overvoltage problem due to the particular way that intermittent ground faults operate an ungrounded system. This capacitance becomes part of our circuit that already includes cable inductance and resistance.

In an ungrounded system, there is no intentional connection between the system conductors and ground. However, there always exists a capacitive coupling between one system conductor and another, and also between system conductors and ground. Consequently, the so-called ungrounded system is in reality a capacitance grounded system, by virtue of the distributed capacitance from the system conductors to ground. Since the capacitance between phases has little effect on the grounding characteristics of the system, it will be disregarded. For simplicity, the distributed capacitive reactance to ground is assumed to be balanced. In an un-faulted condition, with balanced three-phase voltages applied to the lines, the capacitive charging current, *I*co, in phase will be equal and displaced 120° from one another. The phase voltages to ground will also be equal and displaced 120° from one another. Since the neutral of the distributed capacitances is at earth potential, it follows that the neutral of the transformer is also at earth potential, being held there by the capacitance to ground.

If the phase conductor C for example, faults to ground, current flow through that capacitance to ground will cease, since no potential difference across it now exists.

The voltage across the remaining two distributed capacitors to ground will, however, increase from line to neutral to line to line. The capacitive charging current, *I*co, in the two healthy phases will therefore increase by the square root of 3 and the line-to-ground voltages are no longer 120°, but 60° apart.

Hence, the vectorial sum of the capacitive charging current to ground is no longer zero, but is 3 *I*co or three times the original charging current per phase. The fault current, *I*g, flowing from the faulted conductor to ground, leads the original line-to-neutral voltage by approximately 90°.

In an ungrounded system, it is possible for destructive transient over voltages to occur throughout the system during re-striking ground faults. These over voltages, which can be several times normal in magnitude, result from a resonant condition being established between the inductive reactance of the system and the distributed capacitance to ground.

In the isolated system, it is possible for destructive transient over voltages to occur throughout the system during re-striking ground faults. These over voltages, which can be several times normal in magnitude, result from a resonant condition being established between the inductive reactance of the system and the distributed capacitance to ground. Experience has proved that these over voltages may cause failure of insulation at multiple locations in the system, particularly at motors. Transient over voltages from re-striking ground faults are the main reason why ungrounded systems are no longer recommended and grounded systems of some form are the predominant choice. To reduce transient over voltages during re-striking ground faults, one should ground the system using either solid or impedance grounding. Various detection schemes are used to detect the presence of a single line-to-ground fault.

Earthed Systems

Solid Grounding

Solid grounding refers to the connection of a system conductor as given in (Figure 14), usually the neutral of a generator, power transformer, or grounding transformer directly to ground, without any intentional intervening impedance. However, both the impedance of the source and the unintentional impedance in the connection to ground must be considered when evaluating the grounding. To assess the benefits of a solid connection to ground, it is necessary to determine the degree of grounding provided in the system. A good guide in answering this question is the magnitude of ground-fault current as compared to the system three-phase fault current. The higher the ground-fault current in relation to the three-phase fault current, the greater the degree of grounding in the system. Effectively grounded systems will have a line-to ground short-circuit current of at least 60% of the three-phase, short-circuit value.

This is rarely a problem in typical industrial and commercial power systems. A Sufficiently low resistance to earth may be difficult to achieve, but the "ground" to which faults occur will be the bonded conductive electrical enclosures.

The zero-sequence impedance of most generators used in these systems is much lower than the positive sequence impedance of these generators. The zero-sequence impedance of a delta-wye transformer will not exceed the transformer's positive-sequence impedance. There are, however, conditions under which relatively high zero-sequence impedance may occur. One of these conditions is a power system fed by several generators and/or transformers in parallel. If the neutral of only one source is grounded, it is possible for the zero-sequence impedance of the grounded source to exceed the effective positive-sequence impedance of the several sources in parallel.

Another such condition may occur where power is distributed to remote facilities by an overhead line without a metallic ground-return path. In this case, the return path for ground-fault current is through the earth, and even though both the neutral of the source and the non-conducting parts at the load may be grounded with well-made electrodes, the ground-return path includes the impedance of both of these ground electrodes. This impedance may be significant. Another significant source of zero-sequence impedance is the large line-to-ground spacing of the overhead line.

Solid grounding is generally recommended for the following:

1. Low-voltage systems (600 V and below) where automatic isolation of a faulted circuit can be tolerated or where capability is lacking to isolate a ground fault in a high-resistance grounded system.
2. Medium- or high-voltage systems (above 15 kV) in order to permit the use of equipment with insulation levels to ground rated for less than line to line voltage.
3. Medium- or high-voltage applications where the desire for a higher magnitude of ground-fault current in order to be able to provide selective ground-fault detection on lengthy distribution feeders outweighs concerns about arc flash and potential gradients as personnel hazards in a workplace setting.

Resistance Grounding

In a resistance-grounded system, the neutral of the transformer or generator is connected to ground through a resistor. A typical resistance-grounded neutral system is shown in (Figure 14). As commonly installed, the resistance has a considerably higher Ohmic magnitude than the system reactance at the resistor location. Consequently, the line-to ground fault current is primarily limited by the resistor itself.

The reasons for limiting the current by resistance grounding include the following:

1. To reduce burning and melting effects in faulted electric equipment, such as switchgear, transformers, cables, and rotating machines.
2. To reduce mechanical stresses in circuits and apparatus carrying fault currents.
3. To reduce electric-shock hazards to personnel caused by stray ground-fault currents in the ground-return path.
4. To reduce the arc blast or flash hazard to personnel who may have accidentally caused or happen to be in close proximity to the ground fault.
5. To reduce the momentary line-voltage dip occasioned by the occurrence and clearing of a ground fault.
6. To secure control of transient over voltages while at the same time avoiding the shutdown of a faulted circuit on the occurrence of the first ground fault (high resistance grounding). Resistance grounding may be

Figure 14. Resistance-grounded systems

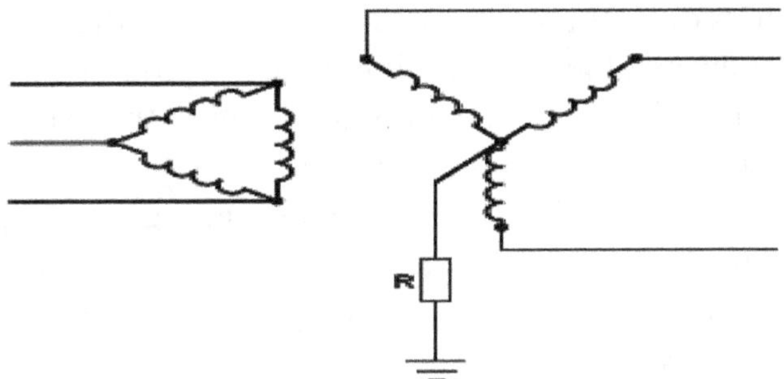

either of two classes, high resistance or low resistance, distinguished by the magnitude of ground-fault current permitted to flow. Although there are no recognized standards for the levels of ground-fault current that define these two classes, in practice there is a clear difference.

High-resistance grounding usually does not require immediate clearing of a ground fault since the fault current is limited to a very low level. The protective scheme associated with high-resistance grounding is usually detection and alarm rather than immediate trip out. A typical scheme for detecting a ground fault in a high-resistance grounded system is shown in (Figure 15). Under normal operation, the neutral point of the transformer is at zero potential. When a single line-to-ground fault occurs, the neutral point is raised to approximately line-to-neutral voltage. This rise in voltage

Figure 15. Scheme for detecting a ground fault on a high-resistance grounded system

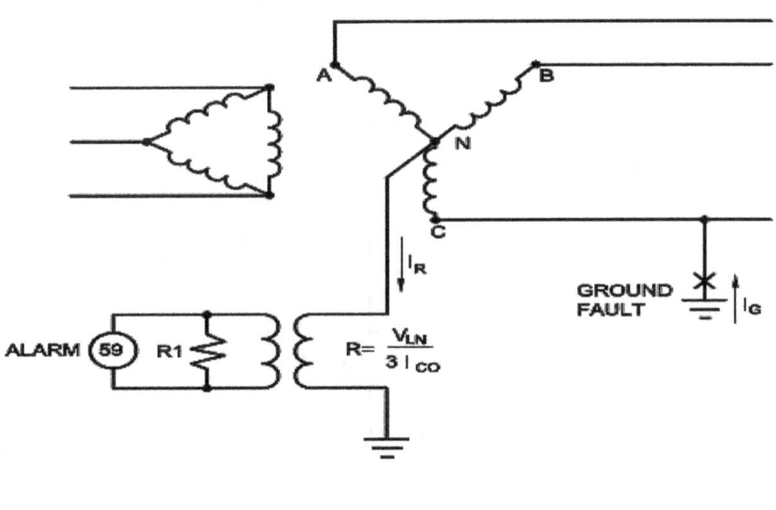

is then detected using an over voltage relay, 59. A step-down transformer is typically used to reduce the line to neutral voltage of the system to a level (usually 120 V) acceptable to the relay. Since a ground fault may persist for an indefinite length of time, a continuous (rather than short term) rating should be imposed on the grounding resistor.

On the other side, the low-resistance grounding is designed to limit ground-fault current to a range between 100 A, and 1000 A, with 400 A being typical. The neutral resistor, *R*, shown in figure16, is selected according to $R = V_{LN} / I_o$, where V_{LN}, is the system line to neutral voltage and I_o is the desired ground-fault current. (Figure 15) illustrates the flow of currents for a single line-to-ground fault on a low-resistance grounded system. Since the combined effects of charging current and system source impedance will affect the ground-current value less than 0.5% in the typical range of utility supplied systems, it is permissible to ignore these effects in calculating the ground-fault resistance value. The general practice is to consider that the full system line-to-neutral voltage appears across the grounding resistor. Only in the case of systems supplied by small generators should departure from this general practice be considered.

Low-resistance grounding has the advantage of facilitating the immediate and selective clearing of a grounded circuit. This requires that the minimum ground-fault current be large enough to positively actuate the applied ground-

Figure 16. Low-resistance grounded system

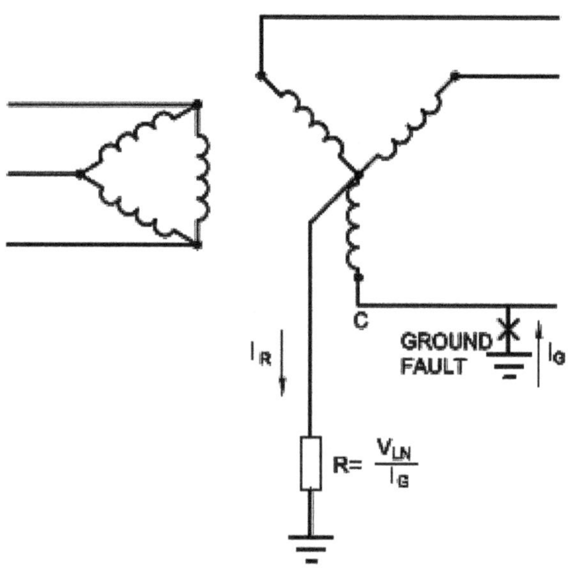

fault relay. One method of detecting the presence of a ground fault uses an over current relay, 51G. This method is presented in figure 17. When a ground fault occurs, the neutral potential is raised to approximately line-to-neutral voltage, resulting in current flow through the resistor. A typical turn's ratio for the current transformer is indicated. Upon indication that a ground fault has occurred, action would be initiated to disconnect the transformer from the secondary circuit.

Normal practice for automatic protection in low resistance grounding is to rate it for 10 s or 30 s, depending upon the degree of security appropriate for the application. In cases of faults that are not, or cannot be, disconnected by secondary breakers, the ability for prompt and automatic disconnection of the primary source is required. Suitable relaying and switching devices for this purpose are an integral part of the low-resistance system design as shown in (Figure 17).

Reactance Grounding

The term reactance grounding describes the case in which a reactor is connected between the system neutral and ground, as shown in (Figure 18). Since the ground fault that may flow in a reactance-grounded system is a function of

Figure 17. Scheme for detecting a ground fault on a low-resistance grounded system

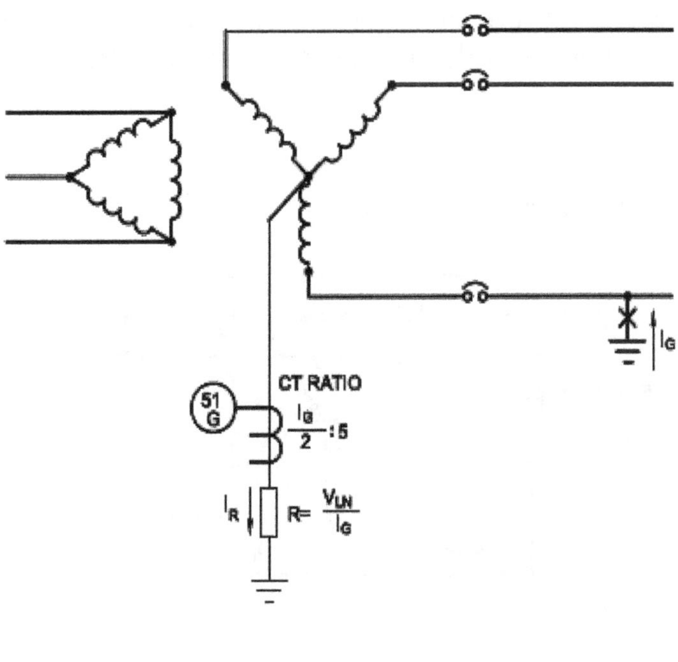

the neutral reactance, the magnitude of the ground-fault current is often used as a criterion for describing the degree of grounding. In a reactance-grounded system, the available ground-fault current should be at least 25% ($X0 = 10X1$) and preferably 60% ($X0 = 3X1$) of the three-phase fault current to prevent serious transient over voltages. The term $X0$, as used, is the sum of the source zero sequence reactance, $X0$, plus three times the grounding reactance, $3Xn$, ($X0 = X0$ source $+ 3Xn$). This is considerably higher than the level of fault current desirable in a resistance grounded system, and therefore reactance grounding is usually not considered an alternative to low-resistance grounding.

Reactance grounding is typically reserved for applications where there is a desire to limit the ground-fault duty to a magnitude that is relatively close to the magnitude of a three-phase fault. Use of neutral grounding reactors to provide this fault limitation will often be found to be a less expensive application than use of grounding resistors if the desired current magnitude is several thousand amperes.

These circumstances may arise in one of two possible instances. One potential setting is where a large substation feeds a medium-voltage distribution system, and the total zero-sequence impedance of the step-down transformers in the station causes the single line- to-ground-fault current to greatly exceed the magnitude of a three-phase fault, and ground-fault limitation is desired to keep the total fault current within the reasonable limits. These conditions tend to occur most often in electric utility distribution practice.

In this system, instead of resistance, a reactance is connected between neutral and earth with ratio of reactance more than 3. The system is represented in the (Figure 18) . Let the earth fault occurring on phase B.

Figure 18. Reactance grounding system

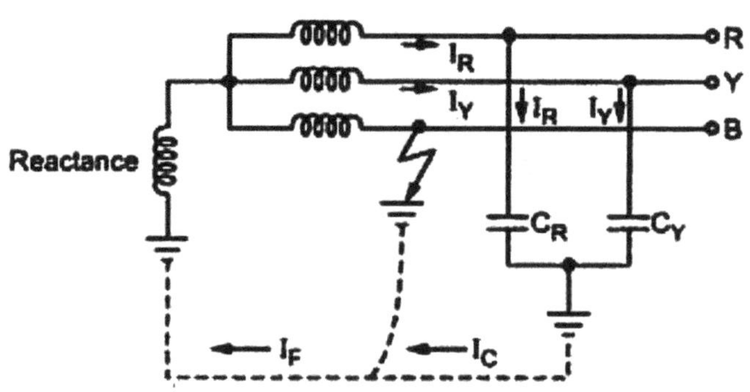

In addition to zero sequence impedance of the source and faulty phase up to point of fault, the fault current is dependent on reactance. By changing the value of reactance, the fault current can be varied. In practice, this method is employed to give characteristics similar to solidly earth system.

There is drawback of this system. With increase in reactance, there is increase in transient voltages resulting from arcing. Hence it is not commonly employed though it ensures satisfactory relaying, partial grading of equipment insulation, less interference with neighboring circuits, and intermediate cost.

For reactance grounding, it is necessary that the magnitude of fault current should be at least 25% of the three-phase fault current. This is higher than the requirement on resistance grounding and thus it can be seen that the resistance grounding and the reactance grounding are not similar. The second instance is where there is a desire to serve single-line-to-neutral-connected load directly at the terminal voltage of generators, i.e., without an intervening generator isolation transformer. In this instance, a current will flow in the generator neutral as a result of unbalance between the loads on the three phases. A resistor in the neutral circuit of the generator will limit the flow of this unbalance, thereby limiting the ability of the system to carry unbalanced single-phase load. Medium-voltage generators are typically not designed to withstand the unbalanced mechanical forces associated with supplying ground-fault currents that exceed the magnitude of current that the machine would produce to a three-phase fault at its terminals, thereby making solid grounding of the neutral undesirable. Use of low-reactance grounding to limit the ground-fault magnitude to a level slightly lower than the three-phase level is a way to resolve these application constraints. The conditions that favor low-reactance grounding of generators are relatively rare, so this practice is somewhat obscure.

Resonant Grounding (Ground-Fault Neutralizer)

A ground-fault neutralizer is a reactor connected between the neutral of a system and ground. The reactor, X_{LN}, is specially selected, or tuned, to resonate with the distributed capacitance, V_{CN} of the system so that a resulting ground-fault current is resistive and low in magnitude. A resistance, *r*, is shown depicting reactor losses. The resulting ground-fault current is in phase with the line to neutral voltage so that current zero and voltage zero occur simultaneously. If the ground fault is in air, such as an insulator flashover, it may be self-extinguishing.

This method of grounding formerly was occasionally seen in high-voltage transmission practice. A key requirement is that because the resonant circuit must be retuned if the distributed parameters of the associated circuit are changed, the ideal application is one that does not involve frequent circuit switching or reconfiguration.

ZIGZAG GROUNDING TRANSFORMER

One type of grounding transformer commonly used is a three-phase zigzag transformer with no secondary winding. The internal connection of the transformer is illustrated in (Figure 19).

The impedance of the transformer to balanced three-phase voltages is high so that when there is no fault on the system, only a small magnetizing current flows in the transformer winding. The transformer impedance to zero-sequence voltages, however, is low so that it allows high ground-fault currents to flow. The transformer divides the ground-fault current into three equal components; these currents are in phase with each other and flow in the three windings of the grounding transformer. The method of winding is seen from (Figure 19) to be such that when these three equal currents flow, the current in one section of the winding of each leg of the core is in a direction opposite to that in the other section of the winding on that leg. This tends to force the ground-fault current to have equal division in the three lines and accounts for the low impedance of the transformer-to-ground currents.

A zigzag transformer may be used for effective grounding, or impedance can be inserted between the derived neutral of the zigzag transformer and ground to obtain the desired method of grounding. This transformer is seldom employed for medium-voltage, high-resistance grounding. An example of low-resistance grounding is shown in (Figure 20). The over current relay, 51G, is used to sense neutral current that only flows during a line-to-ground fault.

WYE-DELTA GROUNDING TRANSFORMER

A wye-delta connected three-phase transformer or transformer bank can also be utilized for system grounding, as shown in (Figure 21 (a)) and (Figure 21 (b)). As in the case of the zigzag transformer, it can be used for effective grounding or to accomplish resistance-type grounding of an existing ungrounded system. The delta connection must be closed to provide a path

Figure 19. Zigzag grounding transformer: (a) core windings, (b) system connection

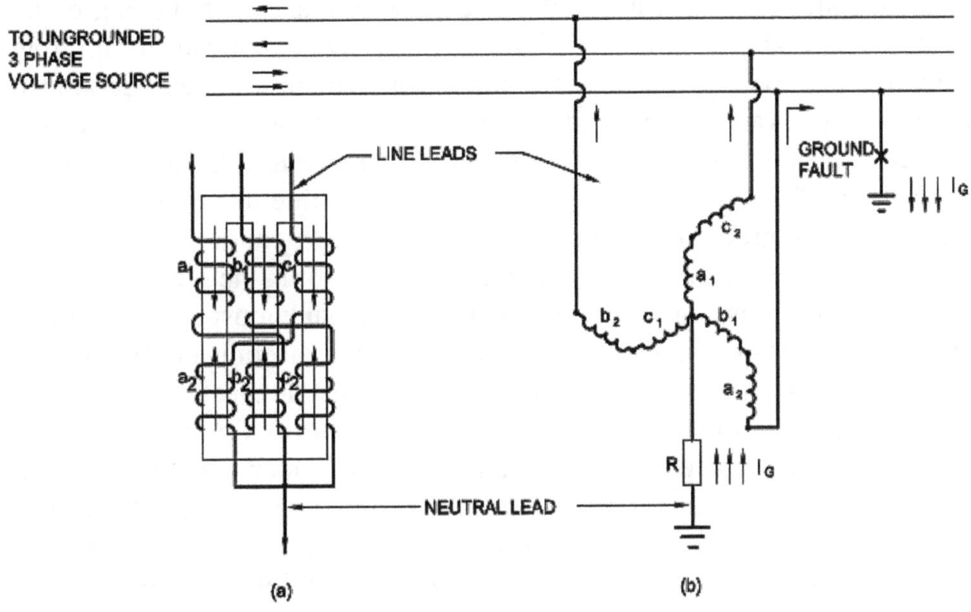

Figure 20. Low-resistance grounding of system through a zigzag grounding transformer with neutral sensing current relay

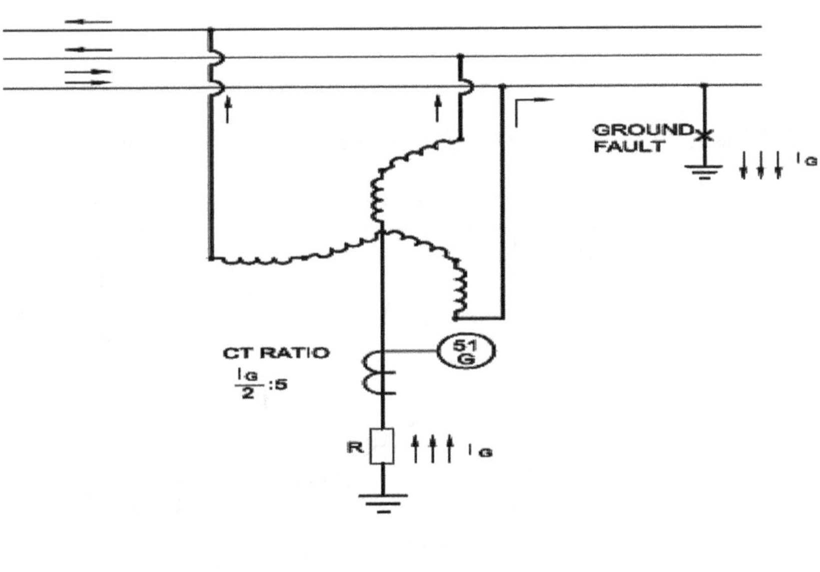

for the zero-sequence current, and the delta voltage rating is selected for any standard value. A resistor inserted between the primary neutral and ground, as shown in (Figure 21 (a)) and (Figure 21(b)) provides a means for limiting ground-fault current to a level satisfying the criteria for resistance-grounded systems. For this arrangement, the voltage rating of the wye winding need not be greater than the normal line-to-neutral system voltage. A neutral sensing current relay, 51G, is shown for detection of a single line-to-ground fault. For high-resistance grounding it is sometimes more practical or economical, as illustrated in (Figure 22), to apply the limiting resistor in the secondary delta connection. For this configuration, the grounding bank must consist of three single-phase transformers with the primary wye neutral connected directly to ground. The secondary delta is closed through a resistor that effectively limits the primary ground-fault current to the desired low level. For this alternative application, the voltage rating of each of the transformer windings forming the wye primary should not be less than the system line-to-line voltage. The rating of a three-phase grounding transformer or bank, in kilovolt ampere (kVA), is equal to the rated line-to-neutral voltage in kilovolts times the rated neutral current. Most grounding transformers are designed to carry their rated current for a limited time only, such as 10 s or 1 min. Consequently, they are much smaller in size than an ordinary three-phase continuously rated transformer with the same rating.

Figure 21. Wye-delta grounding transformer showing current flow

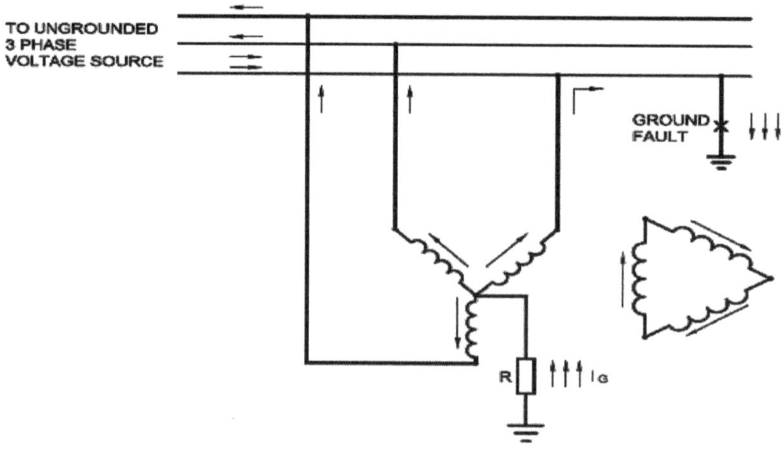

Figure 22. Low-resistance grounding of system through a wye-delta grounding transformer with neutral sensing current relay

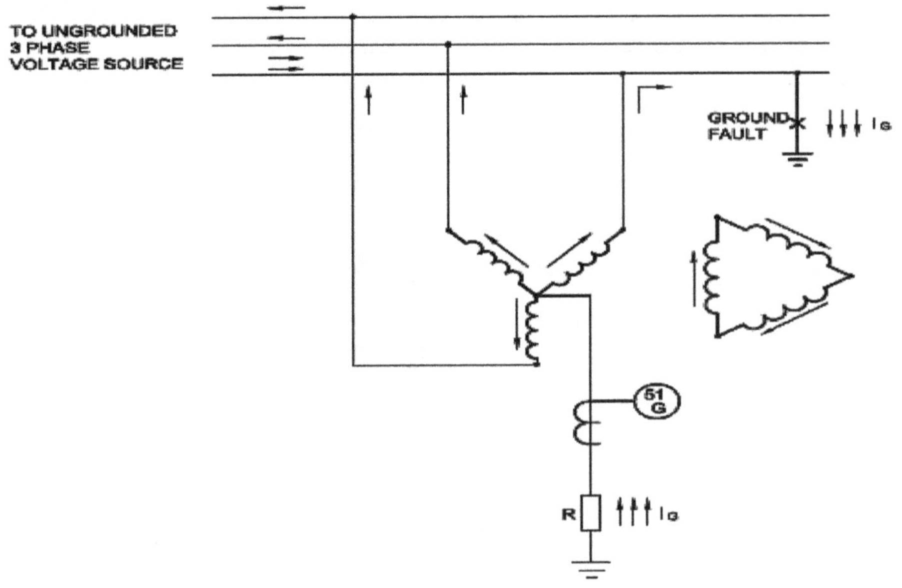

It is generally desirable to connect a grounding transformer directly to the main bus of a power system, without intervening circuit breakers or fuses, to prevent the transformer from being inadvertently taken out of service by the operation of the intervening devices. (In this case the transformer is considered part of the bus and is protected by the relaying applied for bus protection.) Alternatively, the grounding transformer should be served by a dedicated feeder circuit breaker, as shown in part a) of (Figure 21), or connected between the main transformer and the main switchgear, as illustrated in part b) of (Figure 21). If the grounding transformer is connected as shown in part b) of (Figure 21), there should be one grounding transformer for each delta-connected bank supplying power to the system, or enough grounding transformers to assure at least one grounding transformer on the system at all times. When the grounding transformer is so connected, it is included in the protective system of the main transformer. (Figure 22) shows high-resistance grounding of system through a wye-broken delta grounding transformer with neutral sensing voltage relay

Figure 23. High-resistance grounding of system through a wye-broken delta grounding transformer with neutral sensing voltage relay

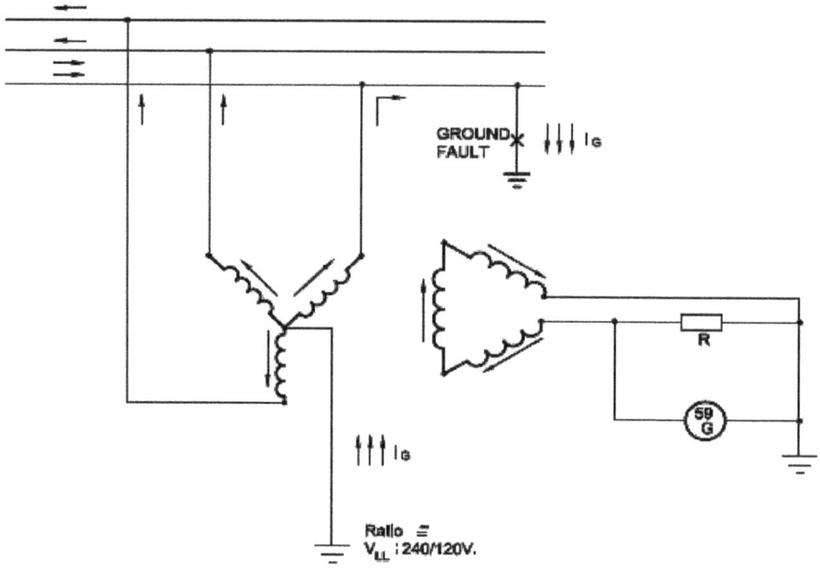

GROUNDING AT POINTS OTHER THAN SYSTEM NEUTRAL

In some cases, low-voltage systems (600 V and below) are grounded at some point other than the system neutral to obtain a grounded electrical system. This is done because delta transformer connections do not provide access to the three-phase system neutral. Two systems are in general use as follows:

- **Corner-of-the-Delta Grounded Systems:** This type of grounding is no longer popular and is not widely used in industrial systems.
- **One Phase of a Delta System Grounded at Midpoint:** This provides neutral grounding for the single-phase 120/240 V and also grounding for the 240 V three-phase system. It is not recommended for voltages over 240 V.

Connection of grounding transformers in delta connected or ungrounded power system to obtain neutral for system grounding: (a) circuit feeder breaker, (b) connected between main transformer and main switch gear

LOCATION OF SYSTEM GROUNDING POINTS SELECTION

Each system as described in this chapter is characterized by its isolation from adjacent grounding systems. The isolation is provided by transformer primary and secondary windings. The NEC defines such a system as "separately derived." A separately derived system is one "...whose power is derived from a generator, transformer, or converter windings and that has no direct electrical connection, including a solidly connected grounded circuit conductor, to supply conductors originating in another system."

Therefore, the new system created by a transformer or generator requires the establishment of a new system ground if it is required or desired that this system to be grounded. The system ground point should always be at the power source as required or permitted by the NEC, including exceptions for multi-source systems.

DELTA WYE TRANSFORMER

In a delta-wye connected transformer, with the load-side neutral grounded, zero-sequence components of current can flow in the secondary wye-connected windings due to a ground fault. Zero-sequence current is then induced into the primary windings of the transformer and circulates in the delta connection. Positive and negative-sequence currents pass through the transformer combining to produce high current in two of the primary phase conductors. A ground fault on the secondary of the delta-wye connected transformer appears as a line-to-line fault on the primary. If the neutral of the wye-connected windings is not grounded, then zero-sequence current cannot flow and the system becomes ungrounded. (Figure 23) shows Zero-sequence impedances of different transformers configurations

Zero-sequence components of current can flow through a wye-wye connected transformer if a neutral path exists on both sides of the transformer. An example is shown in (Figure 24) where a delta-wye connected transformer, T1, supplies power to a wye-wye connected transformer, T2. A fault on the load side of T2 produces zero-sequence current, which flows in the primary and secondary windings of that transformer. Zero-sequence current is permitted to flow in the primary of T2 because a path exists in the delta-wye connected transformer T1. Disconnecting any of transformer neutrals, on either T1 or

Figure 24. Zero-sequence impedances of different transformers configurations

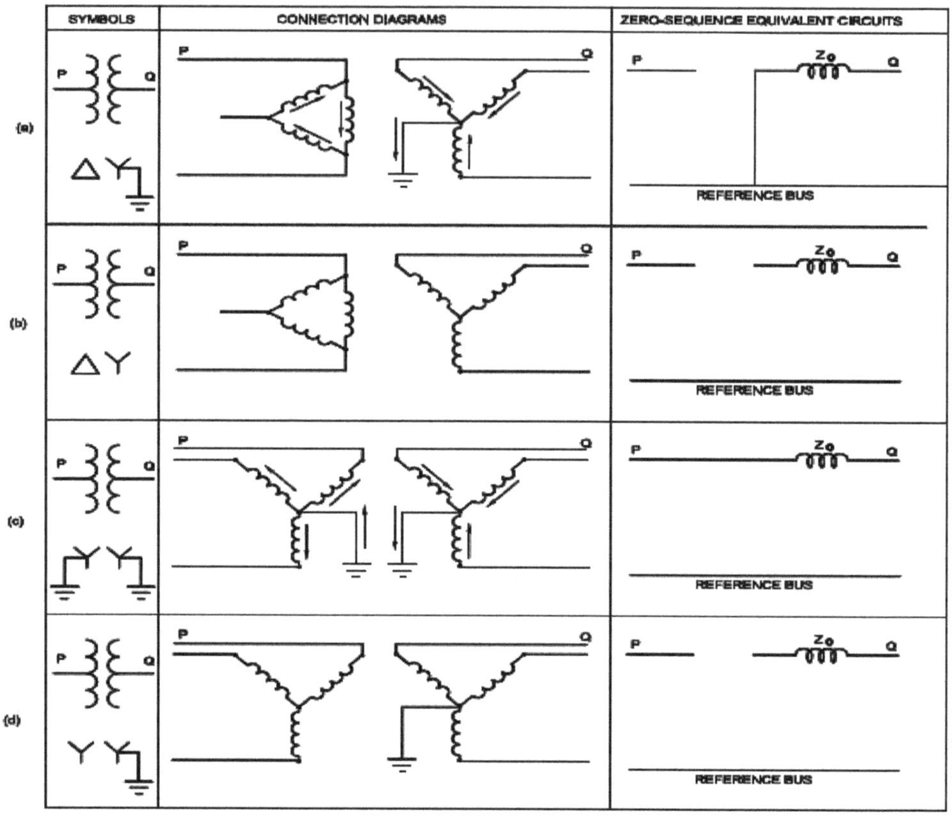

T2, would prevent the flow of zero-sequence current in both transformers, except as allowed by magnetizing reactance.

Depending upon the connections to the transformer, the use of a wye-wye transformer can result in a single system, or its load side may be a separately derived system.

WYE-WYE TRANSFORMERS

A wye-wye transformer, T2, is shown in (Figure 24) with the primary and secondary neutrals interconnected and grounded. This transformer configuration is used on solidly grounded utility distribution systems, particularly underground systems, to prevent Ferro resonance when the supply switches can be operated one pole at a time. The utilities ground the primary

Figure 25. Transformer connections illustrating the flow of zero-sequence current resulting from a line to ground fault

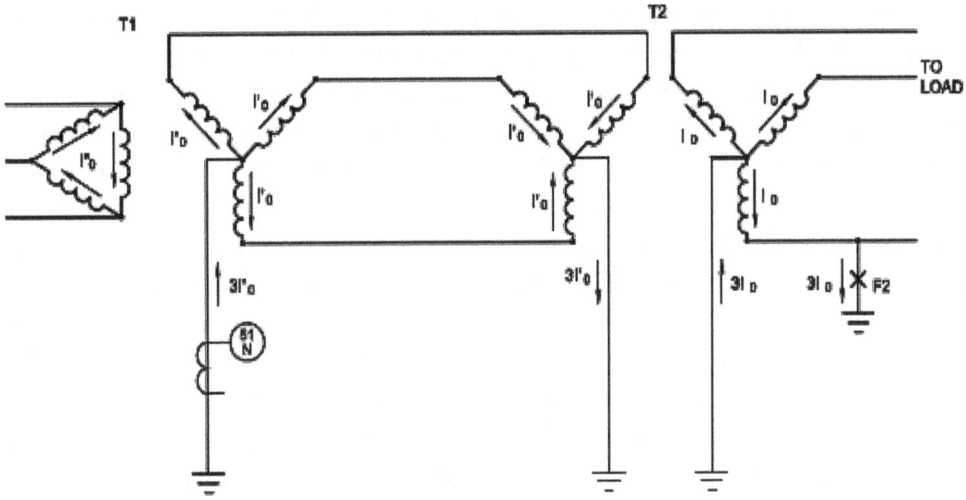

neutral point to minimize the neutral-to-earth voltage throughout the length of the distribution line and by default on underground systems using bare concentric neutral cables. They ground the secondary neutral to provide an effectively grounded low-voltage service.

Note that this multiple grounding of the primary at each transformer is not essential to prevent Ferro resonance or provide secondary grounding as long as the fourth conductor is brought to the primary neutral of the transformer. The neutral-to-transformer case and ground connection minimizes secondary neutral-to ground voltage during a fault between primary and transformer case. In an industrial distribution system, the physical length of the circuit will usually be short enough so that excessive neutral-to-ground voltages will not be present even if there is no ground at the wye-wye transformer common neutral terminals, as shown in (Figure 24). The NEC normally prohibits grounding of the neutral on the load side of the service disconnect, but allows multi grounding of the neutral of an outdoor overhead line or direct burial cable with bare neutral if the circuit voltage is over 1000 V.

As shown in (Figure 24), with a continuous connection from the source neutral to the primary and secondary neutrals of the wye-wye transformer, the output of the transformer would not constitute a separately derived system as defined in the NEC. If the neutral is grounded at the source, T1, the output of the wye-wye transformer will be a continuation of the grounded system, though at the secondary voltage of the transformer. A fault, F2, on the load side of the wye-wye connected transformer, T2, will produce zero-sequence components of current in its primary windings. This zero-sequence current will flow back to the secondary neutral terminal of source transformer, T1. However, this current flowing through 51G cannot determine whether the fault is located before or after the wye-wye transformer, nor can residual or zero-sequence ground detection schemes on the output of T1. The main benefit of this transformer connection is to utilize the standard wye-wye transformer that contains an internal primary-to-secondary neutral connection suitable for utility practice as shown in (Figure 25). The circuit supplied by the wye-wye connected transformer shown in (Figure 26) can be considered a separately derived system, since there are no direct metallic connections between the primary and secondary of the transformer. Primary and secondary ground faults are separately measured and relayed. The secondary of the transformer will not be grounded unless a connection to earth is made. The secondary could be impedance grounded. Secondary neutral grounding will also require a connection from the neutral of the primary source to the primary neutral of the wye-wye transformer to supply zero sequence current. Unlike the delta-wye transformer, the wye-wye transformer itself is not a source of zero-sequence current. Grounding can be achieved without a primary neutral connection if a phase of the secondary rather than the neutral is grounded, since no zero-sequence current is involved. The effect is then identical to corner grounding of a delta-delta transformer.

If a delta tertiary is added to a wye-wye transformer it will not be necessary to supply zero-sequence current from the primary source, since the tertiary

Figure 26. Grounded wye-grounded wye transformer with multi-grounded common neutral

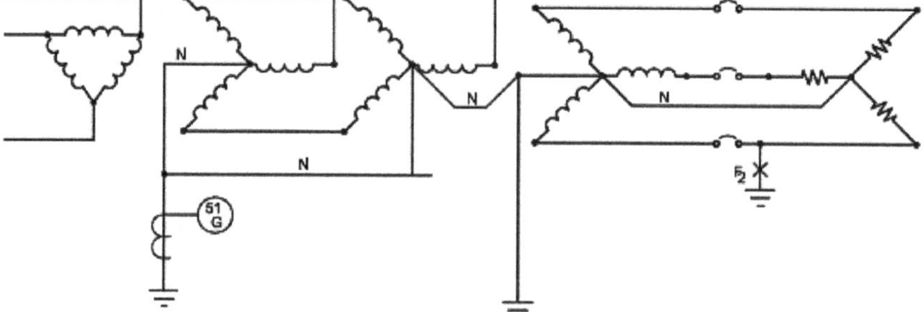

will act as a source of zero-sequence current. Thus, the wye-wye transformer can be considered a part of a single multi-voltage system if the neutrals are interconnected or can be considered to create a separate system if they are not. The symmetry of the wye-wye allows it to provide grounding for its load-side system even though the source and load side may be interchanged at any time.

In a single power source when a system has only one source of power (gen. or transformer), grounding may be accomplished by connecting the source neutral to earth either directly or through a neutral impedance (Figure 27).

Provision of a switch or circuit breaker to open the neutral circuit is not recommended. It is not desirable to operate the system ungrounded by having the ground connection open while the generator or transformer is in service. In the event that some means of disconnecting the ground connection is required for measurement, testing, or repair, a disconnecting link should be used and only opened when the system is de-energized.

In Multiple power sources for installation of interconnected multiple power sources (i.e., generators or power transformers), operated in parallel, system grounding can be accomplished using one of the two following methods:

Figure 27. Grounded wye-grounded wye transformer with separately grounded neutrals

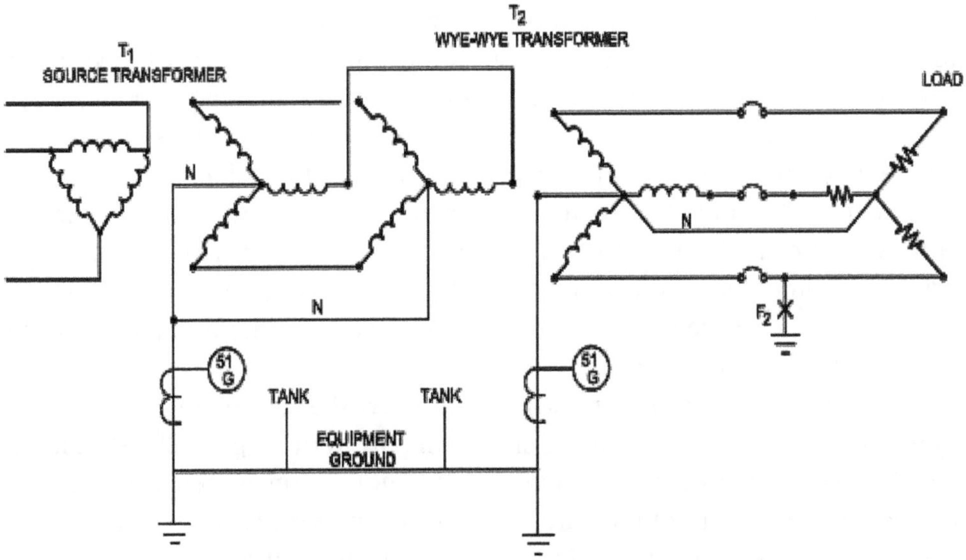

Figure 28. Grounding for systems with one source of power: (a) solidly grounded, (b) R or Z grounded

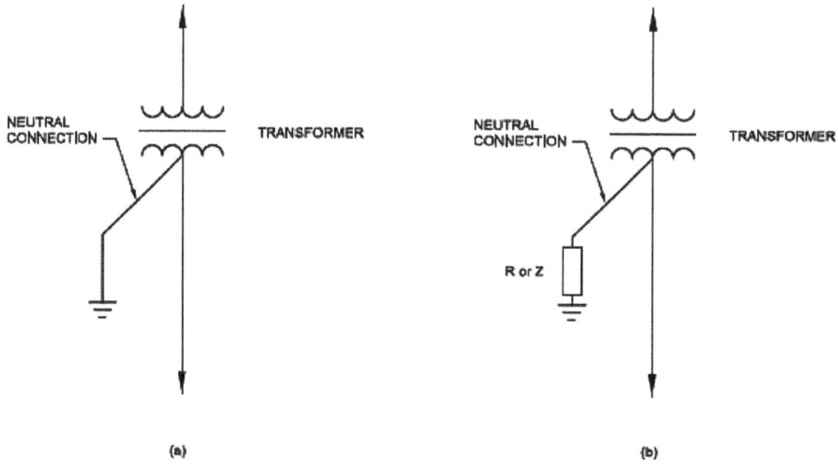

1. Each source grounded, with or without impedance (Figure 28).
2. Each source neutral connected to a common neutral bus, which is then grounded, with or without impedance (Figure 29).

For solidly grounded systems, with multiple sources, where all sources must be solidly grounded, it is acceptable to separately ground each power source as shown in part a) of (Figure 28) unless third harmonics are present or if it results in exceeding the fault capability of the generators. Levels of fault current in systems where generators are paralleled with transformer sources on a four-wire basis must be calculated using symmetrical component sequence values for the sources appropriately combined in the system.

Commercial computer programs are now available that will calculate branch currents for unbalanced faults in systems with both utility and generator sources. There can be a significant increase in the total system ground-fault current as compared to the sum of the fault current available from sources when not in a combined system, while the increase in generator currents can be proportionally even greater.

Where sources are in close proximity, or where the system is four wires, the common neutral or ground bus as shown in part a) of (Figure 29) should be used. In a four-wire system the sources would not be considered as separately derived. If the power sources are not in close proximity, common ground point is not recommended. The impedance in the neutral bus connection may become large enough to prevent effectively grounding the neutral of the source at the remote location. The interconnection may inadvertently become open, allowing the transformer to operate ungrounded. For impedance grounded systems it is acceptable to separately connect each neutral to ground through individual impedance (part b) of (Figure 28). Each impedance rating should allow sufficient current to satisfy the criteria for the grounding system being used. Individual neutral switching devices (automatic or manual) are not recommended, since incorrect operation may allow a power source to operate ungrounded. System relaying is more complex when there are multiple ground-fault sources. The fault current sensed by the feeder is variable, depending on the number of ground-fault current sources that are connected at the time of the fault. When individual source impedances are used for low- or high-resistance grounding, circulation of third harmonic currents between paralleled generators is usually not a problem since the impedance limits the circulating current to tolerable values. When the total ground-fault current from several individual impedances exceeds about 1000 A, a common ground point and single impedance should be considered to provide a single acceptable value of ground-fault current (part b) of (Figure 29). The advantage of this connection is that the maximum fault current is known, and selective relaying can be used to open tie breakers and selectively isolate the faulted bus.

The primary purpose of neutral disconnecting devices in impedance grounded systems, as shown in part b) of (Figure 29), is to isolate the generator or transformer neutral from the neutral bus when the source is taken out of service, because the neutral bus is energized during ground faults.

A generator or transformer disconnected from the power bus, but with an unbroken connection of its neutral to the neutral bus, would have all of its terminals elevated with respect to ground during a ground fault. Disconnecting devices should be metal enclosed and interlocked in such a manner as to prevent their operation except when the transformer primary and secondary switches or generator main and field circuit breakers are open. On low-voltage systems the use of four-pole breakers may provide adequate interlocking. In this case line-to-neutral voltage should not be used for synchronizing.

In the case of multiple transformers, all neutral isolating devices may be normally closed because the presence of delta-connected windings (which are nearly always present on at least one side of each transformer) minimizes the circulation of harmonic current between transformers. Generators that are designed to suppress zero-sequence harmonics, usually by the use of a two-thirds pitch winding, will have negligible circulating currents when

Figure 29. Grounding for systems with multiple power sources (Method 1): (a) solidly grounded, (b) R or Z grounded

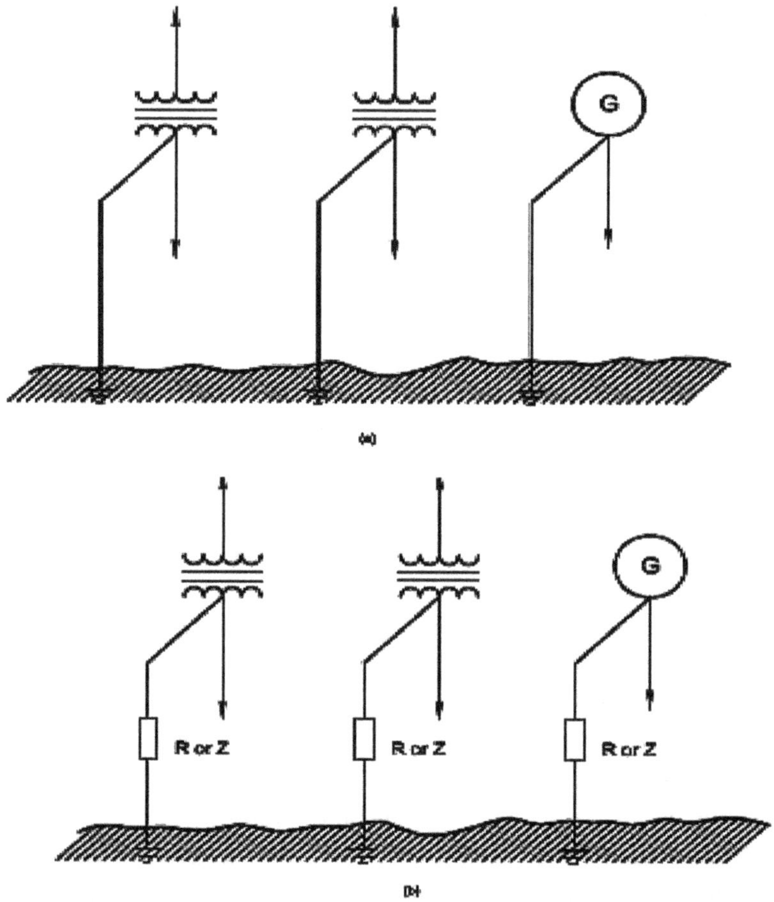

operated in parallel; therefore, it is often found practical to operate these types of generators with the neutral disconnect device closed. This simplifies the operating procedure and increases assurance that the system will be grounded at all times, because interlocking methods can be used.

It is sometimes desirable to operate with only one generator neutral disconnecting device closed at a time to eliminate any circulating harmonic or zero-sequence currents.

In addition, this method provides control over the maximum ground-fault current and simplifies ground relaying. When the generator whose neutral is grounded is to be shut down, another generator is grounded by means of its neutral disconnecting device before the main and neutral disconnecting

Figure 30. Grounding for systems with multiple power sources (Method 2): (a) solidly grounded, (b) R or Z grounded

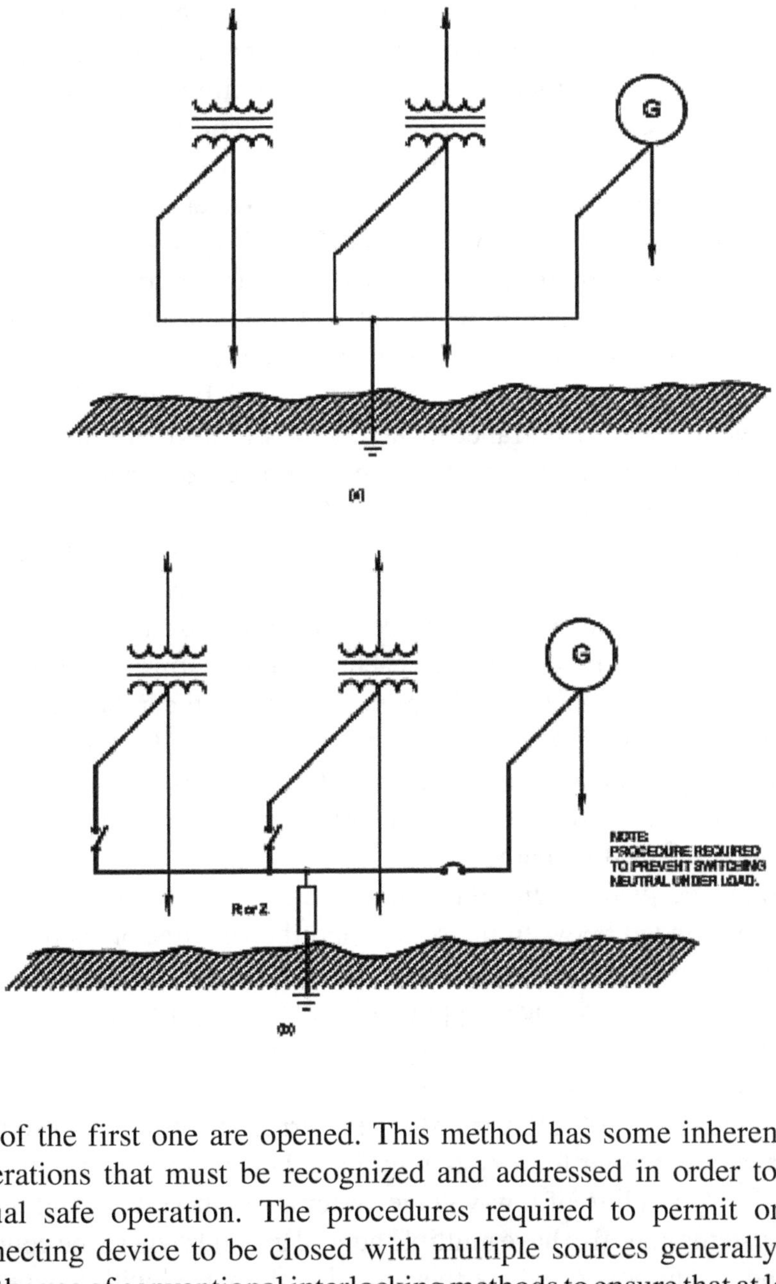

device of the first one are opened. This method has some inherent safety considerations that must be recognized and addressed in order to ensure continual safe operation. The procedures required to permit only one disconnecting device to be closed with multiple sources generally do not permit the use of conventional interlocking methods to ensure that at least one neutral disconnecting device will be closed. Therefore, this method should only be used where strict supervision of operating procedures is assured.

GROUNDING LOCATIONS SPECIFIED BY THE NEC

The following locations for ground connections are required, or permitted, by the NEC for the most common types of power system grounding. This is not intended to be a complete listing of code requirements, and the current edition of the NEC should be consulted for details or recent changes as well as to determine whether grounding is required or prohibited. The purpose is to call attention to location requirements and not to interpret the requirements. On service-supplied systems of 50 V to 1000 V, system grounding, when required or elected, should be made at the service entrance, between the load ends of the service drop or lateral and the neutral land point. If the service is supplied from a transformer external to the building one additional grounding point is required external to the building. If a grounded conductor extends past the service entrance switch, it should have no further grounds on this extension except as noted by the various exceptions in the NEC to this requirement, as follows.

Where dual services feed a double-ended bus, a single ground at the center point of the neutral bus is allowed to replace those previously listed. If more than one building is fed from a single service there should be a system grounding connection made at the entrance to each building. However, if an EGC is run with the load conductors, this ground connection can be eliminated so as to avoid elevating non-current carrying enclosures above ground potential due to load drop in the neutral conductor. Grounding connections should not be located or connected so as to cause objectionable currents in grounding conductors or grounding paths.

Separately derived circuits, if required or elected to have a system ground, should be grounded between the source and the first disconnecting device. System grounding connections downstream of the disconnecting device have the same rules as for service supplied circuits. The point of grounding for systems shall be the neutral or common conductor where one exists; otherwise the point shall be a phase conductor.

On systems over 1000 V, a transformer-derived neutral may also be used as the attachment point for a system ground. This method is not mentioned for effective grounding of low-voltage systems. High-voltage and medium-voltage systems may also have multiple neutral grounds where the conductors are overhead outdoors or where they are directly buried with a bare neutral conductor.

GROUNDING OF INDUSTRIAL AND COMMERICAL GENERATORS

Generators produce slightly no sinusoidal voltages because of saturation and imperfect winding and flux distribution. Industrial generators therefore produce odd harmonic voltages, with the third harmonic voltage being as much as 10%. These harmonic voltages can cause heating from circulating currents in a closed loop. This is one reason why most industrial generators have their internal windings connected in wye rather than delta. The third harmonic voltages produced in the generator's windings are in phase and additive. This would cause third harmonic current to circulate within the delta connected windings, as shown in (Figure 30).

The circulating current would create additional heating within the generator thereby reducing some of its thermal capacity. Generators that operate in a delta connection allow for this in their design. If the generator windings are designed with a two-thirds pitch, the third harmonic voltage can be suppressed, but the zero-sequence impedance will be lowered increasing the ground-fault current.

A grounded generator connected to a delta-wye transformer is shown in (Figure 31). Any third harmonic voltage, *V*3, produced by the generator would be impressed on the primary of the transformer. Since the third harmonic voltages are in phase, the voltage difference across each winding of the transformer's delta will equal zero and no third harmonic, or multiples of the third harmonic, current can be expected to flow.

Any current flowing as a result of a line-to-ground fault on the secondary side of the transformer will appear, as shown in (Figure 32), as a line-to-line fault at the generator output. This type of fault is the most damaging to the generator because of its negative sequence content. There will be no zero-

Figure 31. Circulation of third harmonic current in a delta-connected generator

Figure 32. Analysis of third harmonic current (no zero-sequence loop)

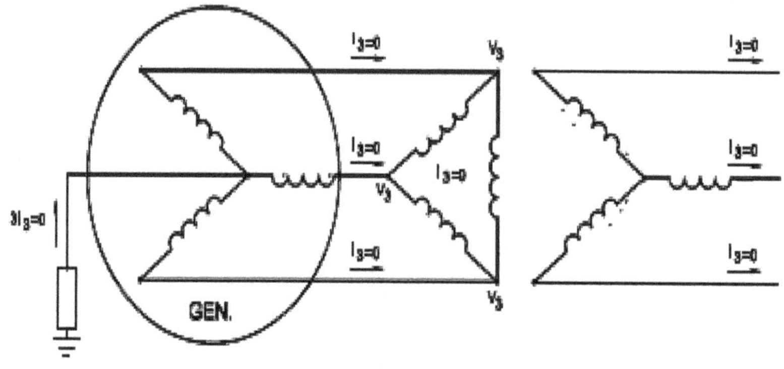

sequence current flow in the generator even though the generator is grounded. Zero-sequence current will circulate in the delta winding of this transformer. The physical limitations imposed by generator construction results in less available insulation thickness, with a resulting reduction in voltage-impulse withstand as compared to no rotating electrical equipment. Thus, special attention should be given to limiting voltage to ground by the grounding of generator neutrals.

Internal ground faults in solidly grounded generators can produce large fault currents.

These currents can damage the laminated core, adding significantly to the time and cost of repair. Such currents persist until the generator voltage decays; since they are not capable of being interrupted by the generator circuit breaker. Both magnitude and duration of these currents should be limited whenever possible.

Figure 33. Zero-sequence currents during a wye side fault

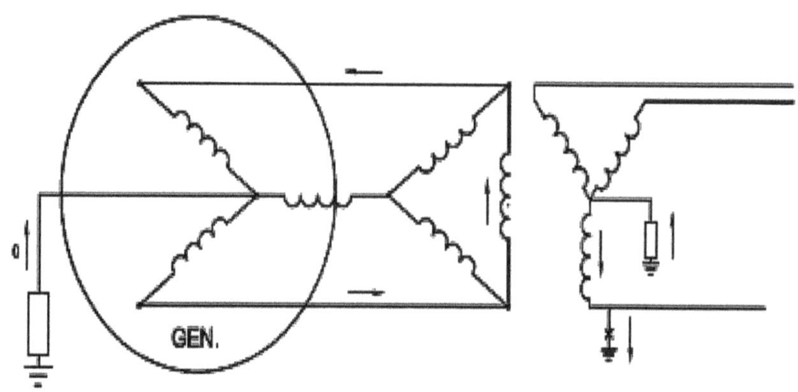

REFERENCES

Schneider. (2016). Characteristics of TT. TN and IT systems. *Wiki Electrical installation*. Retrieved from http://www.electrical-installation.org/enwiki/Characteristics_of_TT,_TN_and_IT_systems

IEE. (2008). Guide to the wiring regulations (17th ed.).

Chapter 4
Grounding System Installation

ABSTRACT

This chapter includes the following points: Methods of installing ground rods, chemical ground rod, chemical treatment of soil lessens seasonal variation of electrodes, soil treatment, trench method of soil treatment, lengthen of the earth electrode in the earth, use multiple rods, Ground Enhancement Material, (GEM), grid installation and conductor materials of grounding system. This chapter draws attention also to the following points: ground rod life expectancy, ground rod annual cost, grounding rods accessories and applications, grounding rods installation methods, chemical grounding rods, copper versus stainless steel electrodes, variation on grounding resistance system by seasonal variation effects and finally how to improve the earthing resistance.

METHODS OF INSTALLING GROUNDING RODS

Grounding Rods

Ground rods are often selected on the basis of their resistance to corrosion. The other major factor is the cost. All too often, the cost of a product is seen as the initial, up front price, but the real cost is determined by the serviceable life of the ground rod.

Galvanized steel rods are one of the cheapest electrodes available. However, they are not the most cost effective since they have a relatively short service life. Solid copper and stainless steel rods have a long service life. However,

DOI: 10.4018/978-1-5225-3853-0.ch004

Figure 1. Copper bonded ground rod versus galvanized ground rod

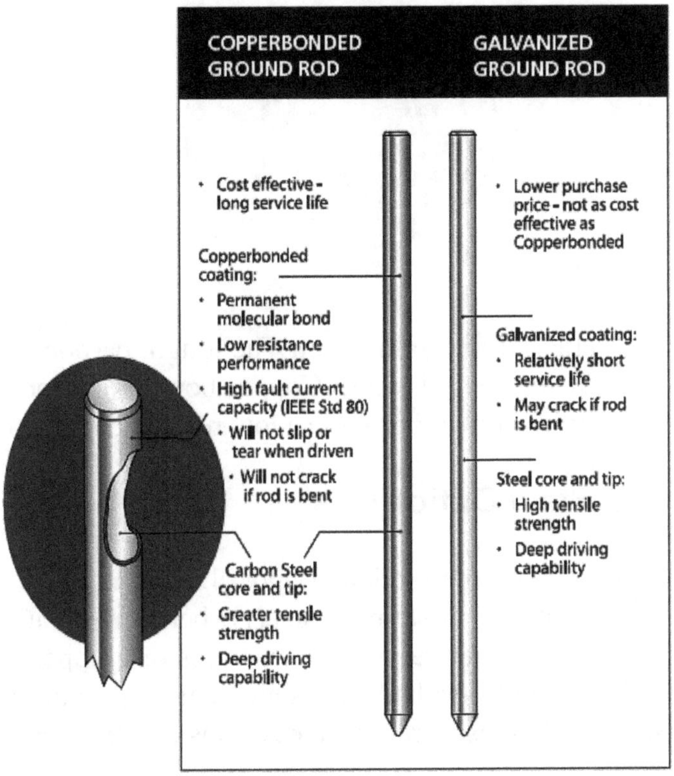

they are considerably more expensive than galvanized steel rods. In addition to this, solid copper rods are not suited to deep driving or even driving short lengths into hard ground, without bending.

As a compromise, steel cord ground rods, swaged in a copper or stainless steel sheath were developed. These ground rods are much less expensive than their solid counterparts. They are capable of being deep driven. However, the sheath of this rod type has been known to slip or tear, particularly the copper version. Once this sheath has been damaged, the integrity of the entire electrode is at risk. (Figure 1) shows copper bonded ground rod versus galvanized ground rod.

The copper bonded ground rod has an electrolytic coating of copper deposited over a layer of nickel. This process ensures a long lasting, molecular bond between the copper layer and the steel core. Recommends copper bonded ground rods because the copper coating will not slip or tear when

Figure 2. Photo of two ground rods subjected to the same pressure load test

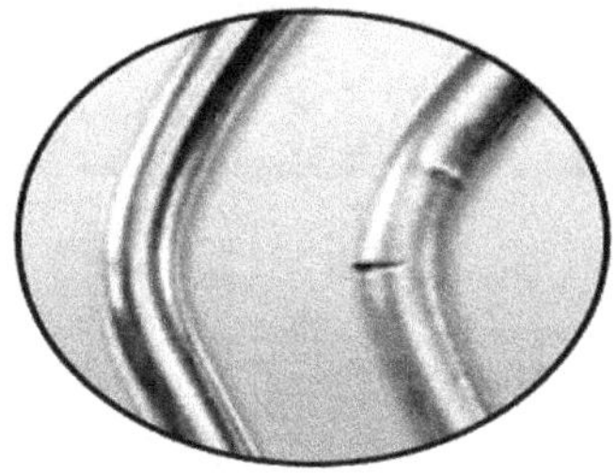

driven nor will it crack if the rod is bent. The tough, carbon steel core has good characteristics for deep driving. Copper bonded ground rods have a high resistance to corrosion and provide a low resistance path to ground.

The Stainless Steel Option

It is important to note that certain soils and land fill areas may not be compatible with copper. In these situations, stainless steel is a better proposition. Stainless steel may also be an alternative, where structures or components, such as steel towers, poles or lead sheathed cables are in close proximity to an array of ground electrodes. In these circumstances, consideration must be given to

Figure 3. Ground rod life expectancy

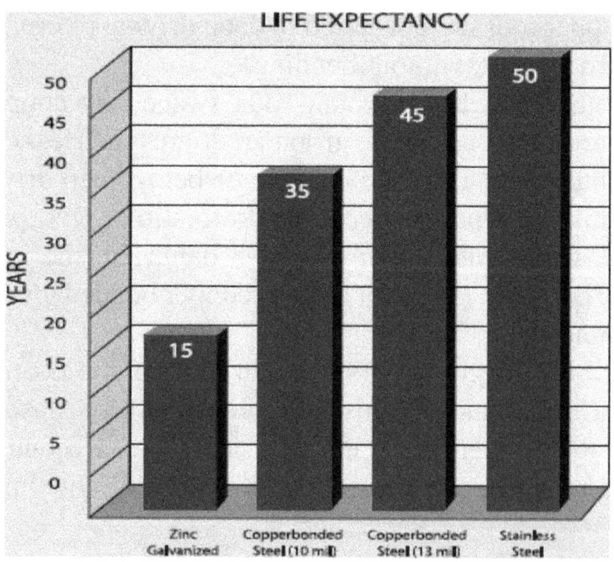

Figure 4. Ground rod annual cost

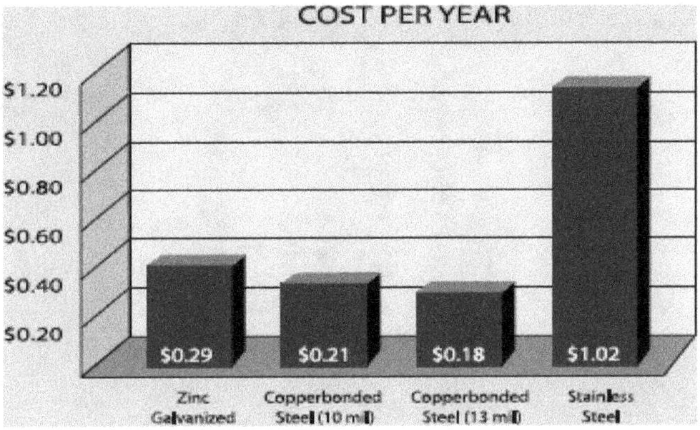

the consequence of galvanic corrosion. The high cost of stainless steel rods prohibits their widespread use. (Figure 2) gives Photo shows two ground rods subjected to the same pressure load test.

The copper bonded ground rod, shown on the left, will bend without tears, cracks or folds, to the outer sheath. The inferior copper clad rod shown on right, has developed cracks and creases to the outer sheath, which will significantly reduce its serviceable life and put the integrity of the entire electrode at risk. (Figure 3) shows the ground rod life expectancy and (Figure 4) shows the ground rod annual cost for different rods types.

GROUNDING ROD ACCESSORIES AND APPLICATION

Driving Sleeves

The driving head fits over the ground rod to protect the rod end from "mushrooming" as the ground rod is driven into earth. See (Figure 5)

Figure 5. Driving sleeves

Figure 6. Coupling

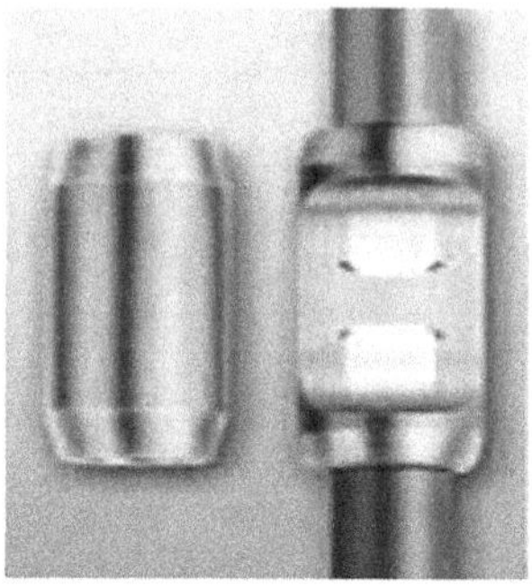

Coupling

Couplings enable ground rods to be driven deep quickly and easily without the risk of rod separation. They are generally tapered so when the rod is driven into the coupling, the two parts compress to form a conductive connection. See (Figure 6)

GROUNDING ROD INSTALLATION

There are two main methods of installing ground rods:

- It is common for the ground rod to be driven into the earth using a hand held ground rod driver or mechanically operated driver.
- Where driving the rod is not possible due to ground conditions, a hole may be drilled to take the ground rod.

Ground Rod Driver

The Ground Rod Driver is a safe, simple and effective way to install longer ground rods. This one tool enables driving of rods to ground level without heavy sledge-hammers or ladders. This saves time and money and dramatically reduces risk of employee injury. See (Figure 7). The Advantages of ground rod drivers are:

- Heavy-duty steel construction provides maximum driving forces, while the soft rubber ergonomic grip provides user comfort.

Figure 7. The ground rod driver provides a safe, simple, and affordable way to install ground rods.

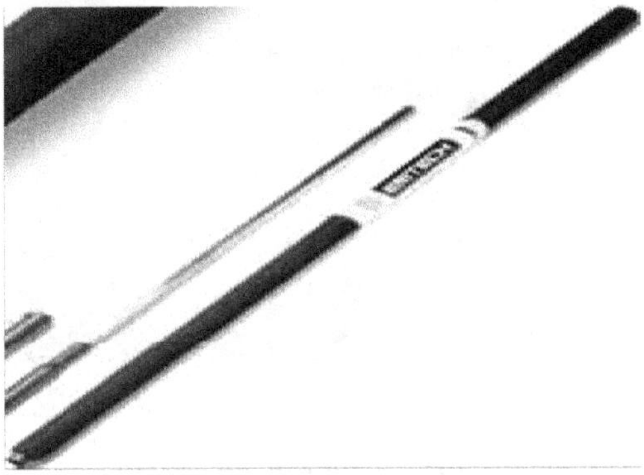

- 5/8" and 3/4" inserts are interchangeable with standard driver body.
- Use on all types of ground rods without deforming ends.
- No heavy tools or ladders required.

The features of the ground rod drivers are:

- Suitable for use on all types of cylindrical ground rods: copper bonded galvanized, stainless steel.
- 5/8" and 3/4" inserts are interchangeable with standard driver body to enable easy driving of standard rods.
- Integral insert prevents driver from slipping off the rods near ground level.
- The driver will not deform the end of the rod – making the connection of the ground conductor quick and easy.
- Convenient retaining collar holds the insert in tool when not in use.
- Self-contained and easy to store.
- Heavy-duty steel construction provides robustness and excellent driving force.
- Soft rubber ergonomic grip provides user comfort.

The steps for the installation of ground rods using the Ground Rod Driver are as follows:

Step One: Remove the insert and slide driver body over ground rod as given in (Figure 8).

Figure 8. Insert and slide driver body over ground rod

Figure 9. Position rod in vertical position and drive rod

Step Two: Position rod in vertical position and drive rod by sliding driver body in an up-and down motion over the ground rod as given in (Figure 9)

Step Three: When the end of the rod reaches approximately 900 mm (30") above ground, slide the body off the rod and place insert on ground rod. Slide the driver body over the insert and drive rod to installed depth as shown in (Figure 10).

The Mechanical Hammer

For adverse soil conditions where it is not suitable to hand drive rods or where a greater depth must be achieved, it may be necessary to use power operated aids. For normal applications we recommend the use of medium tools in the 15 to 25 lb. range, with a stroke of approximately 2" to 4" delivering 2200 blows per minute.

Figure 10. Slide the driver body over the insert and drive rod to installed depth

Drilling Options

Where driving is difficult or progress is inhibited, it is necessary to use purpose-designed drilling equipment, such as electric, pneumatic and petrol driven machinery capable of penetrating rock. Once the hole is drilled, you can use one of two methods to install the ground rod:

- Backfill the hole with Ground Enhancement Material, and add water to the hole as rod is driven to the required depth.
- Insert the rod into the drilled hole, and then add Ground Enhancement Material which has been mixed with water to form slurry.

SAMPLES OF GROUNDING ELECTRODES

Copper Bonded Pointed Grounding Rods

- Copper is molecularly bonded to nickel-sealed high-strength steel cores.
- The company of the ground rods should provide the company name, length, diameter, part number, roll-stamped within 12" of chamfered end and the UL logo and control number where applicable on each rod for easy inspection after installation. The Specifications of Couplings for Copper-bonded Pointed Rods are as follows:

Figure 11. The two types of compression coupling dimensions
A Compression Coupling – Full View
B Compression Coupling – Cut Away

Part No.	Plating Thickness (mil.)	Dia. (in.)	Lgth (ft.)
613852	5	3/8	5
613862	5	3/8	6
613880	10	3/8	8
611255	5	Nom. 1/2	5
611265	5	Nom. 1/2	6
611285	5	Nom. 1/2	8
611380*	10	1/2	8
611205	5	Nom. 1/2	10
611300*	10	1/2	10
615850	10	5/8	5
615860	10	5/8	6
615880*	10	5/8	8
615883*	13 - RUS	5/8	8
615800*	10	5/8	10
615803*	13 - RUS	5/8	10
615812*	10	5/8	12
615815*	10	5/8	15
613460	10	3/4	6
613480*	10	3/4	8
613483*	13 - RUS	3/4	8
613400*	10	3/4	10
613412*	10	3/4	12
613415*	10	3/4	15
614400*	10	1	10

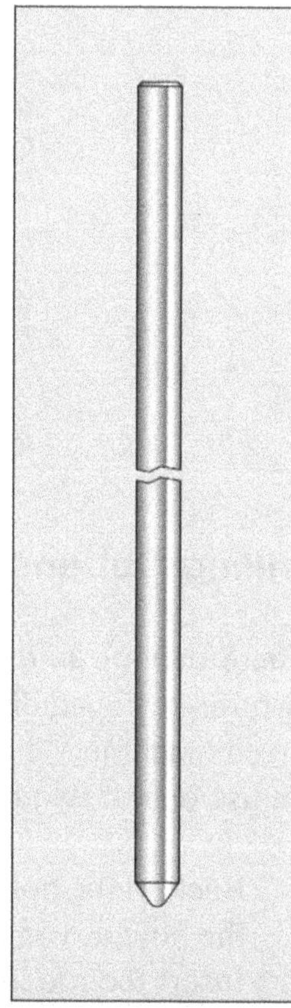

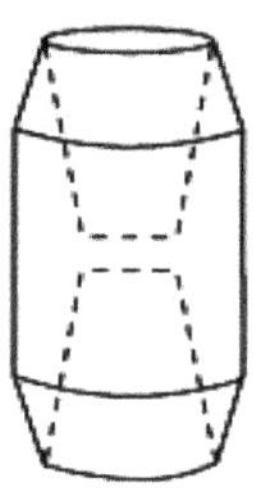

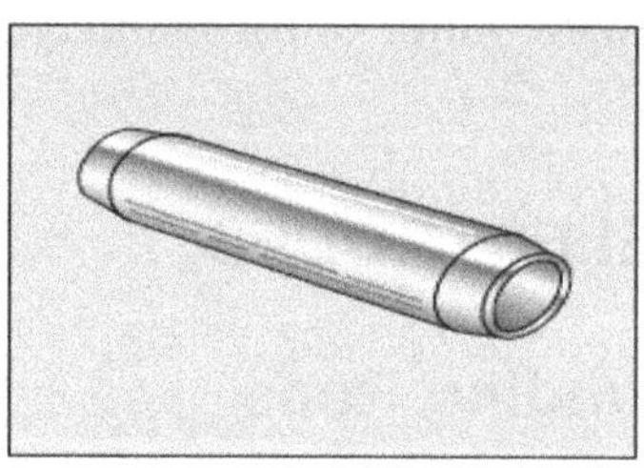

Part No.	Nominal Diameter
CC12	1/2" Nominal
CC12F	1/2" Full
CC58	5/8"
CC34	3/4"

A *Compression Coupling – Full View*
B *Compression Coupling – Cut Away*

- It has to make from high-strength silicon bronze.
- Tapered so when the rod is driven into the coupling, the two parts compress to form a conductive connection. (Figure 11) shows the two types of compression coupling dimensions.

Ground Rod Driving Sleeves

- Slides on top of ground rod to prevent mushrooming while driving in to ground (For plain (unthreaded) ground rods only).
- Sleeves for use in power assisted ground rod drivers

Rods, Couplings and Driving Studs

- The cold-rolled threads with their continuous unbroken grain flows are stronger than cut-threads.

Figure 12. Copper bonded sectional ground

Part No.	Ground Rod Size
B137-14	1/2" Copperbonded or Steel Rod
B137-16	5/8" Copperbonded (.563" dia.)
B137-31	5/8" Steel (.625" dia.)
B137-18	3/4" Copperbonded (.682" dia.)
B137-33	3/4" Steel (.750" dia.)
B137-22	1" Copperbonded (.914" dia.)
B137-37	1" Steel (1.00" dia.)

Part No.	Ground Rod Size (Unthreaded)
DH58	5/8" Copperbonded (.563" dia.)
DH34	3/4" Copperbonded (.682" dia.)

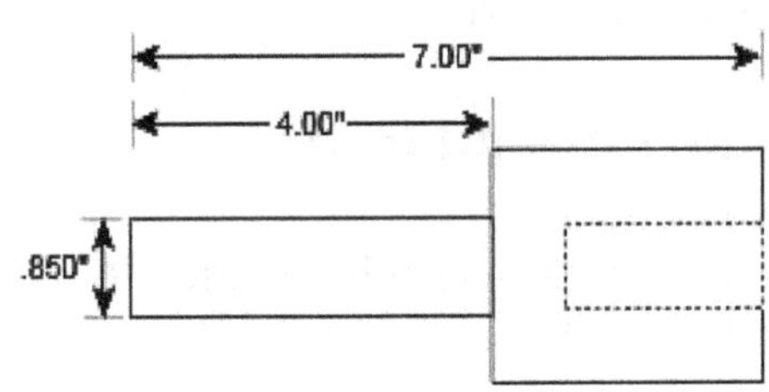

Figure 13. Grounding rods dimensions

Part No.	Plating Thickness (mil)	Dia. (in.)	Length (ft.)
631380*	10	1/2	8
631300*	10	1/2	10
635830	10	5/8	3
635840	10	5/8	4
635850	10	5/8	5
635860	10	5/8	6
635880*	10	5/8	8
635883*	13 - RUS	5/8	8
635800*	10	5/8	10
633480*	10	3/4	8
633400*	10	3/4	10
634400*	10	1	10

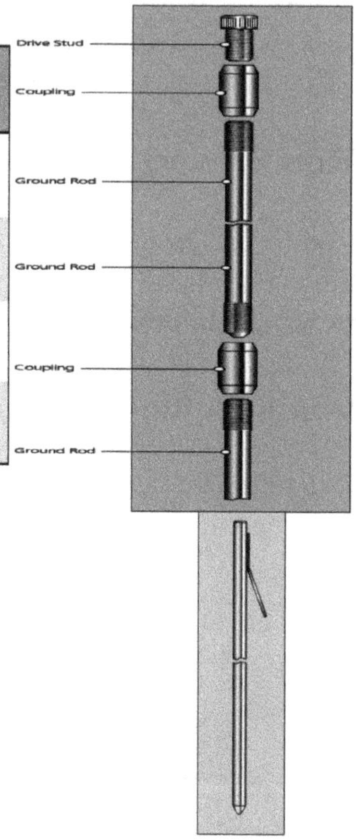

Couplings for Sectional Rods

Part No.	Rod Size
CR-12	1/2" Nominal
CR-12S	1/2" Full Size
CR-58	5/8" Nominal
CR-34	3/4" Nominal
CR-100	1" Nominal

Driving Studs for sectional rod

Part No.	Rod Size
DS12	1/2" Nominal
FDS12	1/2" Full Size
DS58	5/8" Nominal
DS34	3/4" Nominal
DS1	1" Nominal

- High-strength couplings are threaded bronze and chamfered at both ends for easy driving.
- High strength, corrosion-resistant couplings ensure permanent, low resistance copper-to-copper connections. (Figure 12) and (Figure 13) give grounding electrode dimensions

Copper Bonded Pigtail Ground Rods

Pigtails are either copper weld or copper wire welded to rod. Dimensions of such rod are tabulated in (Figure 14).

Figure 14. Dimensions of copper bonded pigtail ground rods

Part No.	Nominal Dia X Length	Cu Thickness (Mil.)	Wire
6138529	3/8" x 5'	5 Mil	#10CW x 18" Lg
6138629	3/8" x 6'	5 Mil	#10CW x 18" Lg
615889	5/8" x 8'	10 Mil	#6 AWG Cu x 48" Lg
615809	5/8" x 10'	10 Mil	#6 AWG Cu x 48" Lg
6158839	5/8" x 8'	13 Mil	#6 AWG Cu x 48" Lg
6158039	5/8" x 10'	13 Mil	#6 AWG Cu x 48" Lg

Figure 15. Dimensions of solid stainless steel ground rods

Part No.	Description	Dia. (in.)	Length (ft.)
681300	304 Pointed	FS 1/2	10
683400	304 Pointed	3/4	10
683400S	304 Sectional	3/4	10
685880	304 Pointed	5/8	8
685800	304 Pointed	5/8	10
685880S	304 Sectional	5/8	8
681300S	304 Sectional	1/2	10

Solid Stainless-Steel Ground Rods

Sectional rods utilize a cut thread for highly corrosive soil. (Figure 15) gives the dimensions of Solid Stainless-Steel Ground Rods

Galvanized Pointed Ground Rods

Such this rod has to meet NEMA, GR1 standards. It has to be Zinc-coated using hot-dip galvanized method for solid protection against corrosion. Its surface has to be rigidly inspected to eliminate seams, slivers and other defects. The dimensions of such this electrode are given in (Figure 16).

Figure 16. Galvanized pointed grounding rods dimensions

Part No.	Rod Size
811250	1/2" x 5'
811260	1/2" x 6'
811280	1/2" x 8'
815860	5/8" x 6'
815880*	5/8" x 8'
815800	5/8" x 10'
813480*	3/4" x 8'
813400	3/4" x 10'
814400	1" x 10'

Figure 17. Copper ground plate electrode with terminated 2 foot welded pigtail connection

Part No.	Dimensions
GPECEAH24 "xx" 24	12"x24"x20 Ga Cu
GPECEAJ18 "xx" 24	18"x18"x20 Ga Cu
GPECEAJ24 "xx" 24	18"x24"x20 Ga Cu
GPECEAJ36 "xx" 24	18"x36"x20 Ga Cu
GPECEAK24 "xx" 24	24"x24"x20 Ga Cu
GPECEAM36 "xx" 24	36"x36"x20 Ga Cu
GPECEBH12 "xx" 24	12"x12"x1/16" Cu
GPECEBH24 "xx" 24	12"x24"x1/16" Cu

Copper Ground Plate Electrode With Terminated 2 Foot Welded Pigtail Connection

As given in (Figure 17).

Copper Ground Plate Electrode With 2 Foot Welded Through Pigtail Connection

As given in (Figure 18).

Figure 18. Copper ground plate electrode with 2 foot welded through pigtail connection

Part No.	Dimensions
GPECCAH24 'xx' 24	12"x24"x20 Ga Cu
GPECCAJ18 "xx" 24	18"x18"x20 Ga Cu
GPECCAJ24 "xx" 24	18"x24"x20 Ga Cu
GPECCAJ36 "xx" 24	18"x36"x20 Ga Cu
GPECCAK24 "xx" 24	24"x24"x20 Ga Cu
GPECCAM36 "xx" 24	36"x36"x20 Ga Cu
GPECCBH12 'xx' 24	12"x12"x1/16" Cu
GPECCBH24 'xx' 24	12"x24"x1/16" Cu

'xx' - Cable code for required conductor type and size

Copper Ground Plates

As given in (Figure 19)

- Usually is making from 20 gauge thick, high conductivity copper sheet.

Figure 19. Copper ground plates

Part No.	Description
LPC753	12" wide by 24" long
LPC754	18" wide by 18" long
LPC755	36" wide by 36" long

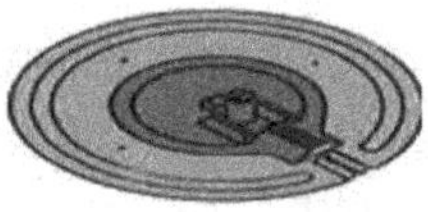

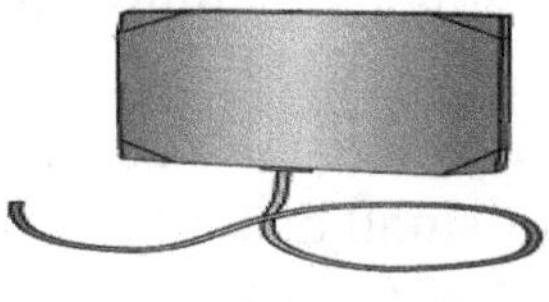

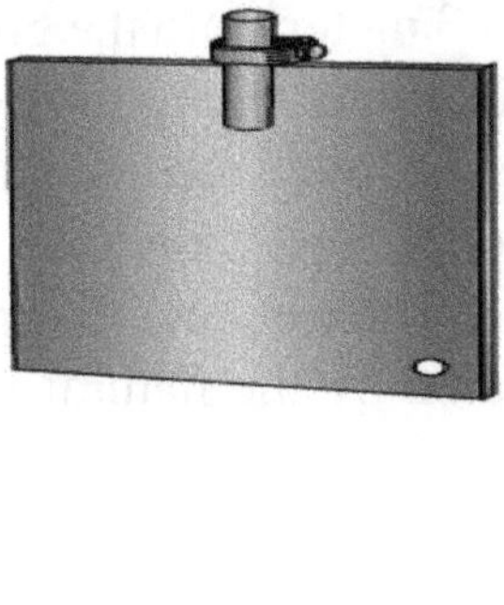

- Two cable attachments LPC535L securely fastened to plate.
- Cable attachments LPC532L securely fastened to plate.

CHEMICAL GROUNDING RODS

Effect of Soil Properties in Grounding

If it is not possible to change the fundamental nature /properties of soil at a given location, the soil resistivity will be reduced by adding an amount of salts to the soil. Local variations of soil conditions occur even in a small area.

When a location for making grounding pit has to be selected, preference should be given to location which is likely to give minimum electrical resistance. The soils have been arranged in ascending order with regard to their electrical resistance as following:

1. Wet marshy lands, or lands containing ashes, cinders or brine waste (average resistivity 2400 ohm/cm)
2. Clay, loamy soil, arable land clay
3. Clay and loam mixed with varying proportion of gravel & sand (average resistivity 15,800 ohm/cm)
4. Damp and wet sands
5. Dry sand
6. Gravel and stones

A location for grounding system, which has water flowing over, is not preferred, because flowing of water normally washes away all conducting salts. When the soil available is of high electrical resistance, then its resistance can be lowered by chemical treatment, the treatment is limited to the area in the neighbor-hood of earth electrode.

Pure water is a poor conductor, but the addition of salts improves the conductivity to a large extent. The presence of small quantity of salt reduces the soil resistance by eighty percent.

Ground electrode should be buried at least 3 meters into the ground 'with charcoal and coke powder mixed with chemical salts like sodium chloride or magnesium chloride put around the electrode. A typical method is to place the chemical in a circular trench around but not touching the electrode. Common salt and Magnesium sulphate are mostly used. The treatment is effective but not permanent, as the chemicals are dissipated by the natural drainage of

Table 1. Ground resistivity for various kinds of the soil and concrete

Type of Soil	Soil Resistivity Ohm.m	Grounding Resistance (1/2)					
		Grounding (m) Depth			Grounding Strip (m)		
	1/2 m	3	6	10	5	10	20
Moist humus soil, moor soil, swamp	30	10	5	3	12	6	3
Farming so, loamy and clay soils	100	33	17	10	40	20	10
Sandy clay soil	150	50	25	15	60	30	15
Moist sandy soil	300	66	33	20	80	40	20
Dry sand soil	1000	330	165	100	400	200	100
Concrete 1:5	400	-	-	-	160	80	40
Moist gravel	500	160	80	48	200	100	50
Dry gravel	1000	330	165	100	400	200	100
Stony soil	30000	1000	500	300	1200	600	300

the soil and must be replaced. A further disadvantage is that corrosion of the rod is accelerated by the chemical treatment. To eliminate the corrosion of electrode to some extent copper is used as the earth electrode since its resistances corrosion is comparatively large to that of the metal iron.

The salts commonly used for' chemical treatment of soil are:

1. Sodium Chloride.
2. Calcium Chloride.
3. Sodium Nitrate.
4. Magnesium Sulpher and Poly Qualifiers mixture of chemicals, which make the soil conductive for a longer period.

Table 1 gives summary of the ground resistivity for various kinds of the soil and concrete

Enhanced Grounding Rods

Chemical Ground Rod sometimes known as: Chemically Activated Ground/earth Electrodes. This solution has a very large contact area with the earth (critical cylinder). In addition, the chem.-rod chemically 'conditions' the surrounding soil in contact with the unit, so that if the moisture drops off it still has the best possible contact via the natural electrolytes/minerals it uses. Typically, one chem.-rod can replace up to 10 standard rods, which makes it the preferred choice in difficult ground conditions or where there isn't a lot of space for deployment.

Simply the Enhanced Ground Rod is a conductive hollow tube ground rod, usually manufactured from copper. It contains special hygroscopic, electrolytic salts. These salts form a saline solution by absorbing moisture out of the atmosphere. This saline solution leaches out of the bottom of the rod, which gradually lowers resistivity of the surrounding soil, forming "electrolytic roots" over time.

The salt mixture is critical. Harger, NEMA (2017) utilizes a special combination of Sodium Chloride and Calcium Chloride. Calcium Chloride is an "active" salt, which continually draws moisture out of the air and forms the solution. Many other providers of this type of electrode utilize salts such as magnesium chloride, some even use common water softener pellets. These salts do not draw moisture out of the air; they must be activated by adding water. This may lower resistivity initially, however, unless water is continually added, the salts dry out over time and resistivity of the electrode goes back up.

To increase the efficacy of the Enhanced Ground Rod, a very low resistance ground enhancement material is placed around the rod. Harger, NEMA (2017) proudly uses Ultra-fill; an ultra- low resistance carbon based material. Harger, NEMA (2017) exothermically attaches a conductor of your choice to the

Figure 20. Enhanced ground rod elements

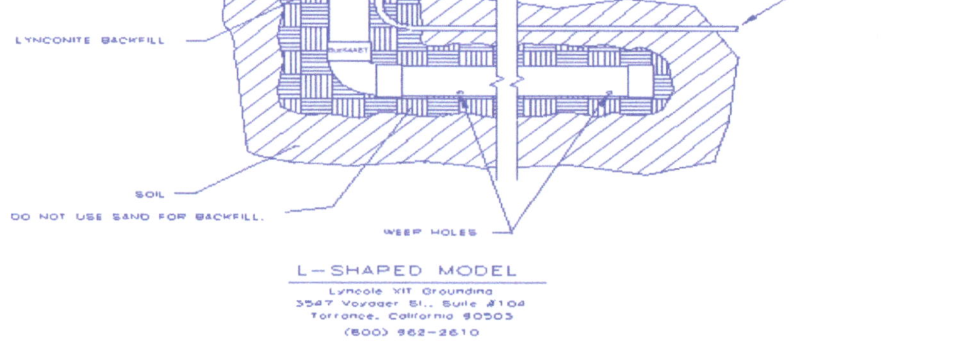

enhanced ground rod. This conductor is called the tail. The tail direction is very important. Harger, NEMA (2017) provides a design which allows the current, either lightning or electrical fault, to maintain a downward sloping path to ground.

Most manufacturers utilize a design which forces lightning to go "uphill" before reaching the ground rod. Some manufactures claim a superior "x" design. Although at first this sounds good, it causes the contractor to make twice as many connections, thus increasing the cost of installation.

Herger offers two basic styles, vertical and horizontal (L-shaped). A variety of lengths, sectionals and different kits to meet the specific requirements are produced by some companies.

The Enhanced Ground Rod Numbering System given in (Figure 21) allows customizing the product to meet specific needs. In the above example, the product specified is a 10' long, L-shaped ground rod with a #2 solid tinned 5' tail, 50 pound bags of Ultra fill and an inspection well.

The vertical ground rods shall be copper clad steel and shall be 3/4 inch or larger diameter, and with a length of 10 to 40 feet. They are to be driven vertically into undisturbed earth, or earth that has been compacted and settled. The tops of the ground rods shall be between 1 and 2 feet below final grade level

Horizontal ground conductors shall consist of No. 4/0 AWG bare copper cables, a minimum of 16 feet in length, and shall be buried a minimum of two feet below final grade. They shall be run perpendicular to an interconnecting cable, such as the perimeter ground cable, or radially from a single interconnection point. Horizontal ground conductor is permitted only when the use of vertical ground rods is not feasible due to geological conditions or other subterranean obstructions at the site.

Figure 21. Enhanced grounding rod numbering system

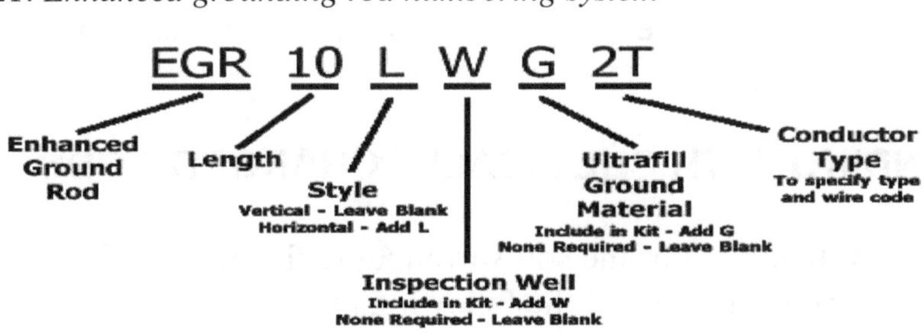

Application notes of Horizontal Ground Conductors are summarized as follows:

- Enhanced grounds are used in high resistivity soil conditions and when low resistance ground electrode systems are critical.
- L-shaped enhanced ground rods are used when proper depth cannot be achieved due to physical conditions such as bedrock.
- Conductor is welded 18" down from the top.
- Outside diameter of copper tube is 2-1/8".
- 35 year warranty.

COPPER VS. STAINLESS STEEL

The permanence of copper in most soils, its resistance to chemical attack, and its inherent low resistance, brings it into widespread use throughout the electrical industry around the world. However, there are certain soils where it is inadvisable to use copper such as in tidal land salt marshes, swamps, and land filled with ashes, coke breeze and like materials. Stainless steel earth rods have a high resistance to both atmospheric and soil corrosion being clad with an austenitic grade stainless steel having chromium content of approximately 17%. Typical of the applications where stainless steel clad earth rods are flavored over copper clad is:

1. Where the chemical composition of the soil reacts more unfavorably than copper - as per conditions described above.
2. Where the earthed item needs to be protected against galvanic attack and corrosion, e.g. Lead sheathed cables, steel poles, etc.
3. Where the tougher sheathing of stainless steel will provide for a more durable and rigid earth rod better suited to hard driving conditions than its copper counterpart. Moreover, the cladding operation imparts an extra toughness to the stainless steel through work hardening.

GROUNDING WITH CHEMICALLY CHARGED RODS

The chemical treatment of the soil surrounding the ground electrodes may be implemented by any one of the following three ways:

Several conductive backfill materials exist on the market that is used to replace poorly conducting soil near the ground electrodes. The impact of putting these materials around the electrodes is significant, since that is where the majority of connection to the earth takes place. Four such materials used for conductive backfills around ground electrodes are described below.

1. Concrete has a resistivity range of 30 to 90 Ohm-meters. Since it is hydroscopic by nature it will tend to absorb moisture when available and keep it up to 30 days, thus maintaining a resistivity lower than the surrounding soil. However, during a long dry season concrete will dry out with a subsequent rise in resistivity. Also, if a substantial amount of fault or lightning current is injected into a concrete encased electrode, the moisture in the concrete may become steam, dramatically increasing in volume and placing a substantial stress on the concrete.
2. Bentonite is naturally occurring clay with a resistivity of about 2.5 Ohm-meters. It used widely as a conductive backfill material. Like concrete it is hydroscopic, which makes it subject to the same drying out concern as concrete. Expansion range of bentonite can reach 300 percent, which means that in dry situations it can shrink away from an electrode, resulting in substantial increase of ground resistance.
3. Clay-based backfill mixtures have generally a resistivity lower than pure bentonite due to the addition of carbon or/and other minerals that provide a greater spectrum of electrically conducting materials. Because these mixtures are clay-based they retain their hydroscopic character to some degree, while the blending of materials dampens the resistance variability with respect to moisture.
4. Carbon-based backfills materials have generally a resistivity lower than clay-based mixtures. Some of these materials can be mixed with concrete to make concrete more conductive. However, these materials tend to be the most expensive and do not retain moisture nearly as well as clay-based materials.

The amount of the backfill material required is determined in most cases by the Interfacing Volume and Critical Cylinder principles. A ground electrode establishes a connection to earth by affecting only a certain volume of earth, called the Interfacing Volume (IV).

For practical purposes for a ground rod the entire connection to earth is contained within an IV whose radius is 2.5 times the length of the rod. Most of the earth connection takes place in a cylinder close to the electrode, called the Critical Cylinder. A study of the influence of soil within the IV demonstrates that six inches of soil along any radial makes up 52 per cent of the connection to earth; a12 inches makes up 68 percent of the connection.

Beyond a diameter of 24 inches there is very little improvement for much larger diameters. Therefore, the recommended diameter for the Critical Cylinder is between 12 and 24 inches, and the calculated amount of the required backfill material is based on that diameter and the length of the ground rod.

The chemically charged ground rods (CCGR) can be used instead of the conventional ground rods. A CCGR is a copper tube of 2-2.5 inches in diameter with several small holes perforated along the length of the tube. The tube is filled with metallic salt evenly distributed along the entire length of the tube. The moisture absorbed from the air and soil form a solution of the contained metallic salt within the CCGR which seeps out through the holes into the surrounding soil, thus lowering the soil's resistivity and increasing the efficiency of the electrode. Figure 1 illustrates the concept of the CCGR.

A combination of the CCGR with backfill can be done. This ground system demonstrates the excellent stability over numbers of years in keeping ground resistance permanently at a low fixed level. As may be seen from that figure, the grounding system based on the implementation of the CCGRs with backfill appears to be the most economical solution in the given conditions of poor conducting soil and low values of the required ground resistance, even without taking into consideration the cost of real estate. Where the appropriate space is not available or too expensive, the CCGRs the only solution in establishing required ground system chemical ground electrodes provide a low impedance ground in locations of high soil resistivity. Used in conjunction with a Bentonite backfill and ERICO's unique GEM material, the chemical ground electrode system, provides a method to improve soil resistivity directly surrounding the electrode.

It maintains a low ground resistance, maintenance-free installation that dissipates lightning energy and other dangerous electrical fault currents, even in sandy or rocky soil conditions. The Chemical Ground Electrode is useful for providing an effective earth in poor soil conditions where spacing for electrodes is limited.

BENFITS AND FEATURES

- Contains natural electrolytic salts, which permeate into the surrounding soil to condition the soil and increase its conductivity.

- Low impedance to effectively dissipate lightning and electrical fault currents.
- Over 70 configurations available, other assemblies made to order.
- Easy connection to ground electrode conductor using the factory provided pigtail (up or down orientation)
- Provides decades of reliable services due to rugged construction, and high quality metals with a 30 year minimum service life
- 2-1/8" OD Type K copper pipe, 0.083" wall contains natural electrolytic salts that permeate into the surrounding soil, improving electrode to soil connection.
- Available up to 20 feet in length. Longer rods can be field assembled using 10 foot sections.
- Optional factory attached radial strips are available to reduce impedance to high frequency lightning energy and to control the direction of the dissipation
- L-shaped rods are available for horizontal installation applications where it is impractical to auger deep vertical holes. The Chemical Ground Electrode is suitable for high soil resistivity, dry soil conditions, as it can replace multiple conventional ground rods, making the system ideal where space limitations apply. Where seasonal variations cause fluctuations in soil resistivity, the Chemical Ground Electrode ensures a constant and reliable low impedance ground.

CHEMICAL GROUNDING ELECTRODE SYSTEM

Chemical Ground Electrodes are most effective when installed as part of a total system that includes high conductivity backfill materials, access/ inspection wells, and permanent, reliable connections. They may be installed either vertically or horizontally, as shown below in (Figure 22) recommends installing the complete system. Chemical electrodes are available in a range of standard and custom configurations and are available individually or as part of a complete kit as reported by ERICO (2017)

(Figure 23) gives the typical vertical installation showing major components required to form an effective ground electrode.

Chemical ground rods can be obtained as individual components or as a complete kit. Kits include the chemical electrode (pre-filled with electrolytic salts), Bentonite, GEM backfill and an access well.

Figure 22. Chemical ground electrode can be installed either vertically or horizontally

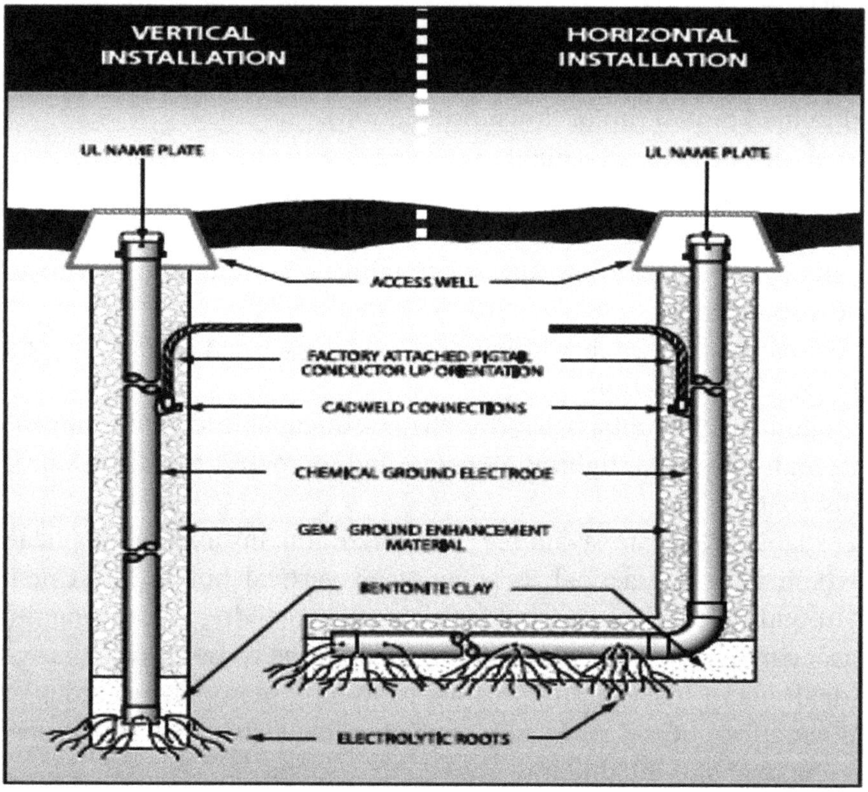

Only rarely do grounding system designers and contractors get to work on a site with good grounding conditions. Even under ideal circumstances, soil structure can vary and make it difficult to achieve uniform, low levels of resistivity across a wide area. Under almost all soil conditions, the use of a ground enhancement material will improve grounding effectiveness. Some are permanent and require no maintenance. When selecting a ground enhancement material it is important to be sure that it is compatible with the ground rod, conductor and connection material.

To improve the conductivity of grounding system recommends using Ground Enhancement Material (GEM).

GEM is a low-resistance, non-corrosive, carbon dust based material that improves grounding effectiveness, especially in areas of poor conductivity. GEM contains cement, which hardens when set to provide a permanent, maintenance-free, low-resistant grounding system that never leaches or washes

Figure 23. Typical vertical installation showing major components required to form an effective ground electrode

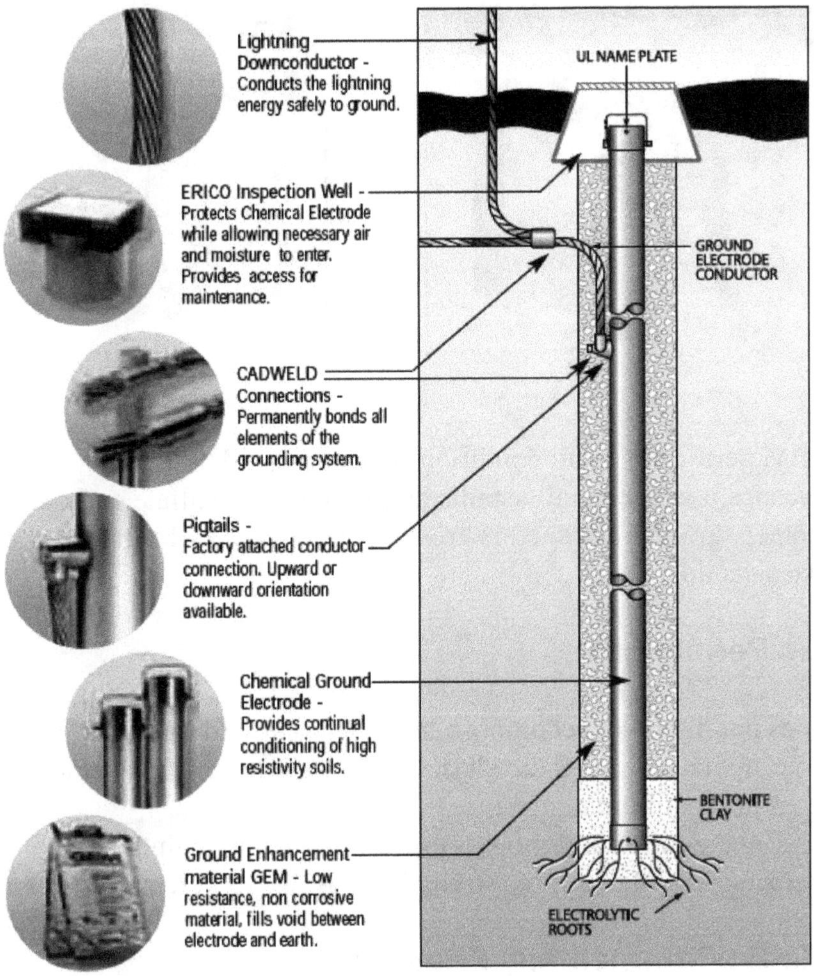

away. This is reported by ERICO (2017). Such this material is complying with IEC 62561-7 standards (2011). (Figure 24) shows that, GEM has a resistivity factor more than 20 times lower than betonies clay.

GEM Is Effective

- GEM can dramatically lower earth resistance and impedance measures
- GEM maintains constant resistances for the life of the system once in set form

Figure 24. The graph clearly shows GEM has a resistivity factor more than 20 times lower than betonies clay.

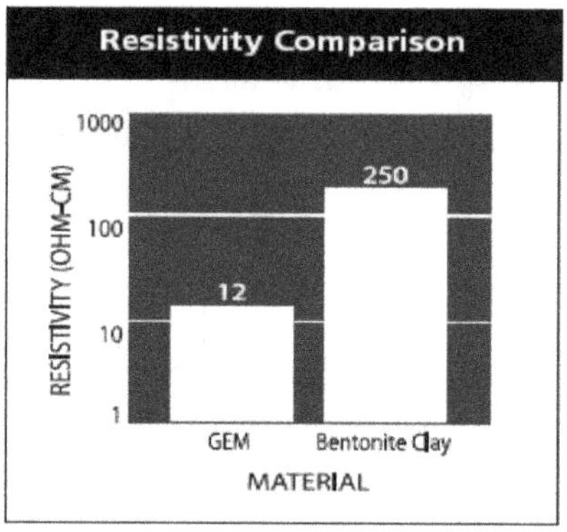

- GEM performs in soil conditions even during dry spells
- Because it is chemically stable and very low in sulfate and chloride, it protects ground conductors from corrosion instead of attacking them like salts do.

GEM Is Permanent

- Does not dissolve, decompose nor leach out with time
- Does not require periodic charging treatments nor replacement
- Does not require maintenance
- Does not require continuous presence of water to maintain effectiveness
- Freezing will increase resistivity by only 10-15 percent

GEM Is Environmentally Friendly

- Does not adversely affect the soil
- Does not leach ions or contaminate ground water
- Meets (USA) Environmental Protection Authority requirements for landfill

Application

GEM improves grounding effectiveness regardless of soil conditions and provides excellent permanent conductivity:

Figure 25. Ground inspection wells

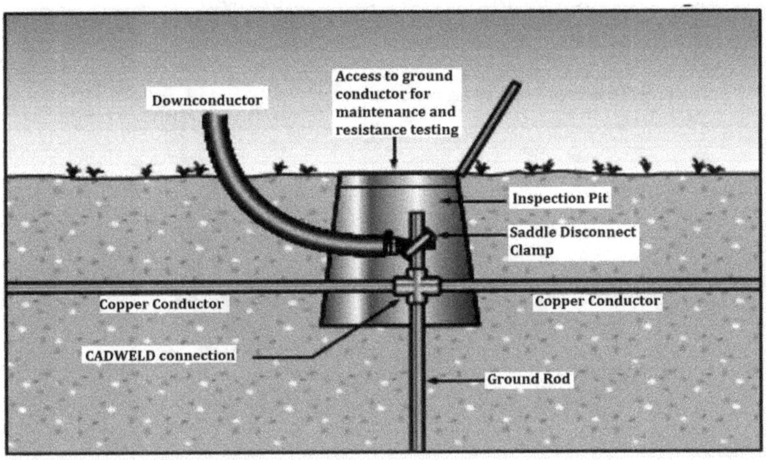

- For areas with high resistance, such as rocky ground, mountain tops, and sandy soil
- As a backfill when you have to drill because the ground is too hard to drive, or where ground rods cannot be driven, or
- Where limited space makes adequate grounding difficult by conventional methods

GROUNDING INSPECTION WELLS

For periodically measuring the electrical resistance of a buried ground system, inspection wells are used as a means of access to the ground conductor. To make electrical resistance measurements, remove the cover and attach a lead from a resistance measuring instrument to the ground conductor. Inspection wells are available in various sizes and materials. Light-duty units are generally used. For areas of high vehicular traffic, you should use heavy-duty inspection wells. See (Figure 25) and (Figure 26).

Polymer Concrete reinforced with heavy weave fiberglass resulting in high strength and minimal weight. Enclosures and covers rated for 10,000 lbs. maximum load. See (Figure 27).

Figure 26. Resistance testing using a ground resistance meter, with access to the ground conductor provided by the ground inspection well

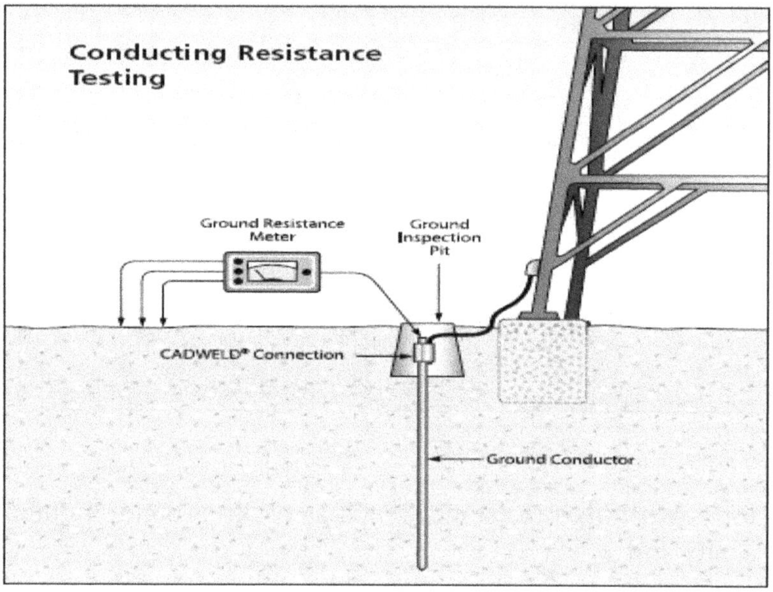

Figure 27. Polymer concrete reinforced

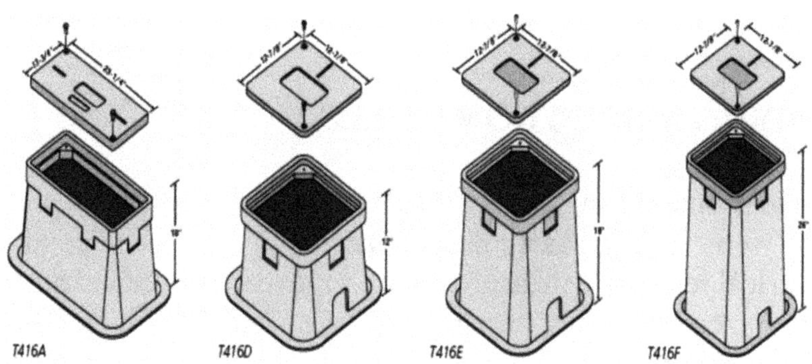

EQUIPMENT GROUNDING BARS

Proper bonding is essential to create an equipotential plane between service grounds and equipment during fault and transient conditions. This equipotential plane provides a near zero voltage differential and serves to protect people and equipment during these events. The most popular bonding product in use today is the ground bar or bonding bar. Ground bars provide a convenient,

Figure 28. Equipment ground bars

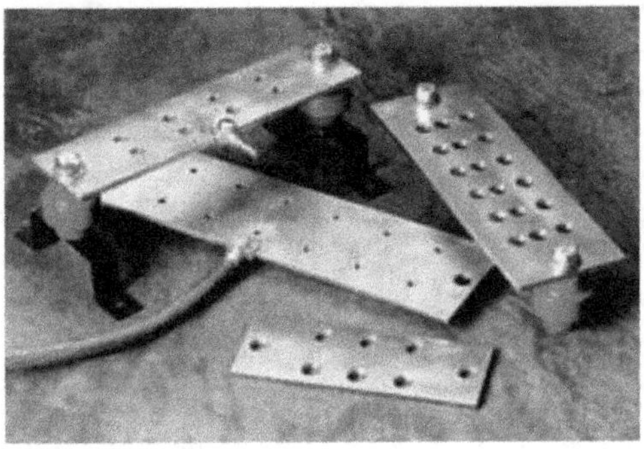

single-point grounding and bonding location. Conductors are welded to the bar or are mechanically fastened by using lugs as given in (Figure 28).

EFFECT OF SEASONAL VARIATION ON GROUNDING RESISTANCE

Because soil resistivity directly relates to moisture content and temperature, it is reasonable to assume that the resistance of any grounding system will

Figure 29. Seasonal variation effects of earth resistance with an electrode of 3/4" pipe on stony clay soil. Depth of electrode in earth is 3 ft for Curve 1 and 10 ft for Curve 2.
As reported by Eaton (1941).

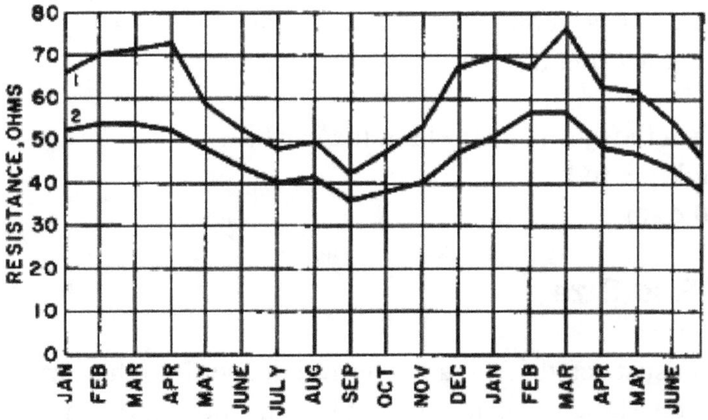

Figure 30. Chemical treatment of soil effects on seasonal variation

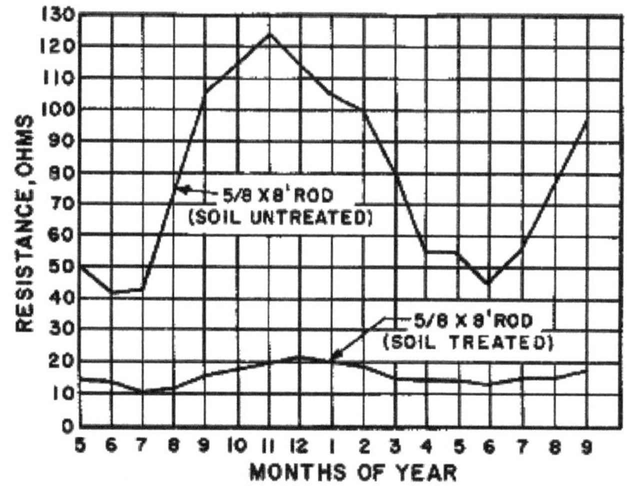

vary throughout the different seasons of the year. Such variations are shown in (Figure 29). Since both temperature and moisture content become more stable at greater distances below the surface of the earth, it follows that a grounding system, to be most effective at all times, should be constructed with the ground rod driven down a considerable distance below the surface of the earth. Best results are obtained if the ground rod reaches the water table. See (Figure 29) and (Figure 30).

HOW TO IMPROVE THE EARTH RESISTANCE

When the earth electrode resistance of your grounding system is not low enough, there are several ways you can improve it as reported by EEP Electrical Engineering (2016):

1. Lengthen the earth electrode in the earth.
2. Use multiple rods.
3. Treat the soil.

Effect of Rod Size

As you might suspect, driving a longer rod deeper into the earth, materially decreases its resistance. In general, doubling the rod length reduces resistance

Figure 31. Earth resistance vs. depth of electrode in earth
Source: Ref 19.

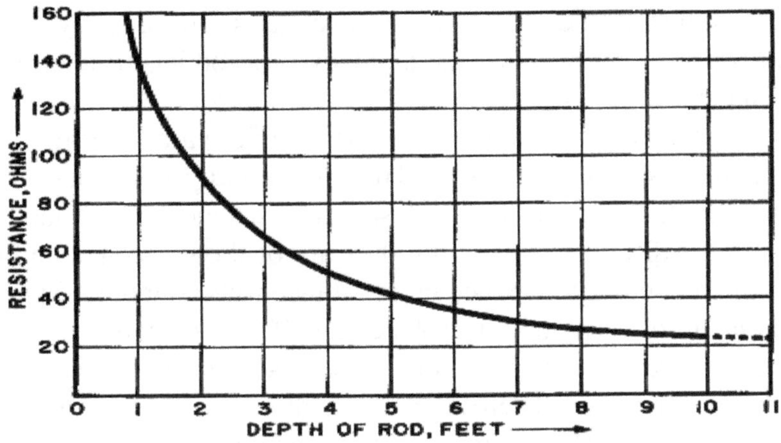

by about 40 percent. The curve of (Figure 31) shows this effect. For example, note that a rod driven 2 ft down has a resistance of 88 Ω; the same rod driven 4 ft down has a resistance of about 50 Ω. Using the 40 percent reduction rule, the reduction in the resistance will be 88 x 0.4 = 35 Ω. By this calculation, a 4-ft deep rod would have a resistance of 88 - 35 or 53 Ω comparing closely with the curve values. You might also think that increasing the electrode diameter would lower the resistance. It does, but only a little. For the same depth, doubling the rod's diameter reduces the resistance only about 10 percent. (Figure 32) shows this relationship. For example, a 10-ft deep rod, 5/8 in. in diameter, has a resistance of 6.33 Ω.; increasing its diameter to

Figure 32. Diameter of a rod has little effect on its earth resistance: curve A, from Ref. 19 given in Megger ERCO (2017), curve B, average of Underwriters Laboratories tests at Chicago, curve C, average of Underwriters Laboratories tests at Pittsburgh

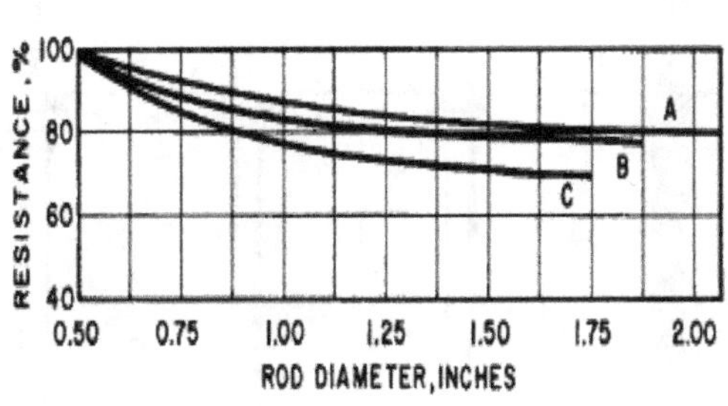

1-1/4 in. lowers the resistance to 5.6 Ω. For this reason, you normally only consider increasing the rod diameter if you have to drive it into hard terrain. (Figure 31) shows that earth resistance decreases with depth of electrode in earth. (Figure 32) shows that the diameter of a rod has little effect on its earth resistance.

Two well-spaced rods driven into the earth provide parallel paths. They are, in effect, two resistances in parallel. The rule for two resistances in parallel does not apply exactly; that is, the resultant resistance is not one-half the individual rod resistances (assuming they are of the same size and depth). Actually, the reduction for two equal resistance rods is about 40 percent. If three rods are used, the reduction is 60 percent; if four, 66 percent, see (Figure 33). When you use multiple rods, they must be spaced apart further than the length of their immersion. There are theoretical reasons for this, but you need only refer to curves such as (Figure 34). For example, if you have two rods in parallel and 10-ft spacing, resistance is lowered about 40 percent. If the spacing is increased to 20 percent, reduction is about 50 percent.

Figure 33. Average results obtained from multiple-rod earth electrodes

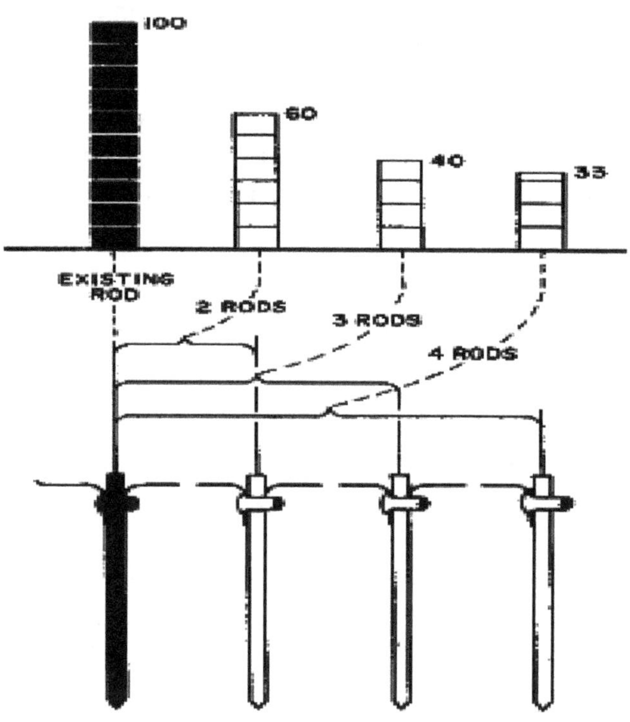

Figure 34. Comparative resistance of multiple-rod earth electrodes, single rod equals 100%

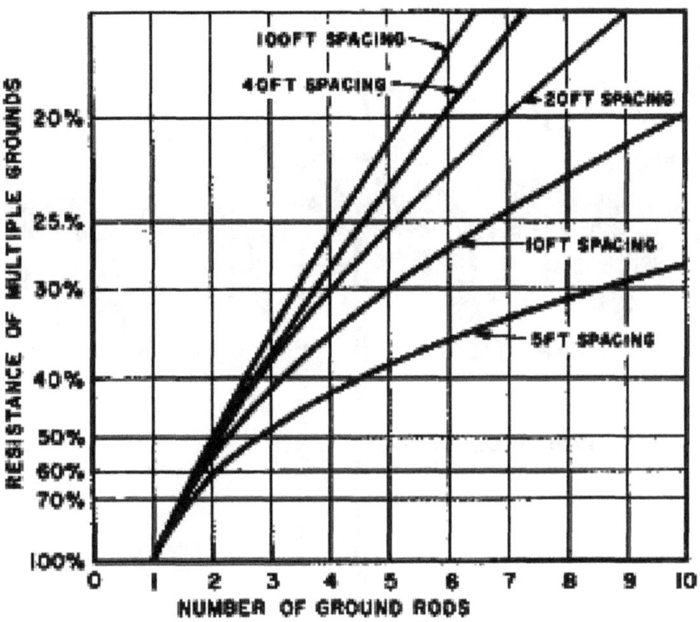

Treatment of the Soil

Chemical treatment of soil is a good way to improve earth electrode resistance when you cannot drive deeper ground rods because of hard underlying rock, for example. It is beyond the scope of this manual to recommend the best treatment chemicals for all situations. You have to consider the possible corrosive effect on the electrode as well as United States Environmental Protection Agency (EPA), (2017) and local environmental regulations. Magnesium sulfate, copper sulfate, and ordinary rock salt are suitable noncorrosive materials. Magnesium sulfate is the least corrosive, but rock salt is cheaper and does the job if applied in a trench dug around the electrode (see Figure 19). It should be noted that soluble sulphates attack concrete, and should be kept away from building foundations. Another popular approach is to backfill around the electrode with a specialized conductive concrete. A number of these products, like Bentonite, are available on the market. Chemical treatment is not a permanent way to improve your earth electrode resistance. The chemicals are gradually washed away by rainfall and natural drainage through the soil. Depending upon the porosity of the soil and the

Figure 35. Trench method of soil treatment

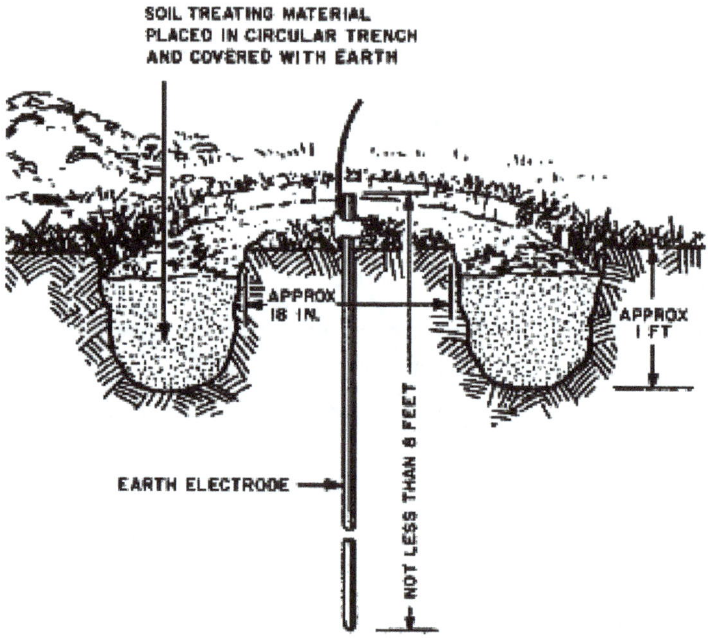

Figure 36. Chemical grounding electrode in the tranche

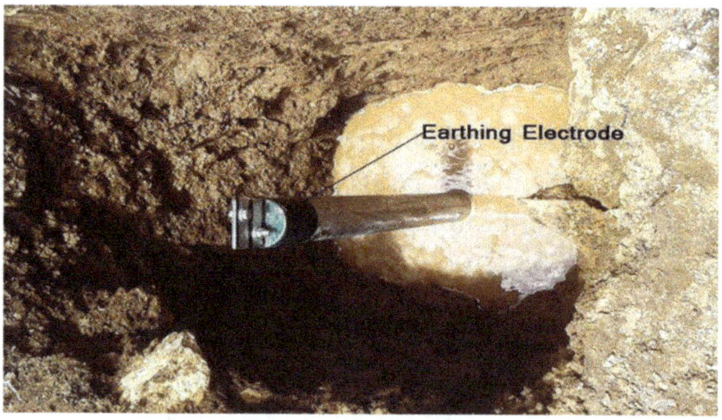

amount of rainfall, the period for replacement varies. It may be several years before another treatment is required. Chemical treatment also has the advantage of reducing the seasonable variation on resistance those results from periodical wetting and drying out of the soil. However, this method should be considered when deep or multiple electrodes are not practical. (Figure 35) shows the trench method of soil treatment. Figure 36 shows chemical grounding electrode in the tranche.

REFERENCES

Eaton, J. R. (1941). *Grounding Electric Circuits Effectively*. General Electric Review.

EEP Electrical Engineering. (2016). Good ways to improve earth electrode resistance. Retrieved from http://electrical-engineering-portal.com/improve-earth-electrode-resistance

EPA. (2017). United States Environmental Protection Agency. Retrieved from https://www.epa.gov/

IEC. (2011). 62561-7 Lightning protection system components (LPSC) - Part 7: Requirements for earthing enhancing compounds.

Megger, E. R. C. O. (2017). A practical guide to earth resistance testing. Retrieved from http://contractconnections.co.nz/gem-standards.html

NEMA. (2014). Harger Lightning and Grounding. Retrieved from https://www.nema.org/Manufacturers/Pages/Harger-Lightning-and-Grounding.aspx, http://www.harger.com/

Chapter 5
Design of Grounding Grid According to IEEE Standards

ABSTRACT

This chapter contains the following points: design procedure of grounding system according to IEEE 80, methods for calculating the grounding grid resistance (Dwight's formula, Laurent and Niemann, Sverak's equation, Schwarz's Formula, Dawalibi, Mukhedkar's Formula, Chow and Salama's Formula, Nahman's Formula and Heppe's Method). It contains also the design of charts of grid earthing system and application of step and mesh potential in safe grounding system design. This chapter draws attention also to the following points: Grounding resistance of grounding system in non-homogeneous soil, calculations of maximum step and mesh voltages, estimation of minimum buried grid conductor length and finally computerized analysis in grounding design.

DESIGN CRITERIA

There are two main design goals to be achieved by any substation ground system under normal as well as fault conditions. These are:

1. To provide means to dissipate electric currents into the earth without exceeding any operating and equipment limits
2. To assure that a person in the vicinity of grounded facilities is not exposed to the danger of critical electric shock.

DOI: 10.4018/978-1-5225-3853-0.ch005

The design procedures described in the following sections are aimed at achieving safety from dangerous step and touch voltages within a substation. It is possible for transferred potentials to exceed the GPR of the substation during fault conditions. The design procedure described here is based on assuring safety from dangerous step and touch voltages within, and immediately outside, the substation fenced area. Since tine mesh voltage is the worst possible touch voltage inside the substation (Excluding transferred potentials).

The mesh voltage will be used as the basis of this design procedure. Since the mesh voltage is the worst possible touch voltage (excluding transferred potentials), the mesh voltage will be used as the basis of design procedure.

Step voltages are inherently less dangerous than mesh voltages. If, however, safety within the grounded area is achieved with the assistance of a high resistively surface layer (crushed rock), which does not extend outside the fence, then step voltages may be dangerous.

In any event, the computed step voltages should be compared with the permissible step voltage after a grid has been designed that satisfies the touch voltage criterion. For equally spaced ground grids, the mesh voltage will increase along meshes from the center to the corner of the grid. The rate of this increase will depend on the size of the grid, number and location of ground rods, spacing of parallel conductors. Diameter and depth of the conductors, and the resistivity profile of the soil. In a computer study of three typical grounding grids in uniform soil resistivity, the data shown in Table (6.1) were obtained. These grids were all symmetrically shaped square grids with no ground rods and equal parallel conductor spacing.

CRITICAL PARAMETERS

The following site-dependent parameters have been found to have substantial impact on the grid design: maximum grid current (I_G), fault duration (t_f), shock duration (t_s), soil resistivity (p), high resistivity surface material (ps), and grid geometry. Several parameters define the geometry of the grid, but the area of the grounding system, the conductor spacing, and the depth of the ground grid have the most impact on the mesh voltage, while parameters such as the conductor diameter and the thickness of the surfacing material have less impact. A brief discussion or review of the critical parameters is given below.

Maximum Grid Current (I_G)

In determining the maximum current I_G consideration should be given to the resistance of the ground grid, division of the ground fault current between the alternate return paths and the grid, the decrement factor, and the future expansion of the power system.

Fault Duration (t_f) and Shock Duration (t_s)

The fault duration and shock duration are normally assumed equal, unless the fault duration is the sum should reflect fast clearing time for transmission and industrial substations. The choices t, and t_s should result in the most pessimistic combination of fault current decrement factor and allowable body current. Typical values for t_f and ts range from 0.25 -1.0 s. Soil resistivity (p)

The grid resistance and the voltage gradients within a substation are directly dependent on the soil resistivity. Since in reality soil resistivity will vary horizontally as well as vertically, sufficient data must be gathered for a substation yard.

The equations can employ only a single value for the resistivity. There is no simple method to determine a value from the field test data that can yield an accurate ground grid analysis using these simplified equations. However, the following points may provide the user with general guidelines.

1. The soil can be considered uniform if the difference between two extremes of all the resistivity values of the field test data is 30% or less.
2. When an equivalent two-layer soil model is determined and the ground system is in the upper layer, the value of *p,* (upper-layer soil resistivity) can be used in the simplified equation for negative values of reflection factor K, the grids designed using uniform soil analysis will have higher step and touch voltages than the grids designed with the equivalent two-layer model, p, if is used as the soil resistivity.

Resistivity of surface layer (ps)

A thin surface layer of crushed rock helps in limiting the body current by adding resistance to the equivalent body resistance. Varies from, 1000-5000 O-m have been used for ps.

INDEX OF DESIGN PARAMETERS

Table 1 contains a summary of the design parameters used in the design procedure.

DESIGN PROCEDURE

The block diagram of (Figure 1) illustrates the sequences of steps to design the ground grid. The parameters shown in the block diagram are identified in the index presented in Table 1.

1. The property map and general location plan of the substation should provide good estimates of the area to be grounded. The soil resistivity test will determine the soil resistivity profile and the soil model needed (that is uniform or two-layer model).
2. The fault current $3I_0$ should be the maximum expected future fault current that will be conducted by any conductor in the grounding system, and the time should reflect the maximum possible clearing time (including back-up).
3. The tolerable touch and step voltages are determined by equations. The choice of time t_s is based on the judgment of the design engineer.
4. The preliminary design should include a conductor loop surrounding the entire grounded area, plus adequate cross conductors to provide convenient access for equipment grounds, etc. The initial estimates of conductor spacing and ground rod locations should be based on the current I_G and the area being grounded.
5. Estimation of the preliminary resistance of the grounding system can be done, more accurate estimation of the resistance may be desired, especially when ground rods are used to reach more conductive sub soils. For this application the effect of different soil resistivity in computing the grid resistance and the rod bed resistance are included by equations. Computer analysis based on modeling the components of the grounding system in detail can compute the resistance with a high degree of accuracy, assuming the soil model is chosen correctly.

6. In order to prevent gross over-design of the grounding system, only that portion of the total fault current $3I_0$ that flows through the grid to remote earth (and contributes to the mesh and step voltages and the GPR) should be used in designing the grid. The current IG should, however, reflect the worst fault type and location, the decrement factor, and any future system expansion.
7. If the GPR of the preliminary design is below the tolerable touch voltage, no access to equipment grounds is necessary.
8. The calculation of the mesh and step voltages for the grid as designed can be done using IEEE 80 (2000) standard formulas.
9. If the computed mesh voltage is below the tolerable touch voltage, the design may be complete [see Step (10)]. If the computed mesh voltage is greater than the tolerable touch voltage, the preliminary design shall be revised [see Step (11)].
10. If both the computed touch and step voltages are below the tolerable voltages, the design needs only the refinements required to provide access to equipment grounds. If not, the preliminary design must be revised [see Step (11)].
11. If either the step or touch tolerable limits are exceeded, revision of the grid design is required. These revisions may include smaller conductor spacings, additional ground rods, etc. More discussion on the revision of the grid design to satisfy the step and touch voltage limits is given
12. After satisfying the step and touch voltage requirements, additional grid conductors and ground rods may be required. The additional grid conductors may be required if the grid design does not include conductors grid conductors be grounded. Additional ground rods may be required at the base of surge arresters, transformer neutrals...etc.

METHODS FOR CALCULATING THE GROUNDING GRID RESISTANCE

Vertical Ground Electrode

The vertical electrode is a simple structure in grounding device. Conformity to the analysis of the electromagnetic field, the grounding resistance of the vertical electrode can be calculated. Not only Sunde and Dwight formula (1968) is used to calculate the grounding resistance but also Tagg and Ollendorf, Zingraff (1964) formulas can be used.

Table 1. Design parameters

Description	Symbol
Symmetrical fault current in substation in Amp	31_0
Maximum grid current that flows between grid and surrounding earth in Amp	I_G
Soil resistivity in Ω m	P
Surface layer resistivity in Ω m	P_s
Surface layer thickness	H_s
Current projection factor for future system growth	C_p
Surface layer resistivity de-rating factor	C_s
Duration of fault current for sizing ground conductor in sec.	t_c
Duration of fault current for determining decrement factor in sec	T_f
Duration of shock for determining allowable body current in sec	t_s
Depth of ground grid in meter	h
Diameter of grid conductor in meter	D
Total area enclosed by ground grid in m²	A
Spacing between parallel conductors in meter	D
Decrement factor for determining IG	D_f
Number of parallel conductors in one direction	n
Spacing factor for mesh voltage	K_m
Spacing factor for step voltage	K_s
Corrected factor for grid geometry	K_i
Corrective weighting factor that adjusts the effects of inner conductors on the corner mesh	K_{ii}
Corrective weighting factor that emphasizes the effects of grid depth	K_h
Total length of grounding system conductors	L
Resistance of grounding system in Ω	R_g
Mesh voltage at the center of the corner mesh	E_m
Step voltage between a point above the outer corner of the grid and appoint 1m diagonally outside the grid	E_s
Tolerable touch voltage for human with 50Kg body weight	$E_{touch\ 50}$
Tolerable touch voltage for human with 70Kg body weight	$E_{touch\ 70}$
Tolerable step voltage for human with 50Kg body weight	E_{step50}
Tolerable step voltage for human with 70Kg body weight	E_{step7}

According to Sunde and Dwight (1968):

$$R_g = \frac{\rho}{2\pi L}\ln\left(\frac{4L}{r} - 1\right) \tag{1}$$

Figure 1. Design procedure block diagram as reported by IEEE 80 (2000) standard

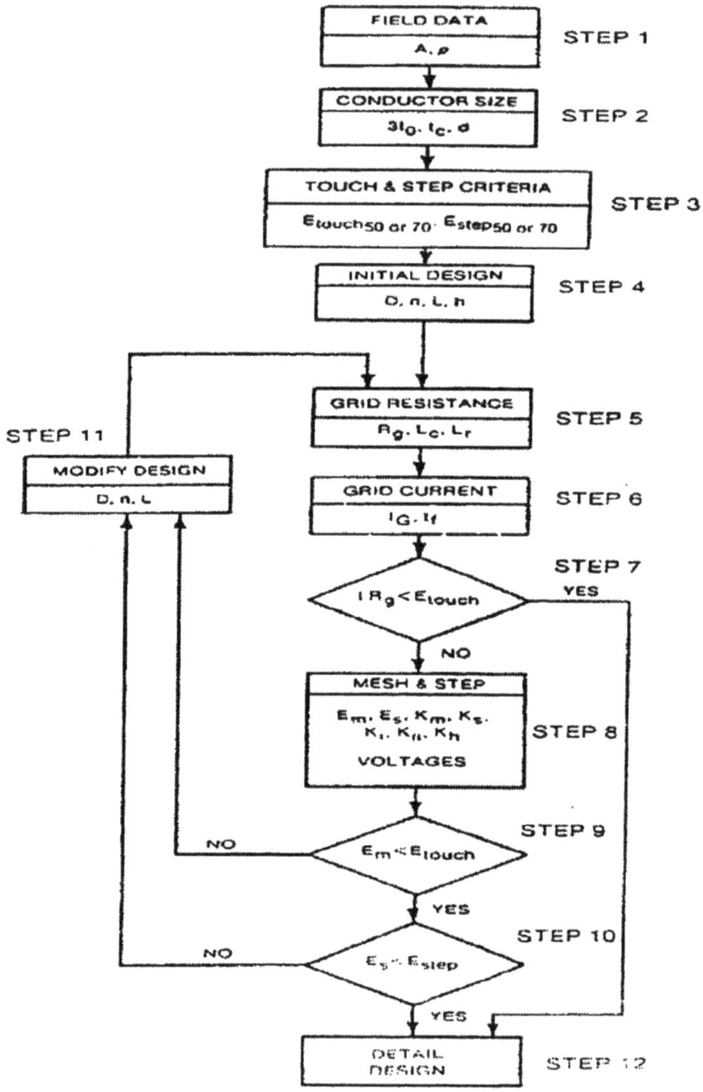

where ρ is the soil resistivity, r is radius of electrode, and L is length of electrode.

And the Tagg, Ollendorf and Zingraff formula (1964) is:

$$R_g = \frac{\rho}{2\pi L}\ln\left(\frac{2L}{r}\right) \quad (2)$$

and the Rudenberg and Datta formula (1986) is:

$$R_g = \frac{\rho}{2\pi L}\ln\left(\frac{L}{r}\right) \tag{3}$$

Horizontal Ground Electrode

For horizontal grounding electrode with length L, radius r, and burial depth h, the grounding resistance formula by Sunde and Schwarz (1968) based on the average potential is:

$$R_g = \frac{\rho}{\pi L}\ln\left(\frac{2L}{\sqrt{2rh}} - 1\right) \tag{4}$$

Formula derived based on the mid-point potential method is:

$$R_g = \frac{\rho}{\pi L}\ln\left(\frac{2L}{\sqrt{2rh}}\right) \tag{5}$$

Rudenberg and Zingraff formula (1964) is:

$$R_g = \frac{\rho}{2\pi L}\ln\frac{L}{r}\left(1 + \frac{\ln\frac{L}{2h}}{\ln\frac{L}{r}}\right) \tag{6}$$

Grounding Resistance of a Grounding Grid

Calculating the total resistance of grounding grid is one of the first steps in set layout and size of grounding grid. The grid grounding resistance depends on foremost on the area to be occupied by the grounding grid, which is known when designing the grounding grid. The value of grid grounding resistance in the uniform soil can be calculated as reported by Xie (1991).

$$R_g = \frac{\rho}{4}\sqrt{\frac{\pi}{A}} \tag{7}$$

where A is area of grid. From the above equation, it is noticed that the area of grounding grid plays an important role in decreasing grounding resistance R_g.

Laurent and Niemann formula as reported by Jinliang He, Rong Zeng, and Bo Zhang (2013) is as follows

$$R_g = \frac{\rho}{4}\sqrt{\frac{\pi}{A}} + \frac{\rho}{L_t} \quad (8)$$

where L_t is the total length of the grounding conductors.

Equation (8) takes into account not only area of grounding grid but also the total length of the conductor. So it gives the maximum possible value of grounding resistance of grid.

Sverak (1984), modified Equation (9) to take into account the effect of grid depth as follows:

$$R_g = \rho\left[\frac{1}{L_t} + \frac{1}{\sqrt{20A}}\left(1 + \frac{1}{1 + h\sqrt{20/A}}\right)\right] \quad (9)$$

Schwarz (1954) developed the next formula from Sunde (1968) and Rüdenberg (1968) to determine the total resistance of grounding grid that consisting horizontal electrode and vertical rod in the homogenous soil.

$$R_g = \frac{R_1 R_2 - R_{12}^2}{R_1 + R_2 - 2R_{12}} \quad (10)$$

where

R_1 = the resistance of grid conductors in Ω

R_2 = the resistance of all ground rods in Ω

R_{12} = the mutual resistance between the group of grid conductors, R_1, and group of ground rods, R_2 in Ω.

Schwarz developed a set of convenient formulas, defining R_1, R_2, and R_{12} in terms of the basic design parameters, assuming uniform soil conditions as reported by IEEE Std. 80. (2000).

However, in practice, it is often desirable to drive ground rods deep into the ground to reach more conductive soils. In the equations below, the expressions for R_2 and R_{12} have been modified for such a possibility.

$$R_1 = \left(\rho_1/\pi l_1\right)\left(\ln\left(2l_1/h'\right) + k_1\left(l_1 / \sqrt{A}\right) - k_2\right) \tag{11}$$

$$R_2 = \left(\rho_a/2n\pi l_2\right)\left(\ln\left(8l_2/d_2\right) - 1 + 2k_1\left(l_2 / \sqrt{A}\right)\left(\sqrt{n} - 1\right)^2\right) \tag{12}$$

$$R_{12} = \left(\rho_a/\pi l_1\right)\left(\ln\left(2l_1/l_2\right) + k_1\left(l_1/\sqrt{A}\right) - k_2 + 1\right) \tag{13}$$

where

ρ_1 = soil resistivity encountered by grid conductors buried at depth h in $\Omega.m$
ρ_a = apparent soil resistivity as seen by a ground rod in $\Omega.m$
H = thickness of the upper layer soil in m
ρ_2 = soil resistivity from depth H downward in $\Omega.m$
l_1 = total length of grid conductors in m
l_2 = average length of a ground rod in m
h = depth of grid burial in m
h'= $\sqrt{d_1 h}$ for conductors buried at depth h, or 0.5 d_1 for conductor at h = 0 (on earth's surface)
A = area covered by a grid of dimensions a• b in m^2
n = number of ground rods placed in area A
K_1, K_2 = constants related to the geometry of the system (Figure 2)
d_1 = diameter of grid conductor in m
d_2 = diameter of ground rods in m
a = short-side grid length in m
b = long-side length in m

The preceding equations (5), (6) and (7) are valid for a two-layer soil environment, with upper layer thickness H, in which ground rods penetrate the more conductive lower layer. In such a case, that is for $\rho_1 \geq \rho_2$ where the grid is buried in the upper layer ρ_1, but the ground rods are partly in ρ_1 and partly in ρ_2, R_2 and R_{12} are calculated with the use of an apparent soil resistivity seen by the ground rods, ρ_a is defined as reported by IEEE Std. 80. (2000):

$$\rho_a = l_2\left(\rho_1\rho_2\right)\left(\rho_2 H + \rho_1\left(l_2 - H\right)\right) \tag{14}$$

For more usual case of the rod's top being in the same laying depth of the grid, the apparent soil resistivity is defined as reported by IEEE Std. 80. (2000)

$$\rho_a = l_2\left(\rho_1\rho_2\right)/\left(\rho_2\left(H - h\right) + \rho_1\left(l_2 + h - H\right)\right) \tag{15}$$

For uniform soils, $\rho_2 = \rho_1$.

If the difference between ρ_1 and ρ_2 is not too great (preferably ρ_2 not lower than 0.2 ρ_1), and the first layer thickness H is at least 0.1 b, the resulting equations are reasonably accurate for most practical calculations and relatively easy to use. Moreover, the ability to work with separate expressions for a grid and a set of rods becomes advantageous in simplified calculations.

A slight problem with the application of these equations was that the coefficients K_1 and K_2 had been originally presented by Schwarz only in a graphical form.

However, given the near-linear character of these curves, it is possible to use a linearized form of $y = px + q$ to obtain K_1 and K_2 within the range of values shown in (Figure 2); or to linearly interpolate between several points taken from the original curve.

Grounding Resistance of Grounding System in Non-Homogeneous Soil

Actually, even there are a lot of equations calculate the grounding resistance in uniform soil, there is a few formula formed to calculate the grounding resistance in non-homogenous soil.

Laurent (1972) suggested equation for calculating resistance of horizontal grounding grid in two soils when top layer thickness is smaller than the size

Figure 2. Coefficients K_1 and K_2: a) coefficient K_1 of Schwarz's Formula; b) coefficient K_2 of Schwarz's Formula
As reported by IEEE Std. 80. (2000).

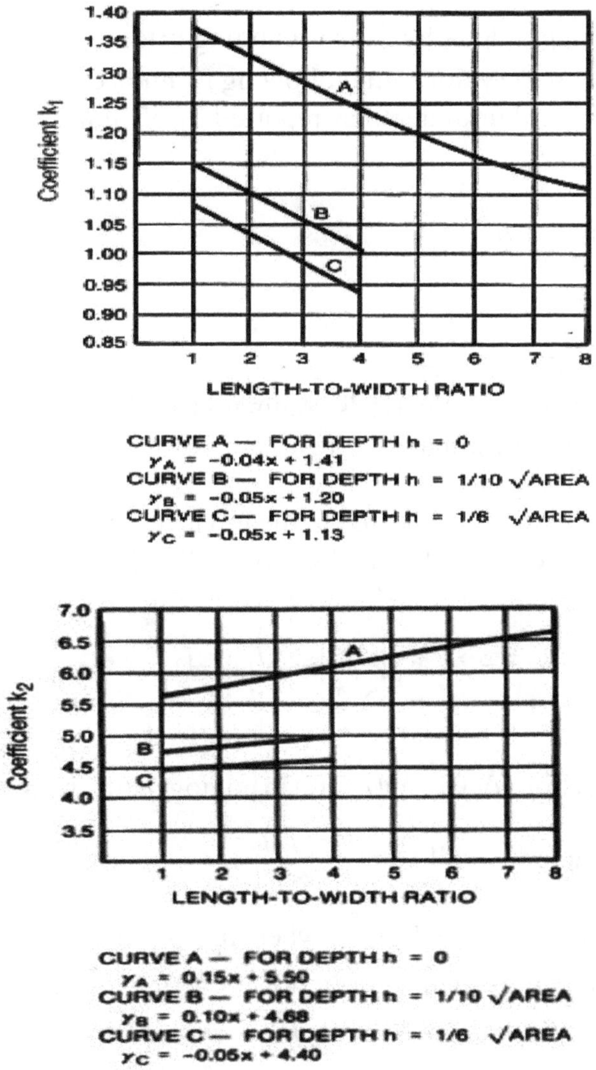

(a)

(b)

of grounding grid and ρ_1 and ρ_2 are the resistivity of top layer and resistivity of bottom layer, respectively.

$$R_g = \frac{\rho_2}{4}\sqrt{\frac{\pi}{A}} + \frac{\rho_1}{L_G} \tag{16}$$

$$R_g = 1.6\frac{\rho_2}{L_P} + 0.6\frac{\rho_1}{L_G} \tag{17}$$

where L_P is the peripheral length of the grounding grid.

If the top soil layer is very thick, the following formula can be adopted to consider the influence of burial depth as reported by Nahman and Salamon (1984):

$$R_g = 1.6\frac{\rho_2}{L_P} + 0.6\frac{\rho_1}{L_G} + \rho_1\frac{H}{A} \tag{18}$$

where H is thickness of top layer.

When grounding grid has vertical rods, Nahman et al. (1986) have modified the Schwarz's equation to estimate the resistance of grounding grid in the two- layered soil.

If all vertical ground rods are located in the top layer, then the grounding resistance of vertical rod can be calculated as reported by IEEE Std. 80. (2000):

$$R_2 = \left(\frac{\rho_1}{2m\pi L_R}\right)\left[\ln\left(\frac{8L_R}{d_2}\right) - 1 + 2K_1K_p\left(\frac{L_R}{\sqrt{A}}\right) - \left(\sqrt{m} - 1\right)^2\right] \tag{19}$$

where K_p is the correction coefficient of soil in homogeneity, which is related to the reflection coefficient of soil $k = {(\rho_2 - \rho_1)}\big/{(\rho_1 + \rho_2)}$ and the soil structural coefficient P. When $\rho_2 > \rho_1$, $= H/\sqrt{A}$; when $\rho_1 > \rho_2$, $P = (H - h)/\sqrt{A}$.

The relationship between the correction coefficient K_p, reflection coefficient K and structural coefficient P is showed in (Figure 3) and (Figure 4).

If vertical ground rods penetrate into the second layer, then the grounding resistance can be calculated as reported by IEEE Std. 80. (2000)

$$R_2 = \frac{\rho_a}{2m\pi L_R}\ln\left(\frac{8L_R}{d_2} - 1\right) + \left(\frac{K_1(0)K'_P\rho_2}{\pi m\sqrt{A}}\right)\left(\sqrt{m} - 1\right)^2 \tag{20}$$

where

$\rho_a = \rho_2 L_R / L_a$, $L_a = L_2 + L_1 \rho_2 / \rho_1$

where L_1 and L_2 are the lengths of the rod parts in the upper and bottom layers, respectively and $\left(L_1 + L_2 = L_R\right)$.

$K_1\left(0\right)$ is the value of correction coefficient K_1 when $h = 0$ and the area is A, and $K_P' = 0$ is the correction coefficient corresponding to the grounding grid buried in the lower bottom layer, as illustrated in (Figure 3).

When the horizontal grounding grid is buried in the top-layer soil, its grounding resistance can be calculated as reported by IEEE Std. 80. (2000)

$$R_1 = \frac{\rho_1}{\pi L_G}\left[K_r \ln\left(\frac{2L_G}{h'} + N - 1\right) + K_P K_1 \left(\frac{L_G}{\sqrt{A}} - \frac{2\left(x+1\right)}{\sqrt{x}}\right)\right] \tag{21}$$

where N the geometry coefficient is defined by Schwarz (1954), as seen in (Figure 5) and K_r is the correction coefficient. When $\rho_2 > \rho_1$, $= H/\sqrt{A}$; when $\rho_1 > \rho_2$, $P = \left(H - h\right)/\sqrt{A}$.

The underlying assumption of the above formula is that all the vertical rods are located inside the region of the horizontal grid. (Figure 6) shows the relationship between the correction coefficient K_r, the soil reflection coefficient K and the structural coefficient P.

The mutual resistance between the horizontal grid and vertical rods as reported by IEEE Std. 80. (2000)

$$R_{12} = \frac{\rho_a}{\pi L_G}\left[K_r \ln\left(\frac{2L_G}{L_e} + N\right) + \frac{\rho_2}{\pi L_G} K_1\left(0\right)\left(\frac{L_G}{\sqrt{A}} - \frac{2\left(x+1\right)}{\sqrt{x}}\right)\right] \tag{22}$$

$$L_e = L_1 + L_2 \rho_1 / \rho_2 \tag{23}$$

Another formula proposed by Nahman and Skuletich calculate the grounding resistance of a grounding grid in two-layered soil as reported by IEEE Std. 80. (2000)

Figure 3. Relationship between the correction coefficient K_p*, reflection coefficient* K *and structural coefficient* P
As reported by IEEE Std. 80. (2000).

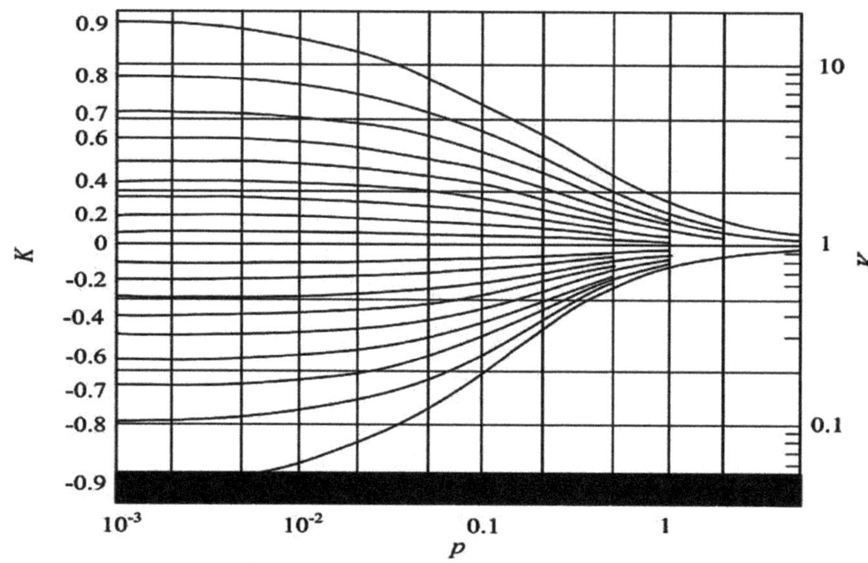

Figure 4. Relationship of K'_p, $P = H/\sqrt{A}$, *and* K, *when the grounding grid is buried in the lower bottom layer*
As reported by IEEE Std. 80. (2000).

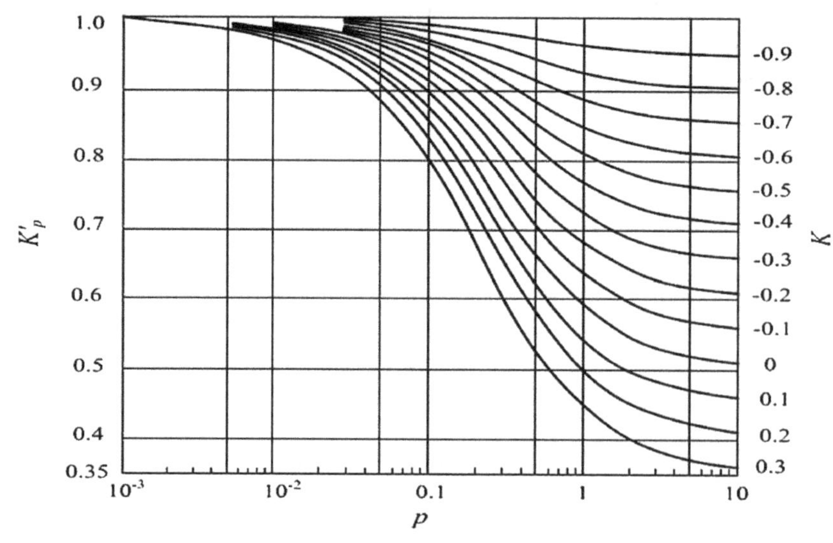

Figure 5. Relationship of geometry coefficient N and the ratio of width to length x of a grounding grid with an area of A
As reported by IEEE Std. 80. (2000).

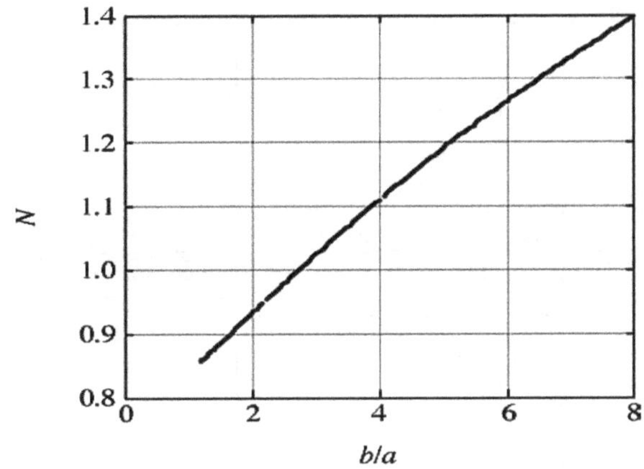

$$R_g = \left[0.53\frac{\rho_2}{\sqrt{A}}\left(\frac{\rho_1}{\rho_2}\right)^g + \frac{\rho_1}{L\sqrt[3]{m'}}\right]\left[1 - 0.8\left(\frac{100hd}{\sqrt{A}}\right)^{0.25}\right] \tag{24}$$

$g = \frac{2H}{\sqrt{A}}, m' = \sqrt{M} + 1$, where M being the number of grid meshes.

Dawalibi-Mukhedkar's Formula

The theoretical basis of a computer program used in grounding resistance calculation performance has been presented by Dalziel, C.F (1972). The ground electrode potentials Vi at different points in earth located at a distance from the center of a conductor are calculated. The resistance of the electrode is then given by:

$$R = \frac{\frac{1}{N}\sum_{I=1}^{N} V_I}{I} \tag{25}$$

where

Figure 6. Relationship between the correction coefficient K_r, the reflection coefficient of soil K and the structural coefficient P
As reported by IEEE Std. 80. (2000).

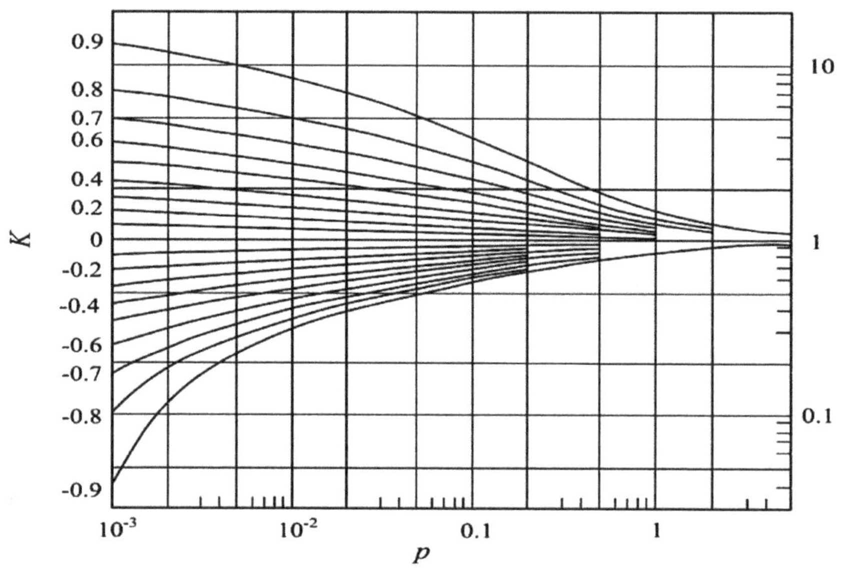

N = the total number of points where the potential has been calculated.
I = the current carried by electrodes.

The theoretical method used by the authors was compared with other existing well known methods currently used to solve grounding problems in uniform earth. No attempt was made to compare the results in a two layer earth structure

Chow and Salama's Formula

A proposed formula to calculate grounding resistance is presented N. H. Malik (1998).

$$R_g = \rho\left[\frac{1}{4}\sqrt{\frac{\pi}{A}} + \frac{1}{N \cdot \Delta L}\left(\frac{1}{\sqrt{2\pi}} \cdot \ln\left(\frac{0.165}{d_o} \cdot \Delta L\right) \cdot \left(1 - \frac{2h}{\sqrt{A}} \cdot 1.128\right)\right)\right] \tag{26}$$

where N is the number of meshes of the grid, ΔL the length of one side of a square mesh in the grid, and d_o is the diameter of the grid conductor

A modification is applied to the method to be used with two layer soil.

$$R_{g\ 2\text{-layer}} = R_g - R_p \tag{27}$$

$$R_p = \frac{\rho_1 \cdot \ln(1-k)}{2\pi(H+h_o)}, \quad k = \frac{\rho_2 - \rho_1}{\rho_2 + \rho_1}$$

Nahman's Formula

The results of the digital analysis of wide set of square and rectangular grounding grids are presented by the basic idea are to develop simple formula for resistance to ground of grounding grids. The formula is developed empirically from the data obtained by digital simulation of a wide set of grounding grids.

The calculations have been performed for various values of grid dimensions from 10m up to 200m, and for three values of burial depth h=0.4725cm, 0.5m and 1 m.

The results obtained by Nahman (1984) supplied sufficient data for developing simple empirical formulas for the grid parameters investigated. A very close approximation of the diagram for grid resistances gives the following formula:

$$R = \rho \cdot \left(\frac{0.53}{\sqrt{A}} + \frac{1.75}{L^3 \sqrt{n}} \right) \cdot \left[1 - 0.8 \cdot \left(\frac{100.h.d}{n\sqrt{A}} \right)^{0.25} \right] \tag{28}$$

where

R = resistance to ground, Ω
ρ = ground resistivity, Ω. m
A = grid area, m^2
L = total lengths of grid conductors, m
h = burial depth, m
n = number of grid conductors in one direction.

Heppe's Method

Equations presented in moat papers on the computation of voltages at the surface of the earth near a buried energized electrode grid of bare wires are based on the assumption that leakage current density is the same at all points along the wire grid. Junyang Ma, Zaijun Wu,et al (2016) present a method for allowing for the effects of variation of leakage current density caused by the proximity of parallel conductors, cross conductor, and angled conductors. The method is relatively simple, straight forward, and easy to be programed, although it is based on the matrix method. For example, solving for the leakage currents in the 112 wire segments of a square 8 x 8 grid, and using these 112 leakage currents to compute the resistance of the grid and the surface voltage at several points, required the solution of only 16 simultaneous equations in 16 unknowns, and took only 4 seconds of CPU time. Finally, this method contains equations permitting one to compute the mutual resistance of orthogonal and angled conductors.

It is assumed that r_{jj} is the sum of the self-resistance of a type j segment and the mutual resistances between the segment and other segments of type j and r_{jk} is the sum of the mutual resistances between a segment of type j and segment of type k. The voltage with respect to far distant ground on the segment is V_j. Currents i_j are the value, of leakage current in the type of segments. The equations state that the sum of the voltage, induced into a segment by its own leakage and the leakage from all other segments must be equal to the applied voltage. Expressed more compactly, in a grid with n types of segments one have:

$$\sum_{k=1}^{n} r_{jk} i_k = v_j \text{ , } j = 1,2,3,.....n \qquad (29)$$

or, still more compactly,

RI = V

where R is the matrix of self and mutual resistance values, I is the column vector of leakage currents in the n types of segments, and V is the list of voltages on the n segments. It will be assumed herein that all of the segments are at the same voltage.

CACULATION OF MAXIMUM STEP AND MESH VOLTAGES

Mesh Voltage

Mesh voltage is the maximum touch voltage inside a mesh. The mesh voltage in the preliminary design is as reported by IEEE Std. 80. (2000)

$$U_m = \left(\rho I_G K_i K_m\right) / L_m \tag{30}$$

where K_m is the geometric correction coefficient of mesh voltage, K_i is the correction coefficient of an irregular grid structure for considering the error introduced in the deduction of K_m (this coefficient only considers the peripheral meshes in which the worst step and touch voltages take place), L_m is the effective burial length of the grounding system, and I_G is the maximum grid current that flows between the grounding grid and the surrounding earth.

The geometric correction coefficient K_m is as reported by IEEE Std. 80. (2000)

$$K_m = \frac{1}{2\pi}\left[\ln\left(\frac{D^2}{16hd} + \frac{\left(D+2h\right)^2}{8Dd} - \frac{h}{4d}\right) + \frac{K_{ii}}{K_h}\ln\frac{8}{\pi\left(2n-1\right)}\right] \tag{31}$$

where D is the spacing between the parallel horizontal grounding conductors, d is the diameter of the grid conductor, and K_{ii} is the corrective weighting coefficient adjusting the influence of inner conductors on the corner mesh, where the equivalent diameter of flat steel $d = b/2$, where b is the width of flat steel. The equivalent diameter of equilateral angle steel $d = 0.84b$, where b is the width of equilateral angle steel. The equivalent diameter of non-equilateral angle steel $d = 0.71\sqrt[4]{b_1 b_2\left(b_1^2 + b_2^2\right)}$ where b_1 and b_2 are the widths of two sides of an equilateral angle steel.

The above equations right when the burial depth is between 0.25 and 2.50 m. When vertical rod are added at the corners or across the entire region of the grounding grid, $K_{ii} = 1$.

When there is no vertical rod, or when there are only a few vertical rods but they are not added at the periphery, then

$$K_{ii} = 1 / \left(2n\right)^{2/n} \tag{32}$$

$$K_h = \sqrt{1 + h / h_0} \tag{33}$$

where n is the number of parallel conductors in the rectangle or equivalent rectangle grounding grid. The following equation can take into account the irregularity of the grounding grid as reported by IEEE Std. 80. (2000)

$$n = n_a n_b n_c n_d \tag{34}$$

For a square grounding grid, $n_b = 1$. For square and rectangle grounding grids, $n_c = 1$. For square, rectangle and L-shaped grounding grids, $n_d = 1$. In other cases

$$n_a = \frac{2L_c}{L_P},\ n_b = \sqrt{\frac{L_P}{4\sqrt{A}}},\ n_c = \left(\frac{L_x L_y}{A}\right)^{\frac{0.7A}{L_x L_y}},\ n_d = \frac{D_m}{\sqrt{L_x^2 + L_y^2}}$$

where, L_c is the total conductor length of grounding grid (m), L_P is the peripheral length of the grounding grid (m), A is the area of the grid (m^2), L_x is the maximum length of the grid in the x direction (m), L_y is maximum length of the grid in the y direction (m) and D_m is the maximum distance between any two points on the grid (m).

For the purpose of a simple calculated of K_m and K_i in the calculation of mesh voltage, it is possible to use $n = \sqrt{n_1 n_2}$, where n_1 and n_2 are the number of conductors in the x and y directions, respectively. But it should be noticed that this equation will fetch an error.

The correction coefficient of an irregular grid structure can be calculated as reported by IEEE Std. 80. (2000):

$$K_i = 0.664 + 0.148n \tag{35}$$

For grounding grids with no vertical rods, or for grids with only a few rods dispersed throughout the grid, but none lying on the corners or along the perimeter of the grid, the effective burial length L_m of the grounding system is:

$$L_m = L_C + L_R \tag{36}$$

where L_R is the total length of all vertical rods (m)

For grids with ground rods on all four corners, as well as along the perimeter and throughout the grid, the effective burial length L_m of the grounding system as reported by IEEE Std. 80. (2000) is:

$$L_m = L_C + \left[1.55 + 1.22\left(\frac{L_r}{\sqrt{L_x^2 + L_y^2}}\right)\right]L_R \tag{37}$$

where, L_r is the length of each vertical rod (m). All rods are assumed to be the same length.

Step Voltages

The step voltage is related to the geometric correction coefficient K_s, the corrective coefficient K_i, the soil resistivity ρ and the average current through per unit length grounding conductors as reported by IEEE Std. 80. (2000)

$$V_s = \frac{\rho I_G K_s K_i}{L_s} \tag{38}$$

where, I_G is the maximum grid current

For grounding grids with or without ground vertical rods, the effective burial conductor length L_s is

$$L_s = 0.75L_c + 0.85L_R$$

The maximum step voltage occurs over a distance of 1 m, beginning at the point above the grounding grid corner along the extension line of the diagonal line outside the corner. For the usual burial depth of 0.25 m $< h <$ 2.5 m, K_s is as reported by IEEE Std. 80. (2000)

$$K_s = \frac{1}{\pi}\left(\frac{1}{2h} + \frac{1}{D+h} + \frac{1+0.5^{n-2}}{D}\right) \tag{39}$$

The formulas that explained it above are based on the assumption in one soil and even spacing of grounding conductors. The area of grounding grids ranges from 6.25 to 10000 m^2, the number of meshes ranges in one direction from 1 to 40, and the mesh size ranges from 2.5 to 22.5 m^2

ESTIMATION OF MINIMUM BURIED CONDUCTOR LENGTH

A simple equation can be developed to permit a preliminary determination of buried grid conductor necessary to keep the maximum touch voltage within the grounded area below the safe limits established by equations and this done as shown below as reported by IEEE Std. 80. (2000)

For E_m < E touch 50, this means that

$$K_m \cdot K_i \cdot \rho \cdot \frac{i_G}{L_m} \langle \left(1000 + 1.5C\left(h_s, k\right)\rho_s\right) \cdot \frac{0.116}{\sqrt{t}} \tag{40}$$

Similarly, for $E_m > E_{touch70,}$

Where the conductor length derived from length is too great to be economically feasible, some of the suggestions mentioned in equation 37 may be considered where soil resistivity and total ground current are very low. The length obtained from equation may give a total length of conductor that is too small to properly connect all equipment to be grounded. In this case, more conductors may be required Even though it is not necessary for safety reasons.

REFINEMENT OF PRELIMINARY DESIGN

If calculations based on the preliminary design indicate that dangerous potential differences can exist within the station, the following possible remedies should be studies and applied where appropriate.

1. Decrease in total grid resistance will decrease the maximum ground grid potential rise and hence the maximum transferred potential. The

most effective way to decrease ground grid resistance is by increasing the area occupied by the grid. Deep driven rods or wells may be used if the available area is limited. Decrease in station resistance may or may not decrease appreciably the local gradients, depending on the method used.

2. Improvement of gradient control. By employing closer spacing of grid conductors, the condition of the continuous plate can be approached more closely. Dangerous potentials within the station can thus be eliminated at a cost. The problem at the perimeter may be more difficult, especially at a small station where earth resistivity is high. However, it is usually possible, by burying the grid perimeter ground conductor outside the y fence line, to ensure that the steeper gradients B immediately outside this grid perimeter do not contribute way to control per meter gradient's is to increase the density of ground rods at the approach to controlling perimeter gradients and step potentials is to bury two or more parallel conductors around the perimeter at successively greater depth as distance from the station is increased.
3. Diverting a greater part of the fault current to other paths, for example, by connecting overhead ground wires of transmission lines or by decreasing the tower footing resistances in the vicinity of the substation. In connection with the latter, however, the effect on fault gradient near tower footings should be weighed.
4. Limiting of short-circuit currents flowing in the ground mat to lower values. If feasible, this will, of course, decrease the total rise in ground mat voltage and all gradients in proportion. Other factors, however, will usually make this impractical. Moreover, if accomplished at the expense of greater fault clearing time, the danger may be increased rather than diminished.
5. Barring of access to limited areas where it may be impractical to eliminate possibility of excessive potential differences during a fault.

By using one or more of the above methods where necessary, designs can be completed for construction purposes. These should be reasonably liberal, as grounding facilities can usually be installed more cheaply if all go in as part of the general construction job, without the necessity of making additions later.

LIMITATIONS OF SIMPLIFIED EQUATIONS

Several simplifying assumptions are made in deriving the equations for E_m and $E_{S,}$ these assumptions may result in inaccurate results, for some cases, in comparison with the results from more rigorous computer analysis or scale model tests. The inclusion of correction factors into the equations of 6.5 practically eliminates the inaccuracy (within certain ranges for the various parameters).

When using the equations of 6.5 the following limits are recommended for square grids, or for rectangular grids having the same number of conductors in both directions:

$n \leq 25$

$0.25\ m \leq h \leq 2.5\ m$

$d < 0.25\ h$

$D > 2.5\ m$

Although the equations 35 and 36 have been tested for n greater than 25 and found to be sufficiently accurate, the tests were not extensive enough to form solid conclusions. Thus, caution should be exercised before exceeding the limits given above.

Furthermore, for equally spaced rectangular grids (that is, with square meshes), the value of n for use on determining the mesh voltage factor K_m and the irregularity factor K_i should be the geometric mean of the number of conductors in either direction. That is,

$N = n_A n_B$ for calculating Em

When n_A and n_B are the number of conductors in Each direction. The value of n for use in determining the step voltage factor K_s and the irregularity factor Kj should be the maximum of HA and n_B, $n = \max(n_A, n_B)$ for calculating E_s

USING COMPUTER ANALYSIS IN GROUNDING DESIGN

There are several reasons that may justify the use of more accurate computer algorithms in designing the grounding system. These reasons include:

1. One or more of the geometric parameters exceed the limits described above.
2. A two-layer soil model is required due to significant variations in soil resistivity.
3. An unsymmetrical grid (that is, L-shaped, with Projections, etc.) makes it impractical to predetermine the location of the worst touch voltage.
4. Uneven grid conductor or ground rod spacing cannot be analyzed using the approximate methods
5. More flexibility in determining local danger points may be desired.

PERMISSIBLE POTENTIAL DIFFERENCE TYPICAL SHOCK SITUATION

Ground potential rise (GPR) is the maximum voltage that a station grounding grid may attain relative to a distant grounding grid may attain relative to a distant grounding point assumed to be at the potential of remote earth.

Under normal conditions, the grounded electrical equipment operates at near zero ground potential, that is, the potential of a grounded neutral conductor is nearly identical to the potential of remote earth. During a ground fault, the portion of fault current that is conducted by a station grounding grid into the earth causes the rise of the grid potential with respect to remote earth. This voltage rise, GPR, is proportional to the magnitude of the grid current, and to the grid resistance.

As it is known the step voltage is the difference in surface potential experienced by a person bridging a distance of 1 m with his feet without contacting any other grounded object.

The touch voltage is the potential difference between the ground potential rise (GPR) and the surface potential at the point where a person is standing, while at the same time having his hands in contact with a grounded structure.

In a conventional substation, the worst touch voltage is usually found to be the potential difference between a hand and the feet at a point of maximum reach distance. However, in the case of a metal-to-metal contact from hand-to-hand. Or from hand-to-feet, which is of concern in the gas insulated substations, both situations should be investigated for the possible worst reach condition including both hands. Mesh voltage is defined as the maximum touch voltage to be found to be found within a mesh of a ground grid. The transferred voltage is known as a special case of the touch voltage where a voltage is transferred into or out of the substation

Typically, the case of transferred voltage occurs when a person standing within the station area touches a conductor grounded at a remote point or a person standing at a remote point touches a conductor connected to the station grounding grid. During fault conditions, the resulting potential to ground may equal or exceed the full GPR of a grounding grid discharging the fault current, rather than the fraction of this total voltage encountered in the ordinary touch contact situations.

In fact, the transferred voltage may exceed the sum of the GPR's of substations, due to induced voltages on communication circuits, static or neutral weirs, pipes, etc. It is impractical, and often impossible, to design a ground grid based on the touch voltage caused by external transferred voltages. Hazards from these external transferred voltages are best avoided by using isolating or neutralizing devices and by treating and clearly labeling these circuits, pipes, etc., as being equivalent to live lines.

Figure 7. Touch voltage limits for metal-to-metal contact and a typical range of enclosure voltages to ground

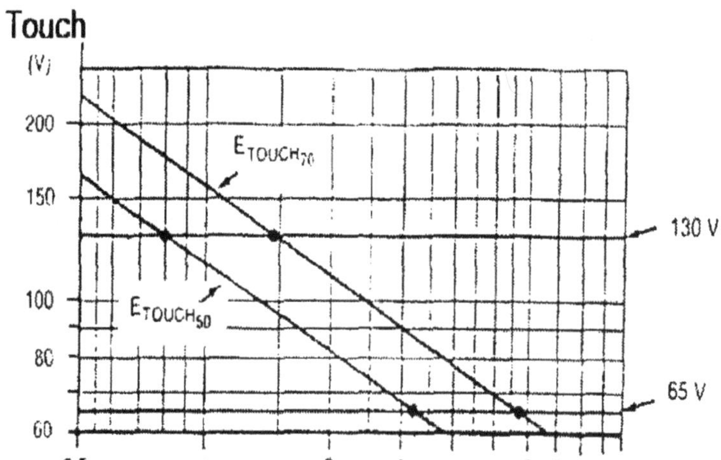

The safety of a person depends on preventing the critical amount of shock energy from being absorbed before the fault is cleared and the system de-energized. The maximum driving voltage of any accidental circuit should not exceed the limits

(Figure 7) shows the touch voltage limits for metal-to-metal contact and a typical range of enclosure voltages to ground.

The actual touch voltage, mesh voltage, or transferred voltage should be less than the maximum allowable touch voltage, E_{t0}uch, to ensure safety.

REFERENCES

He, J., Zeng, R., & Zhang, B. (2013). *Methodology and Technology for Power System Grounding*. Singapore: John Wiley and Sons.

IEEE. (2000). IEEE Guide for Safety in AC Substation Grounding (standard 80).

Laurent, P. G. (1972). Guide sur le calcul l' ´ex´ecution et la mesure des prises de terre. *Rev. Gen. Electr.*, *53*, 455–467, 563–572.

Nahman, J., & Salamon, D. (1984). Analytical expressions for the resistance of grounding grids in non-uniform soil. *IEEE Transactions on Power Apparatus and Systems*, *103*(4), 880–885. doi:10.1109/TPAS.1984.318367

Nahman, J., & Salamon, D. (1986). Analytical expressions for the resistance of rod beds and of combined grounding systems in non-uniform soil. *IEEE Transactions on Power Delivery*, *1*(3), 90–96. doi:10.1109/TPWRD.1986.4307978

Rudenberg, R. (1968). *Electrical Shock Waves in Power Systems*. Cambridge, MA: Harvard University. doi:10.4159/harvard.9780674432390

Schwarz, S. J. (1954). Analytical expression for resistance of grounding systems. AIEE Transactions on Power Apparatus and Systems, 73(13 Pt. 3B), 1011–1016.

Sunde, E. D. (1968). *Earth Conduction Effects in Transmission Systems*. New York: Dover Publications.

Sverak, J. G. (1984). Simplified analysis of electrical gradients above a ground grid; Part I—How good is the present IEEE method. *IEEE Transactions on Power Apparatus and Systems*, *PAS-103*(1), 7–25. doi:10.1109/TPAS.1984.318567

Tagg, G. F. (1964). *Earth Resistances*. London: George News.

Xie, G. R. (1991). *Power System Grounding Technology*. Beijing: China Hydraulic and Electrical Engineering Press.

Chapter 6
Step and Touch Voltages and Grounding Potential Rise (IEEE and IEC)

ABSTRACT

This chapter contains safety criteria according to IEEE Standards, step and touch voltage criteria, reduction factor, simplified equations for mesh and step voltage, application of step and mesh potential in safe earthing system design, safety criteria, IEC 479-1 STD, permissible touch and step voltage according to IEC 479-1, Optimum Compression Ratio (OCR), test and verify, approximated analysis to calculate voltage profile using apparent soil resistivity, Scaling factor, test and verify the approximated method—Ground Potential Rise (GPR) of faulty substations having equal and unequal spacing grounding grids conductors. This chapter draws attention also to the following points: Infinite Series Potential (I.S.M.) expressions in three dimensions and calculations of three dimensions potential rise.

INTRODUCTION

Since the early days of the electric power industry, safety of personnel in and around electric power installations have been a prime concern.

Over the years and after many investigations on the effects of electric current on humans, safe limits have been established and standards have been developed which provide permissible values of body currents to avoid

DOI: 10.4018/978-1-5225-3853-0.ch006

electrocution. Two such standards are IEEE 80 (1961) and the IEC60 479-1(1984). The purpose of both standards is to establish safe (permissible) body currents.

Since IEEE Standards 80 publication in 1961, Guide For Safety In AC Substation Grounding, has been favorably received by the industry and gained a broad acceptance worldwide, this chapter presents the improvements in a broader context of the Guide's development to this date.

This chapter is organized as follows: first the safety criteria as stated in the two documents, IEEE Standards (2000) and IEC 60479-1 (1984), are described in detail and evolution of IEEE standard. A common basis for comparison is developed in terms of permissible touch and step voltages.

A detailed analysis of potential profile at a surface above energized unequally spaced grounding grid has been carried out, for soils with multiple layers. Exact and approximated algorithms are used to obtain the voltage profile and then to calculate the step and touch voltages.

SAFETY CRITERIA ACCORDING TO IEEE 80 STANDARDS

Step and Touch Voltage Criteria

Allowable Body Current Limit

An electric shock accident is determined by the magnitude, duration, and frequency of the current passing through the heart. When a current passes through the heart it causes ventricular fibrillation, and when the current flows through the nerve center it suppresses the respiratory as reported by IEEE 80 (2000).

Therefore, safety design of grounding system takes into account the critical value of an electric shock current causing ventricular fibrillation. Many Researchers have studied the permissible body current limit as reported by Schon. J., H., (1998) and ESBTSD, (1994), for an AC current with a frequency of 50 Hz, it can be classified into three categories according to human responses:

- **Sensory Current:** This is the minimum current a human can feel. For a male adult it is 1.1 mA, and for a female adult, it is 0.7 mA.

- **Extrication Current:** This is the maximum current that can be extricated by human after electroshock. It is also related to gender, for a male adult it is 16 mA and for a female adult, it is 10.5 mA.
- **Fatal Current:** This is the minimum current that threatens life in the shortest time, by causing ventricular fibrillation or asphyxiation as reported by Schon. J.H (1998).

It is concluded that the value of current play an important role in electric shock on the human body, and also the duration of the current passing through the body is considered. If the duration is very short, the current that a human can tolerate and which does not cause ventricular fibrillation as reported by Dalziel, C. F. and Massogilia, F. P, (1956) is:

$$I_K^2 t = K \tag{1}$$

where the unit of I_K is A, t is the time passing through human in second, and K is an energy coefficient related to the weight of the human body. The energy coefficients of 50 and 70 kg humans are $K_{50} = 0.0135$ and $K_{70} = 0.0247$, respectively. Therefore, we can get the allowable body current I_{K50} for a 50 kg human

$$I_{K50} = 0.116 / \sqrt{t} \tag{2}$$

According to the result from Equation 2, the tolerable current corresponding to 1 s duration is 116 mA, while that corresponding to 0.1 s is 367 mA and that corresponding to 0.05 s is 519 mA.

The allowable body current I_{K70} for a 70 kg human is

$$I_{K70} = 0.157 / \sqrt{t} \tag{3}$$

Allowable Body Voltage

Once allowable body current I_K is known the allowable body voltage calculated by the multiplying the allowable body current I_K and the body resistance R_B is obtained:

$$U_{SV} = I_K R_B \tag{4}$$

By setting Equation 3 into Equation 4, the short-time allowable body voltage for a 50 kg human is:

$$U_{SV} = 0.116 R_B / \sqrt{t} \tag{5}$$

Also, the allowable body voltage for a 70 kg human can be obtained:

$$U_{SV} = 0.157 R_B / \sqrt{t} \qquad (5)$$

Touch and Step Voltage Criteria

The safety of a people in substation depends on preventing the critical amount of shock energy from being absorbed before the fault is cleared by protection system.

The maximum voltage of any accidental circuit should not exceed the limits defined as follows.

The limit of step voltage:

$$E_{Step} = \left(R_B + 2R_f\right).I_B \tag{6}$$

For body weight of 50 kg

$$E_{Step} = \left(1000 + 6C_s \cdot \rho_s\right) \cdot \frac{0.116}{\sqrt{t_s}} \tag{7}$$

For body weight of 70 kg

$$E_{Step} = \left(1000 + 6C_s \cdot \rho_s\right) \cdot \frac{0.157}{\sqrt{t_s}} \tag{8}$$

Similarly, the limit of touch voltage:

$$E_{touch} = \left(R_B + \frac{R_f}{2}\right) \cdot I_B \tag{9}$$

For body weight of 50 kg

$$E_{touch} = \left(1000 + 1.5C_s \cdot \rho_s\right) \cdot \frac{0.116}{\sqrt{t_s}} \tag{10}$$

For body weight of 70 kg

$$E_{touch} = \left(1000 + 1.5C_s \cdot \rho_s\right) \cdot \frac{0.157}{\sqrt{t_s}} \tag{11}$$

where;

E_{step} = the step voltage in V, E_{touch} is the touch voltage in V
ρ_s = the resistivity of the surface material in $\Omega.m$.
t_s = the duration of shock current in seconds, C_s is the correction coefficient of the foot grounding resistance.

The goal of a safe grounding design is to prevent the possibility of situations when a person can be vulnerable to absorbing a dangerous level of electrical shock energy, before the fault is cleared and the system de-energized. Hence,

V_{sc}>= Vc, is always required for safety,

V_{sc} = safe voltage limit;
Vc = voltage of accidental circuit.

The expression (12) is obtained when the depth of soil of resistivity is very large. Since the depth of the crushed rock or gravel layer is finite and its resistivity is more than that of soil underneath, ground resistance of the foot is calculated from (13)

$$R_f = \frac{\rho_s}{4b} \tag{12}$$

$$R_f = \frac{\rho_s}{4b} C_s \tag{13}$$

b = 0.08m

Application of Cs Factor in 1986 IEEE Edition: Without the use of computer, Cs is rather tedious to evaluate, especially when a protective surface layer of small thickness h_s, is covering the switchyard and its resistivity is greater than that of the underlying soil. **A** set of curves, was therefore provided, (Figure 1). The curves were based on the following formula:

$$C_s = \frac{1}{0.96}\left[1 + 2\sum_{n=1}^{\infty} \frac{k^n}{\sqrt{1 + (2nh_s / 0.08)^2}}\right] \tag{14}$$

$$K = \frac{\rho - \rho_s}{\rho + \rho_s}$$

Figure 1. Reduction factor Cs as a function of reflection factor K and crushed rock layer thickness hs

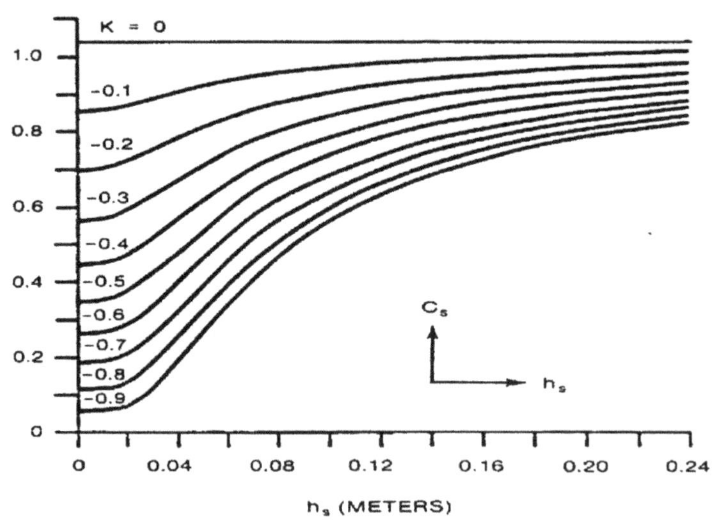

As a footnote option left to designer discretion, the 1986 Guide also suggested a very simple semi-empirical equation for alternative estimate of Cs shown here as C_{sest} in equation 15

$$C_{sest} = 1 - a\left[\frac{1 - \frac{\rho}{\rho_s}}{2h_s + a}\right] \tag{15}$$

a=0.106 m. which avoids the infinite summation series, are also possible;

However, analytical studies done have revealed that equation 14 contains a minor error. The shape of de-rating curves is symptomatic of this oversight: There is an initial convex-to-concave curvature for depths between 0 and 0.04 m, which causes that values of *Cs* are too conservative for hs < 0.15 m.

On the other hand, as also documented by Dawalibi et al. (1975), the seemingly unsophisticated equation 15 which was adapted from Sverak (1984) just for the mentioned "footnote approximation" - has proven itself to yield not only reasonably correct results, but quite unexpectedly, for very thin layers to do so within a smaller error range than other formulas.

Application of Cs Factor in 1996 Edition: In view of the above described findings, it is likely that the following equations will be included in the 1996 edition, giving the value of *Cs* with very good accuracy as given in equation 16:

$$C_s = C_1 - C_2 \tan^{-1}\left(\frac{2h_s}{b}\right) - C_3\left[e^{-7h_s} - e^{-30h_s}\right] \tag{16}$$

$$C_1 = {}^{(1+k)}\!\big/_{(1-k)}, C_2 = 4k / \left[\pi(1-k)\right],$$

$C_3 = 0.21k^2$, hs and b in meters

Equation 16 has been derived as a polynomial fit by Thapar et al. (1991), to give values of *Cs,* very close to the results obtained with their accurate but rather complex analytical method. The error is less than *5%*. This equation is

Figure 2. Reduction factor Cs as a function of reflection factor K and crushed rock layer thickness hs

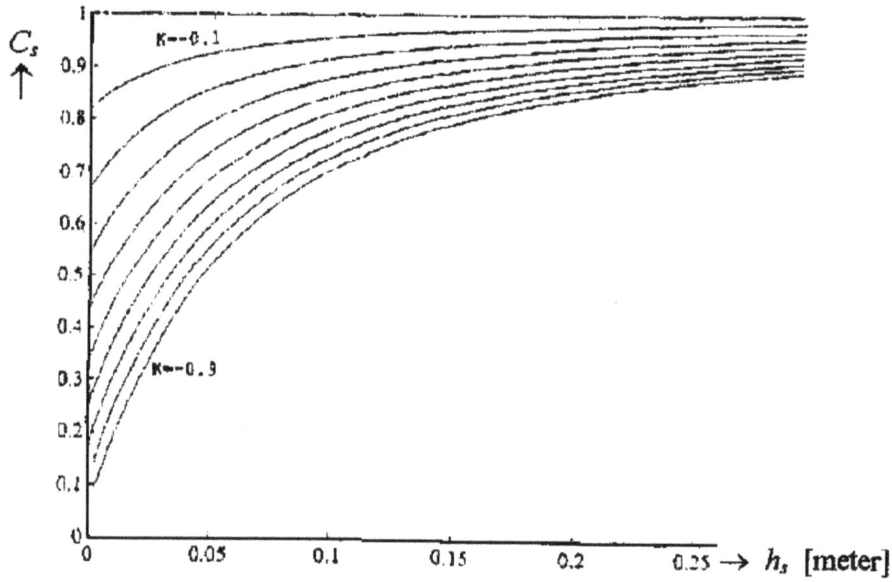

applicable for hs varying from 0 to 0.3 m and K varying from 0 to -0.98. **A** set of Cs curves for b = 0.08 m will also be provided, (Figure 2), replacing (Figure 1).

The older equation 15 is more accurate for very thin surface layers between 0.005 m and 0.02 m, and for the rest of parameter range is about 2 -5% more conservative than equation 16.

Ultimately, using the factor 1/0.96 from equation 14 makes its results practically identical to those of equation 16. Thus,

$$C_s = C_{sest} / 0.96 \quad (17)$$

Simplified Equations for Mesh and Step Voltage

In 1959, a small working group led by Stevens [34] developed a practical mathematical model, allows approximating the maximum step and touching voltages in the critical area of a corner mesh. For its time, this was a remarkable achievement. The model, which so far has been used in all editions of the Guide, is based on a simple abstraction of n parallel conductors in depth h. The mesh voltage, in volts, is obtained as.

$$E_m = \frac{\rho.K_m.K_i.I_G}{L_m} \tag{18}$$

An analogous formula for calculation of the maximum step voltage, Es in volts, is

$$E_s = \frac{\rho.K_s.K_i.I_G}{L_s} \tag{19}$$

Note: A formal distinction between L=Ls and L=Lm indicates that the effective buried conductor length may be defined differently for the step and the touch calculations.

In either formula, the calculated voltage is obtained as a product of the soil resistivity ρ, of an average current density per unit of the effective buried length, IG/Lm or IG/Ls, and of two factors:

Km or Ks characterizing the grid geometry, and Ki, an irregularity factor, which accounts for some of the errors introduced by the assumptions made in deriving Km and Ks

The effective length of buried conductors includes ground rods connected to the grid. The entire grounding system is energized and conducts current IG into the earth. A homogeneous soil of uniform resistivity is assumed.

AIEE No. 80 (1961) and IEEE 80 (1976) Editions reported equations to calculate the constants *Km, Ks* and *Ki* for square grid consisting of *n* parallel conductors evenly spaced *D* as follows:

$$K_m = \frac{1}{2\pi}.\left[\ln\left(\frac{D^2}{16 \cdot h \cdot d}\right)\right] + \frac{1}{\pi}\ln(\lambda) \tag{20}$$

$$K_s = \frac{1}{\pi}\left[\frac{1}{2h} + \frac{1}{D+h} + \frac{1}{2D} + \frac{1}{3D} + \ldots.. + \frac{1}{(n-1)D}\right] \tag{21}$$

$$K_i = 0.65 + 0.172n \tag{22}$$

n = number of parallel grid conductors in one direction
D = spacing between parallel conductors in m
d = diameter of grid conductor, m
h = depth of grid burial, m
λ = a finite series (3/4)(5/6)(7/8).....((2n-3)/(2n-2))

The effective buried length L, used both in IEEE 80 (2000), was simply defined as

$$L = L_c + L_r \tag{23}$$

L_c = total length of grid conductors, in m,
Lr = total length of all ground rods, in m

Factors *Km, Ks* and *Ki* perform the necessary corrections due to the difference in geometries of a configuration of *n* parallel conductors representing the grid and of the actual grounding system. *Km* simulates the influence of currents flowing in *n* -2 conductors of the *n* conductor model upon the current density in two most outward parallel conductors representing the corner mesh, whereas *Ki* compensates for the difference in performances of the whole model and of a complete grid.

To assist the designer in choosing a convenient square grid representation of rectangular or L-shaped grids and in interpreting the results, a figure was provided for six different grid spacing patterns, depicting the distribution of mesh voltages within each grid. These auxiliary data reflected the outcome of Koch's (1950), measurements on miniature grid models in an electrolytic tank. A complementary example, describing the development of a grounding design for an L-shaped grid from a rough idea to a viable concept, was also provided. Otherwise, implicitly assuming the presence of one's good engineering sense and prudent judgment, no explicit applicability limits were mentioned.

In practical use, however, some deficiencies emerged. It has been found that the application of *Km* as defined by equation 20, causes optimistic errors in the calculated values of the mesh voltage and, for very dense grids with high *n* and *D* approaching the order of *h* parameter, the result becomes negative. In contrast, *Ks* defined by equation 21 often yields overly conservative values for grids buried deeper than *0.25* m below the earth surface.

IEEE 80 Edition (1986) improved the accuracy of calculations for a broad class of rectangular grids with and without ground rods, Sverak. J.G. (1984) reassessed the effect of simplifying assumptions used in AIEE No. 80 (1961) and established a deferent modeling approach, to develop new simplified equations for Km and Ks, which corrected most of the deficiencies of the previous equations.

Given below as equation 24, the resulting equation for Km is more accurate and versatile than the previously used

$$K_m = \frac{1}{2\pi}\cdot\left[\ln\left(\frac{D^2}{16\cdot h\cdot d}+\frac{(D+2h)^2}{8\cdot D\cdot d}-\frac{h}{4\cdot d}\right)+\frac{K_{ii}}{K_h}\ln\frac{8}{\pi(2n-1)}\right] \quad (24)$$

K_{ii}=1 for grids with ground rods along perimeter or with ground rods both in the grid corners and inside the grid area, or, $K_h = \sqrt{1+\frac{h}{h_0}}$, $h_0 = 1$

$K_i = \frac{1}{(2n)^{\frac{2}{n}}}$ for grids with a few inner ground rods, or no rods.

The approximation of a current density distribution pattern includes two weighing terms, K_{ii} and K_h The first term adjusts the influence of n - 2 conductors upon the current density in the first two conductors forming the corner mesh, according to the presence (K_{ii} = 1), or the absence (K_{ii} = 2n (exp)-2h) of peripheral grounding rods. The second term, K_h, corrects for non-zero depth of the remaining n – 2 conductors. The finite series λ = (3/4) (5/6) ... etc., which had been previously used in equation 20 to represent these conductors, is replaced by an asymptotic sum.

Ks Model: For the calculation of step voltages, two versions of factor *Ks* were used:

- The older formula given in equation 21, applicability of which was limited to shallow depths between 0 and 0.25 m; and
- Sverak's formula below, applicable to depths greater than 0.25 m.

$$K_s = \frac{1}{\pi}\left[\frac{1}{2h}+\frac{1}{D+h}+\frac{1}{D}\left(1-0.5^{n-2}\right)\right] \quad (25)$$

Other Changes: Compared to the value given in equation 22, the value of a starting point in the definition of irregularity factor *Ki* was slightly adjusted, yielding exactly *Ki* = 1 for a singular mesh (n = 2);

$$K_i = 0.656 + 0.172n \tag{26}$$

An important novel detail in the application of *Km* and *Ks* was a more precise definition of parameter *n* for rectangular grids, characterized by an actual number of the parallel conductors in each direction, n_A, n_B . For the calculation of mesh voltages,

$$n = \sqrt{n_A n_B} \text{ Rounded to the nearest integer} \tag{27}$$

For the calculation of step voltages,

$$n = Max.\left(n_A, n_B\right) \tag{28}$$

Only one value of the effective buried conductor length *L,* was used both in the general formula for mesh and step voltage, that is, $L_M = Ls = L$. However, it was defined differently for grids with peripheral ground rods and for other grids. For grids with ground rods predominantly around perimeter

$$L = L_c + 1.15L_r \tag{29}$$

The use of older definition presented in equation 23, that is, of *L* = *Lc* + *Lr,* was restricted to grids with no ground rods, or with only a few rods located in inner parts of the grid, away from the perimeter.

Finally, the following, conservative applicability limits were recommended for square and rectangular grids:

$n < 25$ $D > 2.5$ m

d < 0.25 *h* 0.25 m < *h* < 2.5 m

In IEEE Edition (1996) to extend the applicability of simplified calculations to T-shaped, L-shaped and triangular grids, as well as to further improve the overall accuracy of results, it is proposed to introduce a new definition of parameter *n,* which will broaden the applicability of existing mesh voltage and step voltage equations Additional minor refinements concern various empirical co-factors in definitions of the effective conductor length and an adjustment of *Ki* formula., reflecting the result of bench-mark tests performed by the Working Group. The specific refinements are:

1. Equation 24 for calculations of the mesh voltage, retain without change the formulation of *Km* itself, but broaden its applicability by a new definition of parameter *n.*
2. Equation 25 for calculations of the step voltage, retain without change this more recent formulation of *Ks* itself (as given for grids buried in depths >= 0.25 m), but broaden its applicability by the new definition of parameter *n,*
3. Equation 21 abandon this older version of *Ks,* which has been already limited by the 1986 edition to grids in a shallow depth < 0.25 m.
4. Using four grid shape components developed by Thapar et al (1991) the newly extended formulation of parameter *n,* is:

$$n = n_I . n_{II} . n_{III} . n_{IV} \tag{30}$$

where

$n_{II} = 1$ for square grids
$n_{III} = 1$ for square and rectangular grids
$n_{IV} = 1$ for square, rectangular and L-shaped grids

and

$$n_I = \frac{2L_c}{L_p} \; ; \; n_{II} = \sqrt{\frac{L_p}{4\sqrt{A}}} \; ; \; n_{III} = \left[\frac{L_x L_y}{A}\right]^{\frac{0.7A}{L_x L_y}} \; ; \; n_{IV} = \frac{D_m}{\sqrt{L_x^2 + L_y^2}}$$

A = area of grid, m^2
Lc = total length of conductors forming the grid, m
Lp = peripheral length of the grid, m
Lx = maximum length of the grid in the x direction, m
Ly = maximum length of the grid in the y direction, m
Dm = maximum diagonal of the grid, m

5. Equations 18 and 19 modify the definitions of effective buried conductor lengths, $L = Lm$ and $L = Ls$, using different co-factors for the mesh and step voltage formula, as follows:
   a. For the calculation of mesh voltages, the effective buried length, L_M, applicable to grids with ground rods in the corners, as well as along the perimeter and inside the grid, is:

$$L_m = L_c + \left[1.55 + 1.22\left(\frac{L_r}{\sqrt{L_x^2 + L_y^2}}\right)\right] \cdot L_R \tag{31}$$

where

Lc = total length of grid conductors, in m
L_R = total length of all ground rods, in m
L_r = length of each ground rod, in m

   b. For the calculation of step voltages, the effective buried length Ls, applicable to grids with or without ground rods, is:

$$L_s = 0.75L_c + 0.85L_r \tag{32}$$

6. Implement a modification of the irregularity factor Ki, which has been derived by Thapar in conjunction with the extended formulation of parameter n. Compared to previous definitions, there is a slight adjustment both in the slope and in the starting point of the Ki curve:

$$K_i = 0.644 + 0.148n \tag{33}$$

APPLICATION OF STEP AND MESH POTENTIALS IN SAFE GROUNDING SYSTEM DESIGN

In IEEE 80 standard (2000) it is concluded for grounding system design that:

$$K_m \cdot K_i \cdot \rho \cdot \frac{i_G}{L_m} \langle \left(1000 + 1.5C\left(h_s, k\right)\rho_s\right) \cdot \frac{0.116}{\sqrt{t}}$$

Figure 3. Touch potential of the grounding system as function of grounding system length

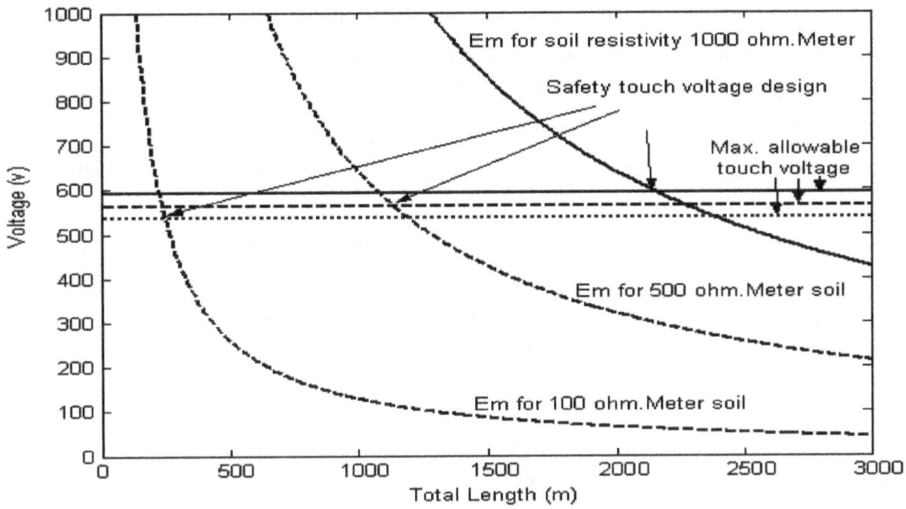

This relation could be used to calculate the safe conductor length to ensure safe step and touch voltages as follow. The influence of soil resistivity on step and touch potential (Figure 3) gives the relation between touch potential of the grounding system and total length of rods and grid conductors, (Figure 4) shows relation between step potential of the grounding system and total length of rods and grid conductors. Grid dimension 50 m.50 m with 11 parallel

Figure 4. Step potential of the grounding system as function of grounding system length

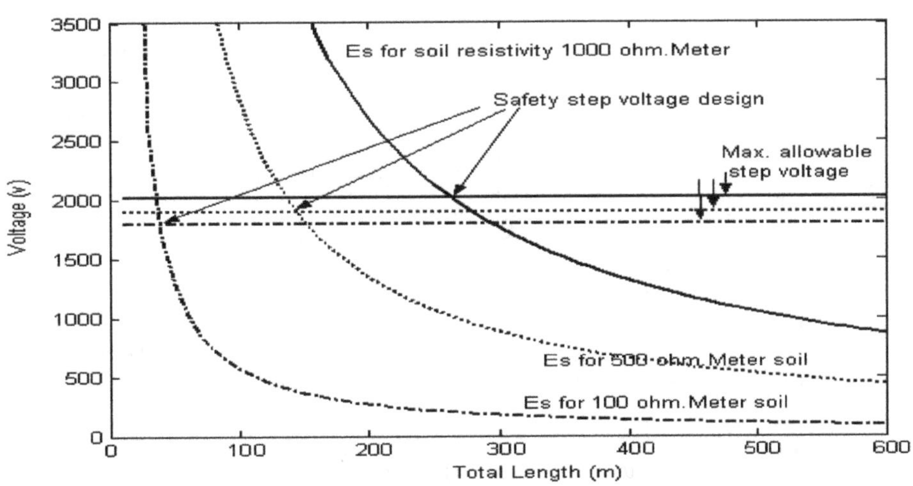

grid conductors at each direction and rods located at perimeter, buried at 0.5 meter depth, crushed rock of 3500 Ω.m resistivity and thickness 10 cm used, grid current assumed to be 1 kA for 1 second fault clearance duration. The calculations are done taking three values of soil resistivity 100,500 and 1000 Ω.m. The safety values (step and touch) are shown also in the figures. The safety design length decreases with decreasing the earth resistivity.

In case of step and mesh potential are very high than the tolerable step and touch potential and the ground system design must be modified, it is possible to increase crushed rock thickness or resistivity as it will affect the maximum allowable step and touch potential. (Figure 5) shows the relation between step voltage and maximum allowable step voltage for different crushed rock resistivity. The influence of crashed rock thickness on touch potential is shown in (Figure 6). This figure shows the relation between mesh potential and maximum allowable touch voltage for different crushed rock thickness (0.1, 0.15 and 0.2 meter).

SAFETY CRITERIA- IEC 60479-1 STD

The parameters of the electric shock model are listed in (Figure 7) showing differences between IEEE and IEC. IEC 60479-1(1984) is less specific than IEEE Std 80 (2000) for design purposes. The standard provides data for body resistances as a function of body voltage, which are illustrated in (Figure 8),

Figure 5. Step potential of the grounding system for different crushed rock resistivity (h_s=0.1m)

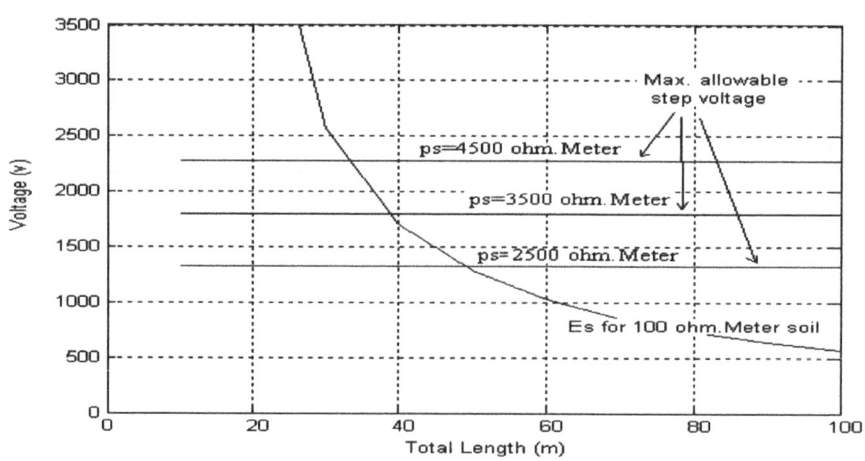

Figure 6. Touch potential of the grounding system for different crushed rock thickness (ρ_s =3500 Ω.m)

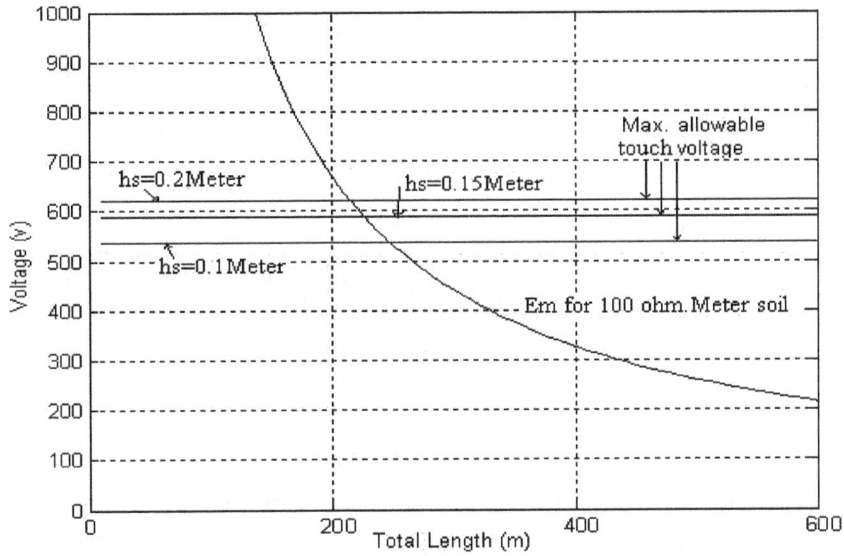

and data of body resistance as a function of path, which are illustrated in (Figure 8). In (Figure 8), the 5% curve indicates body resistance values which were not exceeded by 5% of the population; the 50% curve indicates body resistance values which were not exceeded by 50% of the population, etc. All the values in (Figure 8) are for hand to hand resistance. In (Figure 9), the numbers not in brackets indicate the impedance of several paths in the body

Figure 7. Electrical shock model differences between IEEE 80 Std. and IEC 60479-1

	IEEE Std 80	IEC 479-1
Body Resistance	1000 ohms	Voltage Dependent and Path Dependent
Thevenin Equivalent Resistance	$1.5c_s\rho_s$ for touch voltage $6.0c_s\rho_s$ for step voltage	no guidance
Thevenin Equivalent Voltage	Simplified Equations k_i k_r L I or use of computer models is suggested	no guidance
Permissible Body Current	$0.116A/\sqrt{t}$ for 50 kg person $0.157A/\sqrt{t}$ for 70 kg person	S-curves independent of human size

Figure 8. Human body resistance as a function of body voltage

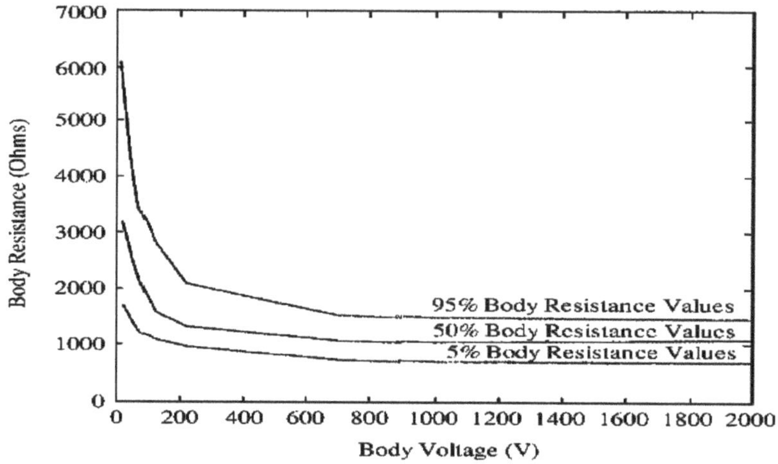

as a percentage of the hand to hand impedance. The numbers in brackets are for current paths between both hands and the corresponding part of the body. IEC 60479-1, (1984) also provides values of permissible body current versus electric shock duration as shown in (Figure 10). In this figure, points on curve C1 represent 0.14% probability of ventricular fibrillation, points on curve C2 represent 5% probability of ventricular fibrillation and points on curve C3 represent 50% probability of ventricular fibrillation. These curves separate the space of body current and shock duration into zones. As an example, Zone 4 represents all the combinations of body current and shock duration which will lead to ventricular fibrillation with probability more than 50%.

The data of IEC 60 479-1 (1984) can be utilized in two ways:

1. Actual body currents can be computed for an individual subjected to touch or step voltage in a specific system and under specific conditions, and
2. Permissible touch and step voltages can be computed for a specific system.

PERMISSIBLE TOUCH AND STEP VOLTAGE IEC 60479-1(1984)

The permissible (or allowable) touch, V_T^a, and step, V_S^a, voltages are computed from the following equations (See Figure 10):

Figure 9. Internal impedance of the human body as a function of the current path

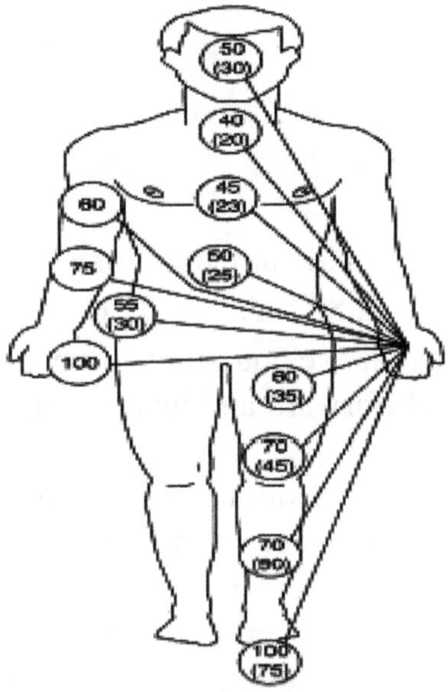

Figure 10. Permissible body current per IEC 60479-1

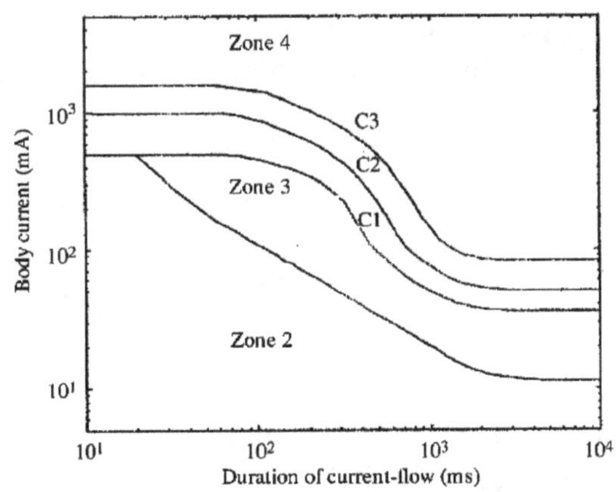

$i_{b,\text{perm}}(t)$	is the permissible body current per IEC 60479-1 for an electric shock duration t. This current is obtained from the data shown in (Figure 8).
R_b^T (*ib*,perm(t))	is the body resistance for the path specified by the touch voltage (typically, hand to two feet) and for a body current equal to $i_{b,\text{perm}(t)}$. This value can be obtained from the data shown in (Figures 8) and (Figure 9).
R_b^S (*ib*,perm(t))	is the body resistance for the path specified by the step voltage (foot

$$V_T^a = i_{b,perm}(t)\left[R_b^T\left(i_{b,perm}(t)\right) + r_{eq,T}\right] \tag{34}$$

$$V_S^a = i_{b,perm}(t)\left[R_b^S\left(i_{b,perm}(t)\right) + r_{eq,S}\right] \tag{35}$$

IEC 60479-1 does not provide any data for $r_{eq,T}$ or $r_{eq,S}$. For this reason, we shall use the data of IEEE 80 Std., $req_s = 6C_s\rho_s$, $req_T = 1.5C_s\rho_s$

Note that application of above equations to obtain the permissible touch and step voltages is straightforward and involves the following steps:

Step 1: For a given electric shock duration t and a given assumed probability of ventricular fibrillation, determine the value of permissible body current, $i_{b,perm(t)}$, from (Figure 10).
Step 2: For the current $i_{b,perm(t)}$, determine the body resistances R_{bT} and R_{bS} from the data shown in (Figure 8) and (Figure 9). For touch voltage, it is expedient to assume that the path will be one hand to two feet (75% of the body resistance given in (Figure 9), and for step voltage that the path is foot to foot (100% of the body resistance given in (Figure 9).
Step 3: Compute $r_{eq,T}$ and $r_{eq,S}$ per IEEE 80 Std.
Step 4: Compute permissible touch and step voltages using Equations 34 and 35.

For purposes of comparison with IEEE Std 80 (2000), the permissible body current $i_{b,perm(t)}$ is computed for probability 0.5% of ventricular fibrillation using proper interpolation between curves C1 and C2 in (Figure 10).

COMPUTATION OF POTENTIAL PROFILE AT A SURFACE ABOVE ENERGIZED UNEQUALLY SPACED GROUNDING GRID

Studying the effect of voltage distribution is an important parameter to design the grounding systems of substations to ensure the safety of people above the substation ground. This can be achieved by equalizing the potential distribution of the ground surface and reducing step and touch potentials. This is reported by Zhongdong, Du., Yao, Zhenyu, Wen, Xishan, Xu, Hua, (2005)

The earth surface potential distribution above the equally spaced grid is extremely non-uniform. In this section, the voltage distribution is calculated for unequally spaced by using the suggested image method and also by using approximated method in case of multi-layer soil. The obtained results are compared with the IEEE method as reported by ANSI/IEEE Std. 80-1986. IEEE guide has proved to be a very popular guide for substation grounding applications. IEEE gradient analysis method allows a recursive point by point integration of surface gradients through all parallel grid conductors. The analytically most significant equations of the IEEE gradient analysis method are given by Dawalibi and Mukhedkar (1975).

In the suggested algorithm, the expressions of potential at any point due to a point current source in multi-layer soils can be obtained by image technique or by solution of Laplace's equations. These are in the form of infinite summation series. The rate of convergence of these series is mainly dependent on the reflection factor between layers of the soil. For a reasonable accuracy, up to a few thousand terms of series may have to be computed. Determination of potential from such expressions, therefore, forms significant part of computational effort in analysis and design of a grounding system in three-layer soil.

A new simplified method for calculating the grounding voltage distribution of grounding grid in three-layer earth structure (multi-layers) has been presented.

Analytical Analysis

(Figure 11) shows the grounding system model under study, which consists of horizontal three layers. The grid is placed in the first layer of grounding system which its resistivity (ρ_1), over a second layer of resistivity (ρ_2), which is over the third layer of resistivity (ρ_3). Thickness of the first layer and second layer are (d_1), and (d_2) respectively. The depth of the buried grid in the first layer is (b), while the distance between the grid and surface of the second layer is (h). The ground voltage profile of the grid can be determined with the use of image technique.

The three boundary planes between air, first layer, second layer, and third layer produce an infinite sequence of images. All these images discharge current into the first layer of its resistivity (ρ_1). Whenever an image is taken on the boundary between air and first layer, the current of the image remains the same. However, when an image is taken on the boundary between the first and second, the current of the image will be reduce by a ratio.

$$k_{12} = \frac{\rho_2 - \rho_1}{\rho_2 + \rho_1} \tag{36}$$

Whenever, an image is taken on the boundary between the second and third, the current of the image reduces by a ratio

$$k_{23} = \frac{\rho_3 - \rho_2}{\rho_3 + \rho_2} \tag{37}$$

where K_{12}: Reflection factor between first and second layer, K_{23}: Reflection factor between second and third layer

Infinite Series Potential (I.S.M.) Expressions in Three Dimensions

(Figure 11) shows the three-layer model under study in case of calculating the voltage distribution on the grid surface where: A: Area of grid, m^2, b: Burial depth of grid, m, For any grounding system in three-layer earth structure, the current in each conductor element produces a potential at a point (x, y,

Figure 11. Grounding system model under study

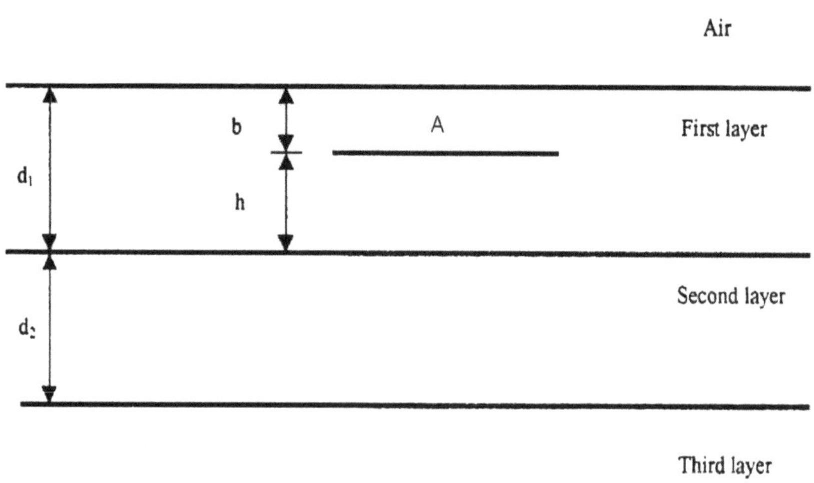

z) in the surrounding medium in three dimension with taking the reflection factors between first, second and third layer in account will be as follows as reported by Heppe, R., J. (1979).

$$v(x,y,z)=\frac{I\rho_1}{4\pi l}\left[\begin{array}{l}\ln f(x,y,z)+\ln f(x,y,2b)+\\ \sum_{n=1}^{\infty}(k_{12}^{n}.\ln f(x,y,z+n2h+(n-1)2b)\\ +2k_{12}^{n}\ln f(x,y,z+n2h+n2b)\\ +k_{12}^{n}\ln f(x,y,z+n2h+(n+1)2b)\\ +\sum_{n=1}^{\infty}(k_{23}^{n}.\ln f(x,y,z+n2(d_2-h)+(n-1)2(d_1+h))\\ +2k_{23}^{n}\ln f(x,y,z+n2(d_2-h)+n2(d_1+h))\\ +k_{12}^{n}\ln f(x,y,z+n2(d_2-h)+(n+1)2(d_1+h))\end{array}\right] \quad (38)$$

where,

$$f\left(x,y,z\right)=\frac{\sqrt{\left(x+l/2\right)^2+y^2+z^2}+x+l/2}{\sqrt{\left(x-l/2\right)^2+y^2+z^2}+x-l/2} \quad (39)$$

l = Length of a grid side, (m),
r = Radius of grid conductor, (m),
n = Number of images

Computer program has been developed based on the I.S.M. The flowchart is given in (Figure 12)

OPTIMUM COMPRESSION RATIO (OCR)

In unequally spaced grid when the conductors are arranged according to an exponent regularity, the span from the center could be calculated by the equations

$$d_n = d_{max}.C^n \quad (40)$$

Figure 12. I.S.M program flowchart

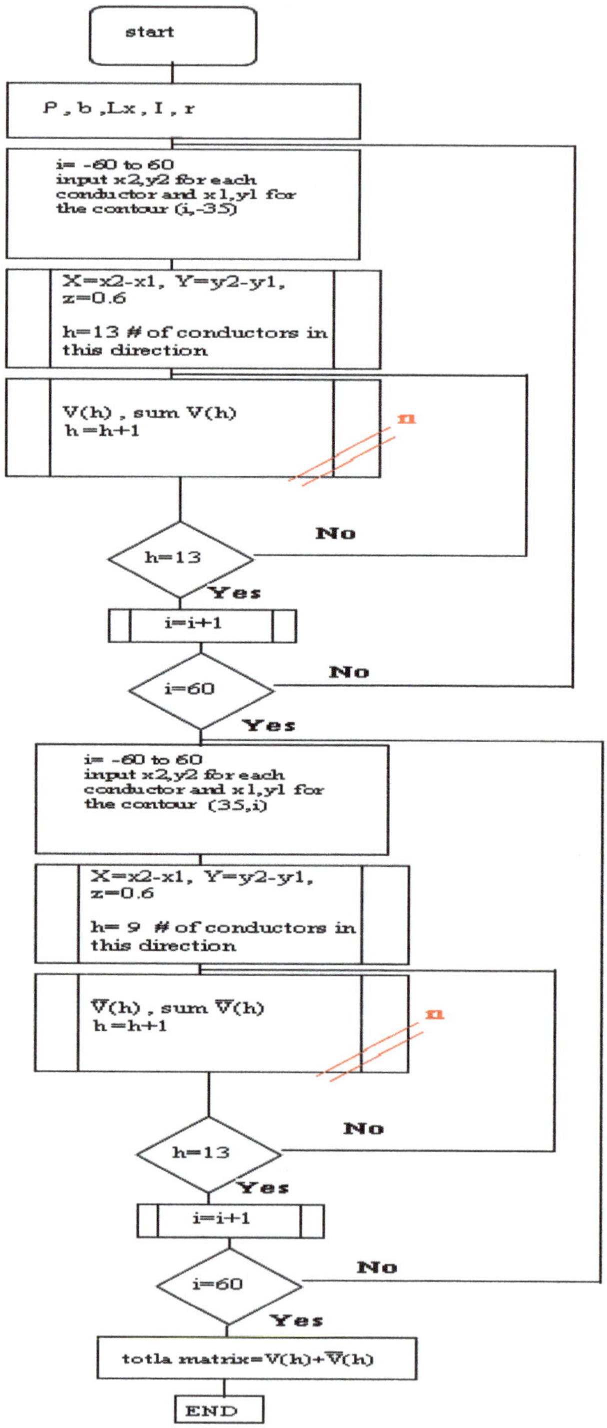

$$d_{max} = \frac{L(1-c)}{1+c-2c^{(N/2+1)}} \text{N even} \tag{41}$$

$$d_{max} = \frac{L(1-c)}{2\left(1-2c^{(N-1)/2}\right)} \text{N odd} \tag{42}$$

where C is the compression ratio, L is the side length of grounding system, N is the conductors number arranged on it and d_{max} is the central span. Empirical expression is obtained by Hung, Chen and Yan (1995) to calculate (OCR).

Testing and Verification

The analytical analysis in three-dimension is applied for a grid grounding system consists of 96 meshes with dimensions given in Table 1 and soil structure given in Table 2 to obtain the voltage profile on the earth surface in the different position along the grid length. The short circuit current of substation 220 kV grounding system is 10 kA. The (I.S.M.) method suggested in this section is applied on example of IEEE 80 (2000) and proves that it is very accurate. The maximum percentage difference is approximately 5%.

Table 1. Grid dimension parameters

Grid area, m^2	120X80
burial depth, m	0.6
Radius of grid conductors, m	0.0175
number of meshes	96

Table 2. Soil structure parameters

thickness of the first layer, m	2
Thickness of the second layer. m	2
resistivity of the first layer Ωm p1=30	30
resistivity of the second layer Ωm p2=40	40
resistivity of the third layer Ωm p3=250	250
Number of images that vary to infinity.	30

(Figure 13) to (Figure 16) show the voltage profile for the ground grid with different compression factor.

Number of images that is used has been chosen to be 30 which will give acceptable result accuracy for different reflection factors (k = 0.1 to 0.9).

Figure 13. Voltage profile along the unequally spaced grid in case of C = 0.4

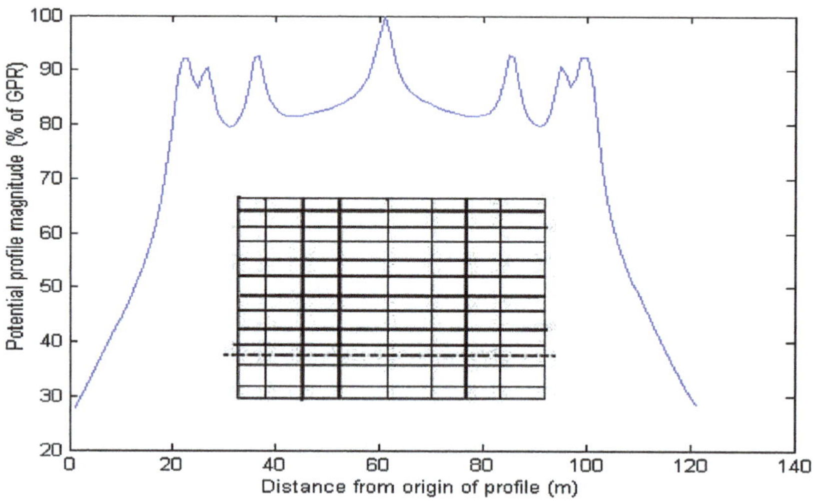

Figure 14. Voltage profile along the unequally spaced grid in case of C = 0.6

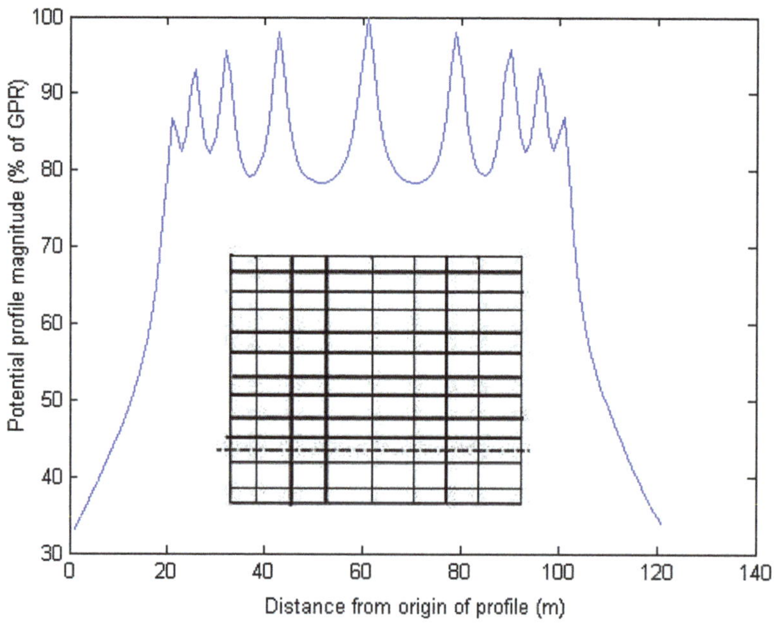

Figure 15. Voltage profile along the unequally spaced grid in case of C = 0.8

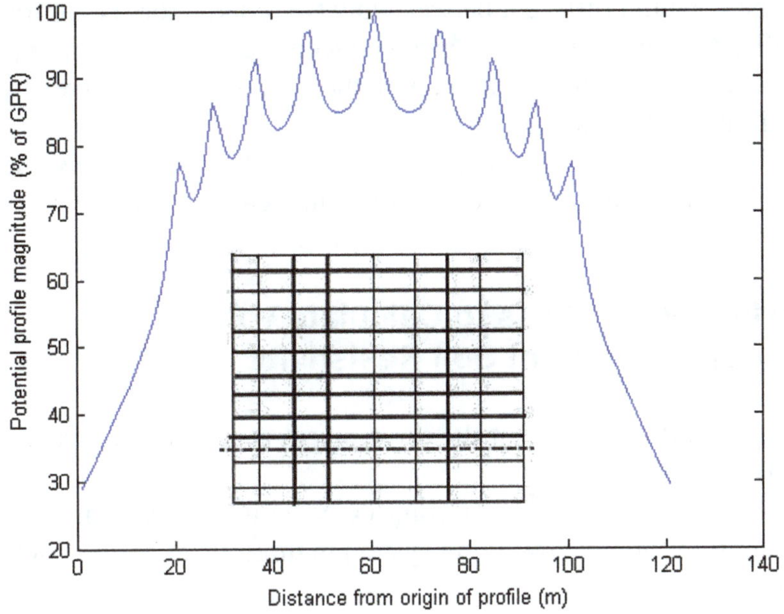

Figure 16. Voltage profile along the unequally spaced grid in case of C = 1.0

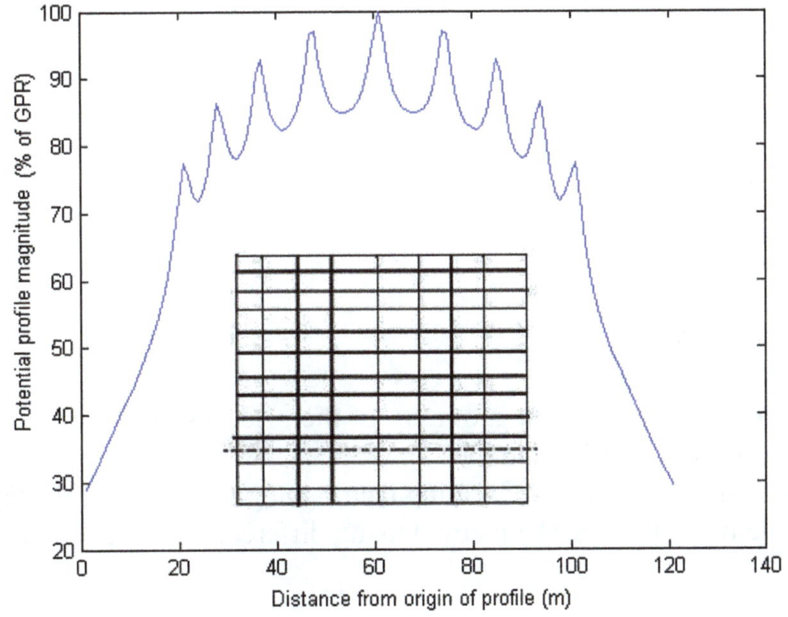

From the obtained results given in (Figures 13 to 16) it is noticed that when C $\approx$ 0.6 the potential different between any two conductors is minimum and this may lead to minimum difference between the maximum and minimum touch potential, this is in agreement with the obtained results given by Hung, Chen and Yan (1995) in calculating OCR.

Calculating the voltage profile by the used method consumes long calculation time to give accurate result for this reason the following method may be used.

Approximated Analysis to Calculate Voltage Profile Using Apparent Soil Resistivity

The apparent soil resistivity of the three layers is calculated by using IEEE 80 (2000) formula.

In this case the potential at a point (x, y, z) in surrounding medium in three-dimension for any grounding system in three-layer earth structure will be as follows:

$$v\left(x,y,z\right)=\delta\frac{I\rho_a}{4\pi l}\left[\ln f\left(x,y,z\right)+\ln f\left(x,y,z+2b\right)\right] \tag{43}$$

where,

$$f(x,y)=\frac{\sqrt{\left(x+l/2\right)^2+y^2+z^2}+x+l/2}{\sqrt{\left(x-l/2\right)^2+y^2+z^2}+x-l/2} \tag{44}$$

δ = Scaling factor

Scaling Factor

Scaling factor δ depends on geometry of the grid, distance from the point of origin and compression ratio C

Table 3 shows the values of scaling factor that gives voltage profile closer to that calculated by (I.S.M.) method using different Compression ratio

Table 3. The influence of scaling factor with compression ratio

Compression Ratio C	Scaling Factor δ
1	4.1
0.8	4.72
0.6	4
0.4	5.571

In order to ascertain the scaling factor when we know the compression ratio, we must find a general law, using the method of least squares to fitting curves.

The fitting equation is as follows:

$$\delta = -76.088C^3 + 165.67C^2 - 115.71C + 30.215 \quad (45)$$

Testing and Verification of the Approximated Method

Effect of Compression Ratio on Voltage Profile

The designed computer program requires various parameters of the grounding system, which are divided as follows:

The same grid and soil parameters given in Tables 1 and 2 are used to calculate the voltage profile by using the approximated method using ρ_a = 220 Ωm. Computer program has been developed based on the approximated method (Figure 17) shows flowchart for the program. The obtained results are given in (Figures 18 to 21).

From the obtained results given in (Figure 21) the maximum difference between the peaks values of the two methods is about 12% in case of compression ratio=0.4 and minimum difference is 0%.

When C = 0.6,0.8 and 1 (Figure 20), (Figure 19) and (Figure 18) the maximum difference between the peak values of the two methods is about 4%, 8% and 13% respectively and the minimum difference in all cases is 0%.

The minimum value of the maximum percentage difference between the I.S.M. and the approximated method occurs at C = 0.6.

Effect of Number of Meshes on Voltage Profile for Square Grid

Four different grids are used for testing and verification CS01, 04, 16 and 64 given in (Figure 22). Table 4 and Table 5 give the parameters of the four

Figure 17. Approximated method program flowchart

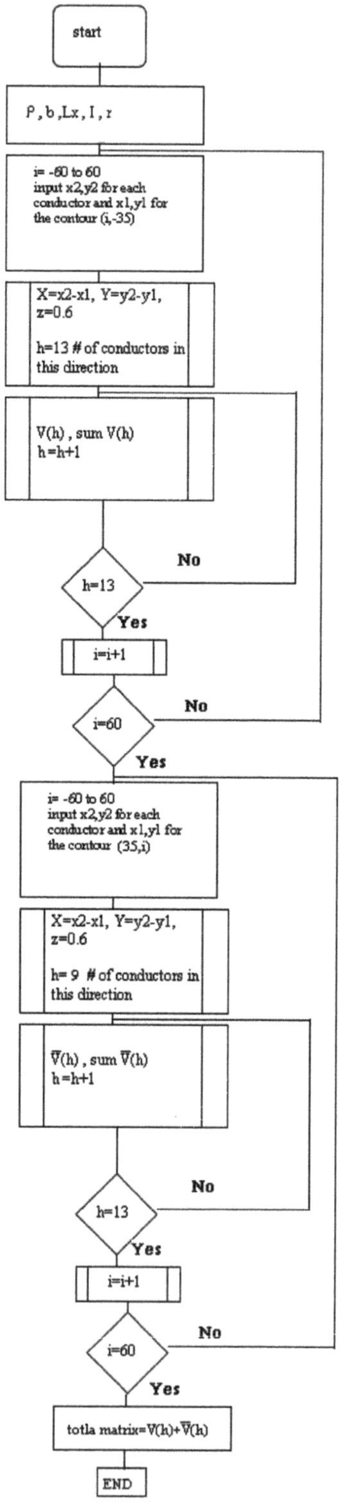

Figure 18. Voltage profile along the unequally spaced grid in case of C = 0.4

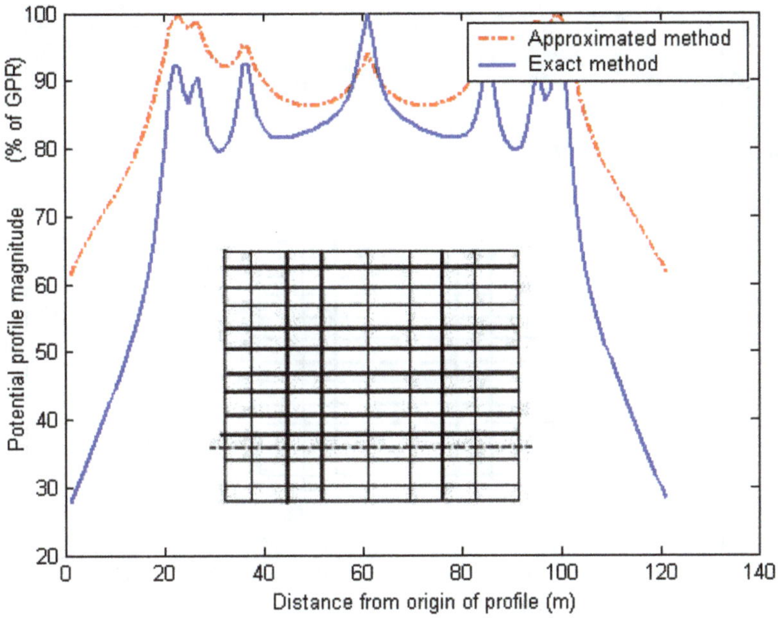

Figure 19. Voltage profile along the unequally spaced grid in case of C = 0.6

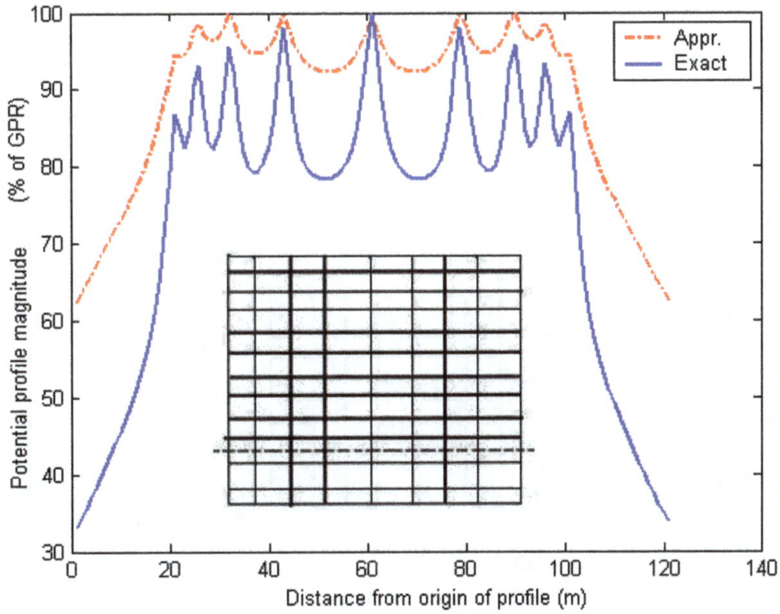

Figure 20. Voltage profile along the unequally spaced grid in case of C = 0.8

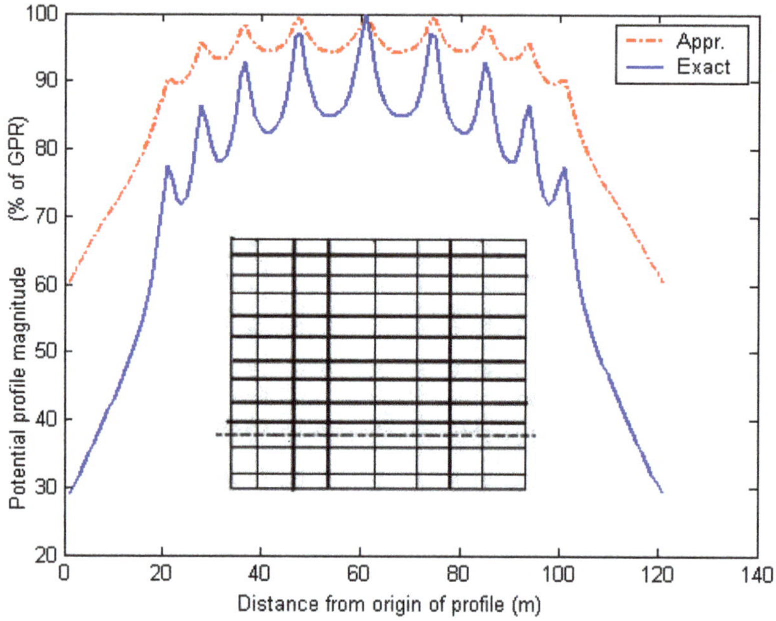

Figure 21. Voltage profile along the unequally spaced grid in case of C = 1

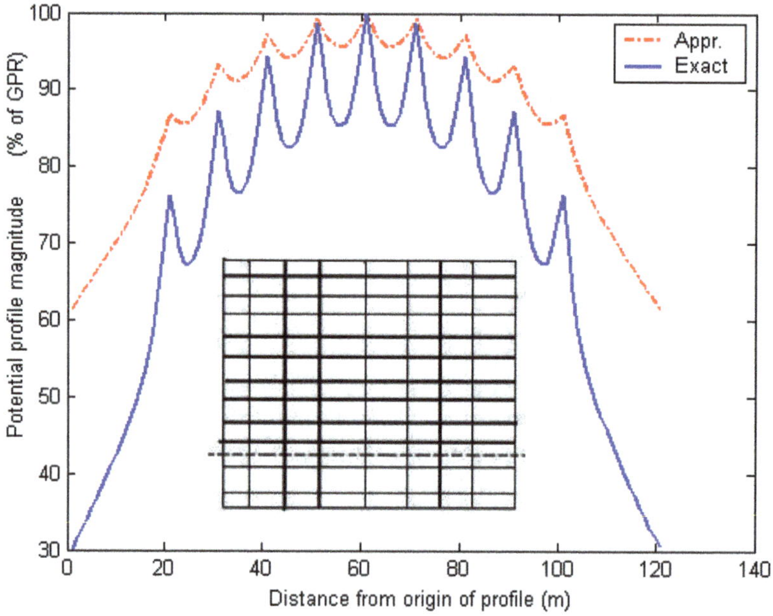

Figure 22. Different numbers of meshes for square grids

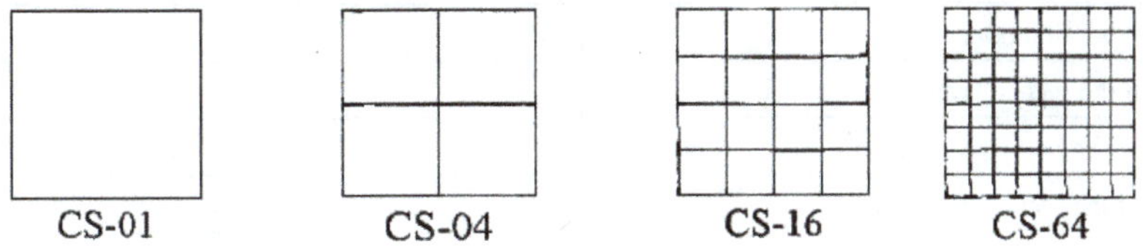

Table 4. Parameters of the four square grids

Case	Area m×m (m^2)	n (Number of Parallel Conductors)	N (Number of Meshes	D (Space Between Conductors)
CS01	10×10	2	1	10
CS04	10×10	3	4	5
CS16	50×50	5	16	12.5
CS64	100×100	9	64	12.5

square grids and soil structure parameters. (Figures 23, 24, 25 and 26) show the voltage profiles for the grids given in (Figure 22) for the purpose of comparison between the exact and approximated calculations.

From the obtained results given in (Figure 23) it is notice that when number of meshes=1 the potential difference of the both methods (I.S.M.

Figure 23. Voltage profile along the CS01 grid

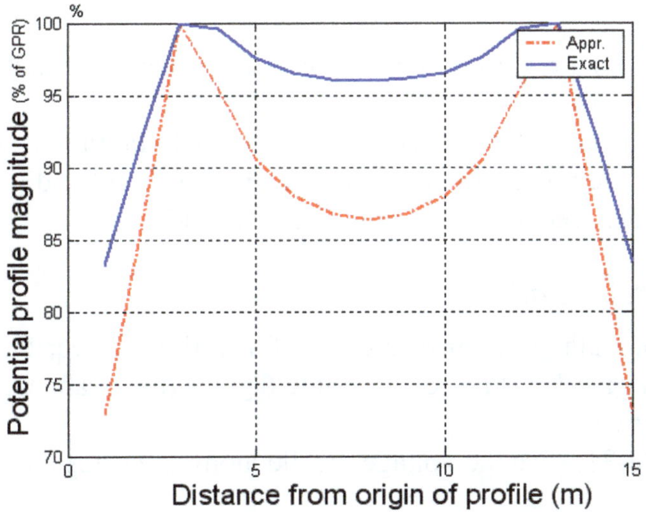

Table 5. Soil structure parameters

Thickness of the first layer, m	2
Thickness of the second layer. m	2
Resistivity of the first layer Ωm	100
Resistivity of the second layer Ωm	300
Resistivity of the third layer Ωm	500
Number of images that vary to infinity.	30

Figure 24. Voltage profile along the CS04 grid

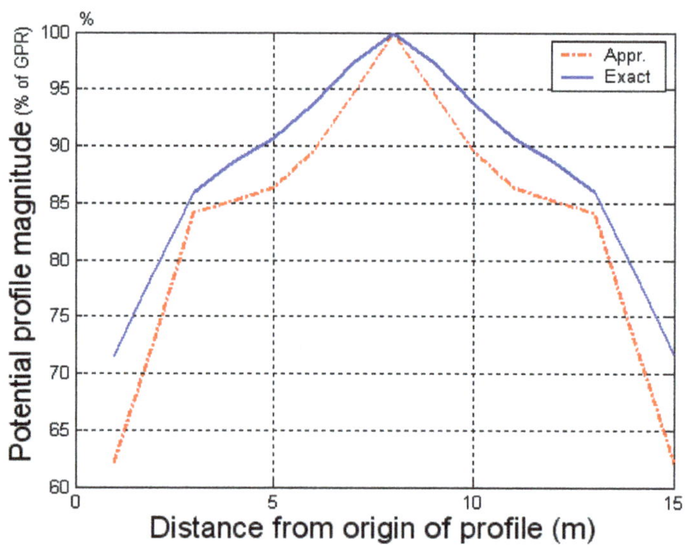

& approximated) are closed to each other at the perimeter and differ at the center of the grid

When numbers of meshes equal 4, 16 and 64 (Figures 24, 25 and 26) both methods are much closed at the center of the grid and with potential difference not exceed 15% at perimeter of the grid

For Rectangular Grid

Four different rectangular grids are used for testing and verification CR01, 04, 16 and 64 given in (Figure 27). Table 6 gives the parameters of the four rectangular grids

(Figures 28-31) give the voltage profile along the grids under study

Figure 25. Voltage profile along the CS16 grid

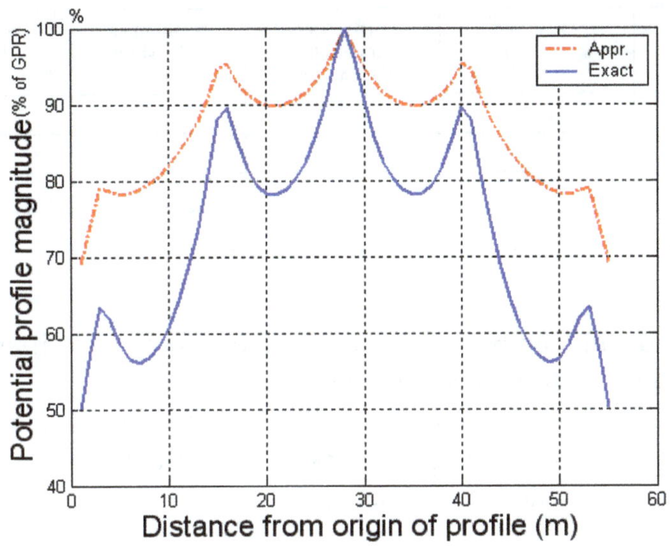

Figure 26. Voltage profile along the CS64 grid

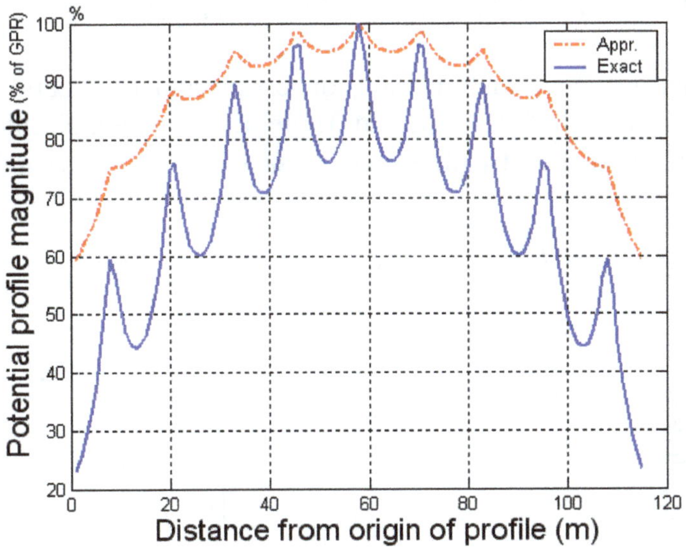

Figure 27. Different number of meshes for rectangular grids

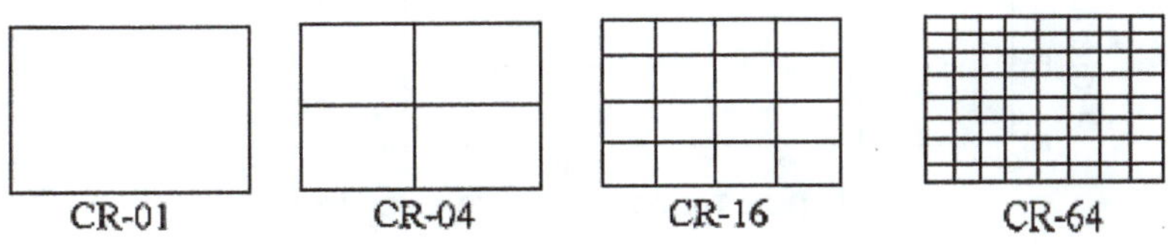

Table 6. Parameters of the four rectangular grids

Case	Area m×m (m^2)	n (Number of Parallel Conductors)	N (Number of Meshes)	D (Space Between Conductors)
CR01	10×20	2	1	14.14
CR04	10×20	3	4	7.07
CR16	50×100	5	16	17.67
CR64	100×200	9	64	17.7

For the grid with number of meshes equal one, potential difference at the center of the grid is difference for the two methods.

The results of the two methods are much closed for number of meshes 4, 16 and 64 and it is noticed that the potential difference between the two methods vary from center to perimeter.

CALCULATION OF THREE DIMENSIONS POTENTIAL USING THE APPROXIMATED

Four different square grids are used for testing and verification of CS01, CS04, CS16 and CS64 described in Table 4, the potential calculated for every meter in both x and y direction and saved in 3D matrix using Matlab

Figure 28. Voltage profile along the CR01 grid

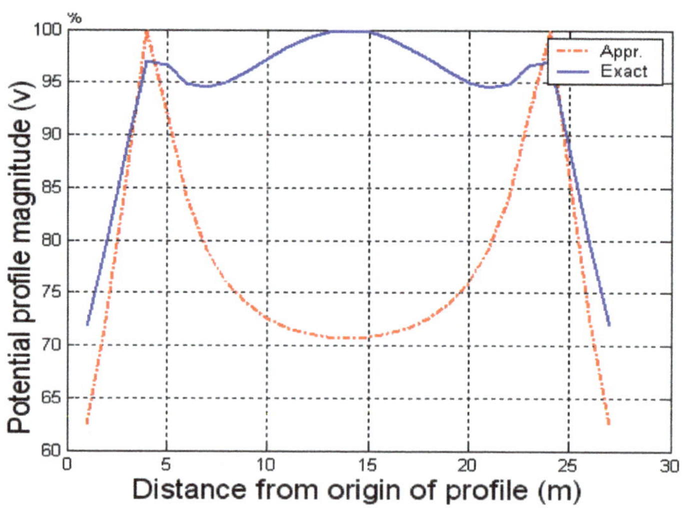

Figure 29. Voltage profile along the CR04 grid

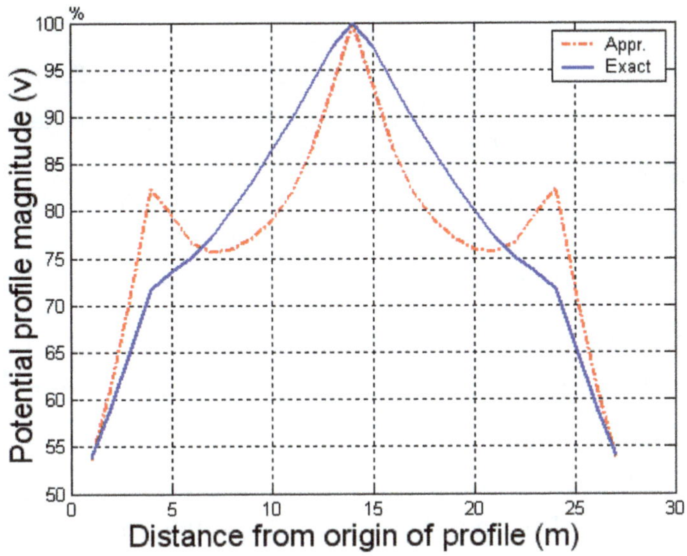

program. Results shown in (Figures 32, 33, 34 and 35 are so closed to that obtained by Bendito, Carmona, Encinas et al. (2000).

For grid consists of one mesh the difference between the center potential and the perimeter potential is 12% and for grid consists of four mesh the difference

Figure 30. Voltage profile along the CR16 grid

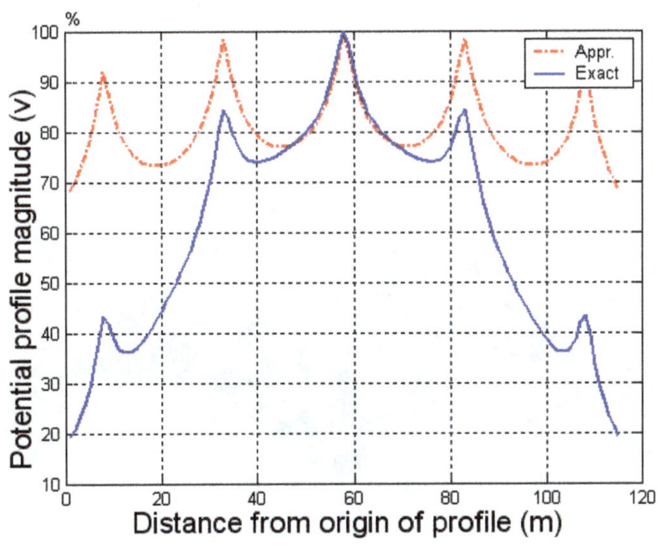

Figure 31. Voltage profile along the CR64 grid

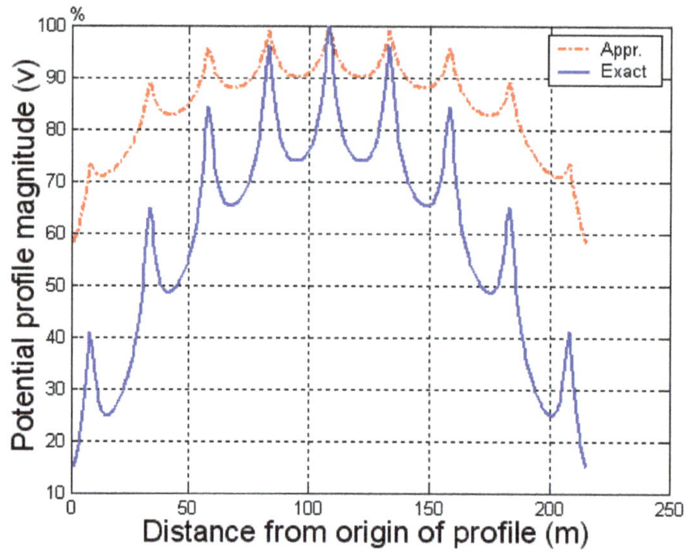

between the center potential and the perimeter potential is 15% and for grid consists of sixteen mesh the difference between the center potential and the perimeter potential is 35% and for grid consists of 64 mesh the difference between the center potential and the perimeter potential is 40%.

Figure 32. 3D Earth surface potential over CS01 grid

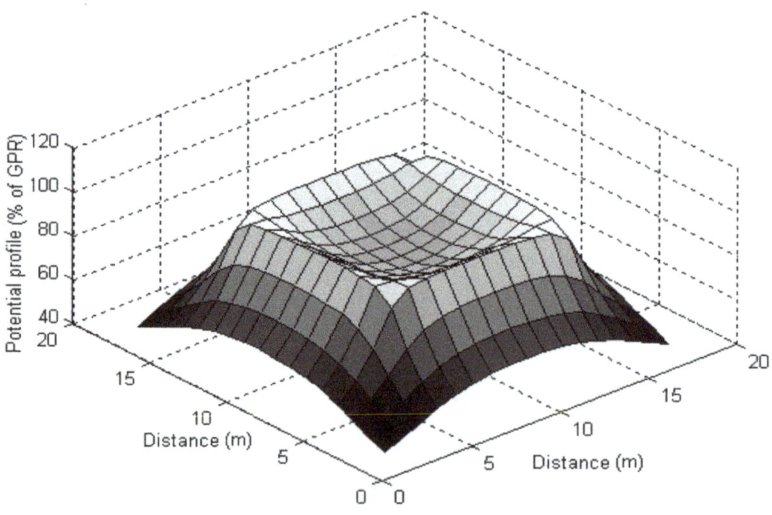

Figure 33. 3D Earth surface potential over CS04 grid

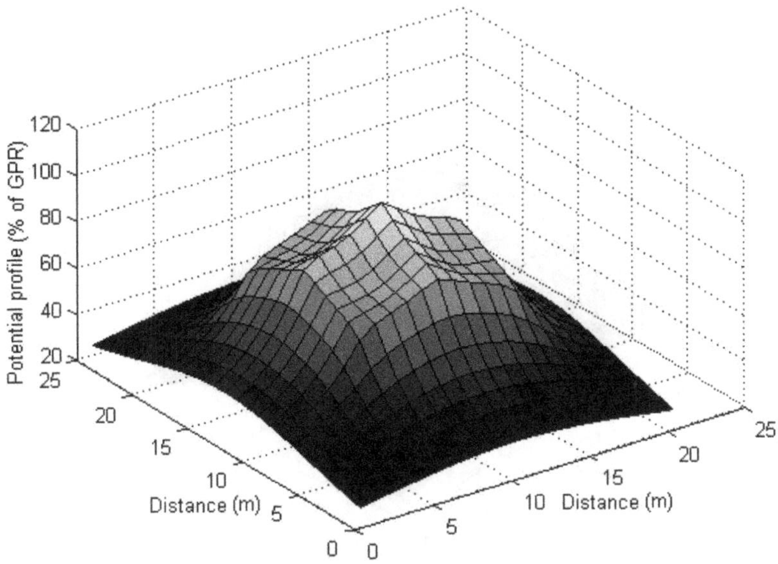

Figure 34. 3D Earth surface potential over CS16 grid

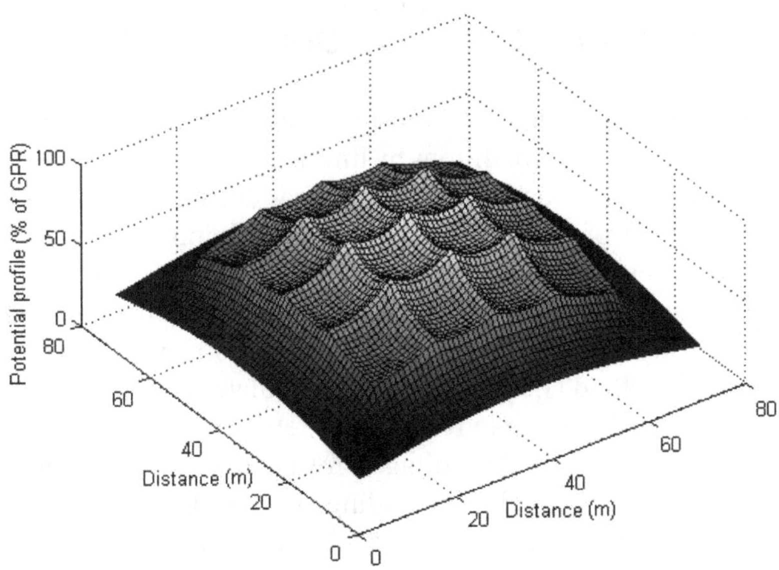

Figure 35. 3D Earth surface potential over CS64 grid

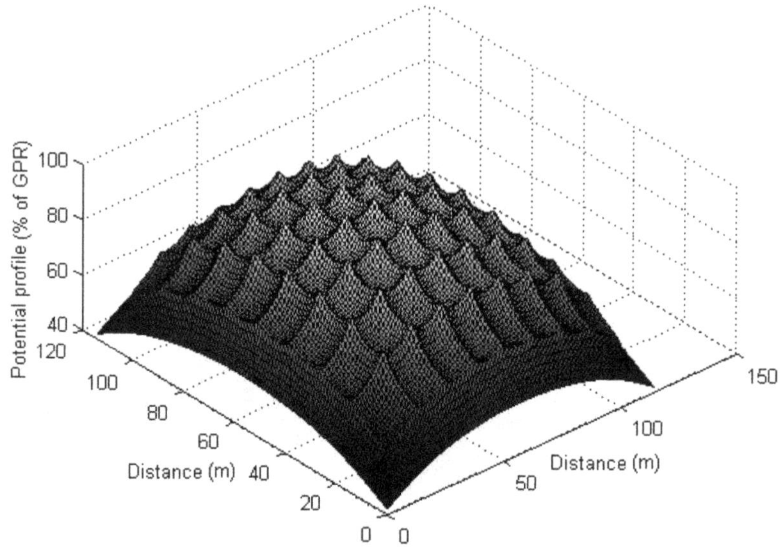

GROUND POTENTIAL RISE (GPR) OF FAULTY SUBSTATIONS HAVING EQUAL AND UNEQUAL SPACING GROUNDING GRIDS CONDUCTORS

Grounding of power substations is to provide safety for personnel during normal and fault to ground conditions by limiting step and touch potentials, to dissipate lightning strokes and to stabilize voltage during transient conditions. One important goal of this paper is to present simplified approaches for the earth surface potentials and ground potential rise (GPR) of substation grounding in case of uniform and non-uniform soil. The equalization of the current density and then decreasing of GPR is achieved. The Infinite Series Potential Method (I.S.M) is used to calculate the GPR and the earth surface potentials with accurate results for the grids buried in uniform and non-uniform soil structure. The voltage profiles have been designed in 3D to make the selection of suitable grounding system easier.

There are two main design goals to be achieved by any substation grounding system under normal as well as faulty cases. One of them is to provide means to dissipate electric currents into the earth without exceeding any operating and equipment limits, and the other is to assure that a person in the vicinity of grounded facilities is not exposed to the danger of critical electric shock. Step voltages are inherently less dangerous than mesh voltages. However, safety within the grounded area is achieved with the assistance of a high resistivity surface layer (surface material), which does not extend outside the fence. In any event, the computed step voltages should be compared with the permissible step voltage after a grid has been designed that satisfies the touch voltage criterion.

Studying the effect of voltage distribution on the grid surface is an important parameter to design the grounding system of substations to ensure the safety of people above the substation ground. This can be achieved by equalizing the potential distribution of the ground surface and reducing step and touch potentials. As it is well known the person's safety is affected by the ground potential rise (GPR) of grounded structures during electric power faults. In order to limit these voltages to safe values, it is necessary to design a right grounding system that effectively connects all metallic structures of the substations to ensure the safety criteria for both human and substations taking into account the economical purposes.

However, the unequally spaced grid can make current density which is emanating from grid conductors more uniform, and therefore remarkably decrease of potential gradients on the earth's surface. Then, the maximum touch voltage for this grid will reach its lowest values and thus the safety conditions for human above the substation earth surface will be ensured. The unequally spaced grid saves about 34% of grounding grid material. It also costs less to install the grid.

Unequally Spaces Grids Design Procedures

The spacing between the grounding grid conductors should be so arranged that the touch and step voltages will be smaller than the safe value for the personnel. The optimal design of grounding systems for substations is to suitably arrange the conductors of grounding grid systems to keep the maximum touch voltages within the safe limit; this would ensure making all grounding grid conductors sufficiently utilized so it can be certified as a safe and economic design method.

The conductor compression ratio "C" is related to the grid dimensions and the maximum conductor separation d_{max} which occurs toward the center of the grid and is given by Zhongdong, Du., Yao et al. (2005):

$$d_{\max} = \frac{L(1-C)}{1+C-2C^{(N/2+1)}}, \text{ with N even number} \tag{46}$$

or,

$$d_{\max} = \frac{L(1-C)}{2\left(1-C^{(N-1)/2}\right)}, \text{ with N odd number} \tag{47}$$

$$d_i = d_{\max} C^i \tag{48}$$

where N is the desired number of conductors perpendicular to grid axis of length L. In unequally spaced grid when the conductors are arranged according to an exponent, the span from the center could be calculated according to equations 46, 47 and 48, where i, is the conductor number.

For the optimal design of grounding grid system, the optimal compression ratio (OCR) could be determined, in which the touch voltage reaches its minimum if the grounding grid is designed using this compression ratio. An empirical expression is suggested by Zhongdong, Du., Yao et al. (2005) to calculate OCR as follows:

$$OCR=a_0+a_1\exp(0.0001h)+a_2\exp(bh) \tag{49}$$

where

$$b=-0.3503-9.6311\exp(-0.03666L) \tag{50}$$

$$a_0=a_{01}+a_{02}k+a_{03}k^2 \tag{51}$$

$$a_1=a_{11}+a_{12}k+a_{13}k^2 \tag{52}$$

$$a_2=a_{21}+a_{22}k \tag{53}$$

k is the number of segments of the conductor in any direction; h is the grid laying depth. The relationship between a_0, a_1... and L is given in Table 7.

Table 7. Relation between a_0, a_1... and grid length

Coefficients	L≤100 m	100 m ≤ L ≤ 175 m	175 m ≤ L ≤ 250 m	L > 250 m
a_{01}	0.44	0.38	-0.51	0.32
a_{02}	-77.43	-50.65	-33.18	-15.44
a_{03}	15.63	13.88	18.49	13.42
a_{11}	0.033	0.19	1.15	0.38
a_{12}	76.9	50.21	32.82	15.16
a_{13}	-15.56	-13.83	-18.44	-13.38
a_{21}	-0.067	-0.037	-0.029	-0.022
a_{22}	0.5	0.41	0.34	0.26

Effect of Conductors' Distributions on Voltage Profile of Grids Buried in Uniform Soil

Infinite Series Potential Method (I.S.M)

Software is designed by the authors to illustrate the effect of equal and unequal grid conductor's distribution on the earth surface potentials above the grid. A square grid of 105×105 m^2 with number of meshes 64 equally and unequally spaces parallel conductors in each direction is taken as an example as shown in (Figure 36). The grid material is copper having 5 mm radius, calculated according to IEEE standard 80.

By using Infinite Series Potential Method (I.S.M), the potential difference in volts can be calculated in two dimensions using the following equation.

$$V(x,y) = \frac{I\rho_1}{4\pi l} \ln f(x,y) \tag{54}$$

where,

$$f(x,y) = \left(\sqrt{\left(x + l/2\right)^2 + y^2} + x + l/2\right) \Big/ \left(\sqrt{\left(x - l/2\right)^2 + y^2} + x - l/2\right)$$

where I is the short circuit current, L is grid side length in meter and ρ_1 is the soil resistivity in Ω.m. The following similar relation is used to obtain the grounding grid voltage profile in three dimensions.

Figure 36. Examples for grid module of N=64 with equally, unequally spaced having denser conductors at the edges and at the middle, (A), (B) and (C) respectively.

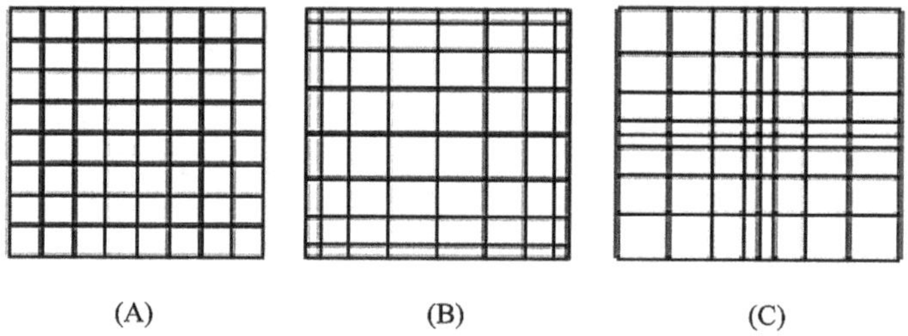

$$V\left(x,y,z\right)=\frac{I\rho_1}{4\pi l}\ln f\left(x,y,z\right) \tag{55}$$

where,

$$f\left(x,y,z\right)=\left(\sqrt{\left(x+l/2\right)^2+y^2+z^2}+x+l/2\right)\Big/\left(\sqrt{\left(x-l/2\right)^2+y^2+z^2}+x-l/2\right)$$

To achieve the effect of grid conductors distribution on the ground potential rise, assuming the current emanating from the faulty substation into the grounding grid at short circuit condition of 220 kV substations is I = 10 kA, grid burial depth is 0.6 m. The earth surface in volts and as percentage of GPR for grid modules A, B and C are calculated. The obtained results are given in Table 8 when the soil resistivity was 100 Ω.m.

From the obtained results, it is concluded that the GPR decreases in unequally spaces grids having denser conductors at the grid edges compared with equally spaces grid conductors and unequally spaces having denser conductors at the grid middle, in general using grounding grids decreases the potential difference between each two conductors and more uniform voltage profile can be obtained. (Figure 37) represents the three dimensions earth surface potentials, for every one meter in both x and y directions, above the mentioned grids of different conductors distribution configurations.

Regarding to grid conductors' distribution with equally and unequally spaces, the percentages of the potential rise at the corner conductors with respect to the grid center potential rise in case of number of meshes 9×9, 13×13 and 17×17, are given in Table 8.

Table 8. Earth surface potentials in volts and as percentages of GPR along center conductor for equally and unequally spaced grid of (A=105×105 m^2)

Equally Spaces Grid Conductors									
x, (m)	-52.5	-39.38	-26.25	-13.13	0	13.1	26.3	39.4	52.5
V, (volt)	4003	4732	5076	5249	5297	5249	5076	4732	4003
V, (% GPR)	0.76	0.89	0.96	0.99	1	0.99	0.96	0.89	0.76
Unequally Spaces Having Denser Conductors at the Grid Edges									
x, (m)	-52.4	-46.8	-35.1	-19.1	0.0	19.1	35.1	46.8	52.4
V, (volt)	4080	4431	4629	4762	4799	4762	4629	4431	4080
V, (% GPR)	0.85	0.92	0.96	0.99	1	0.99	0.96	0.92	0.85
Unequally Spaces Having Denser Conductors at the Grid Middle									
x, (m)	-52.4	-33.3	-17.3	-5.6	0.0	5.6	17.3	33.3	52.4
V, (volt)	4156	5265	5785	6138	6197	6138	5785	5265	4156
V, (% GPR)	0.66	0.85	0.93	0.99	1	0.99	0.93	0.85	0.66

From this table, it is noticed that increasing the number of grid meshes at the same area reduces the grid GPR regardless of grid configuration.

Noteworthy Tables 8 and 9 and (Figure 37) confirm the previous conclusion, that the increasing of unequally grid conductors having denser conductors

Figure 37a. Voltage profile along grid conductors in case of equally spaced grid (grid A), the number of meshes = 64, short circuit current 10 kA, soil resistivity 100 Ω.m

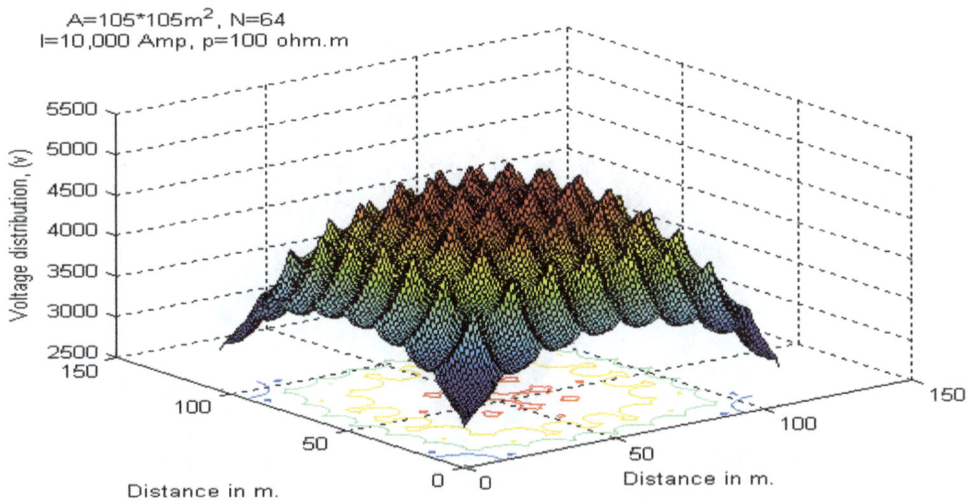

Figure 37b. Voltage profile along grid conductors in case of unequally spaced grid having denser conductors at the middle, where the number of meshes =64, short circuit current 10 kA, and soil resistivity 100 Ω.m

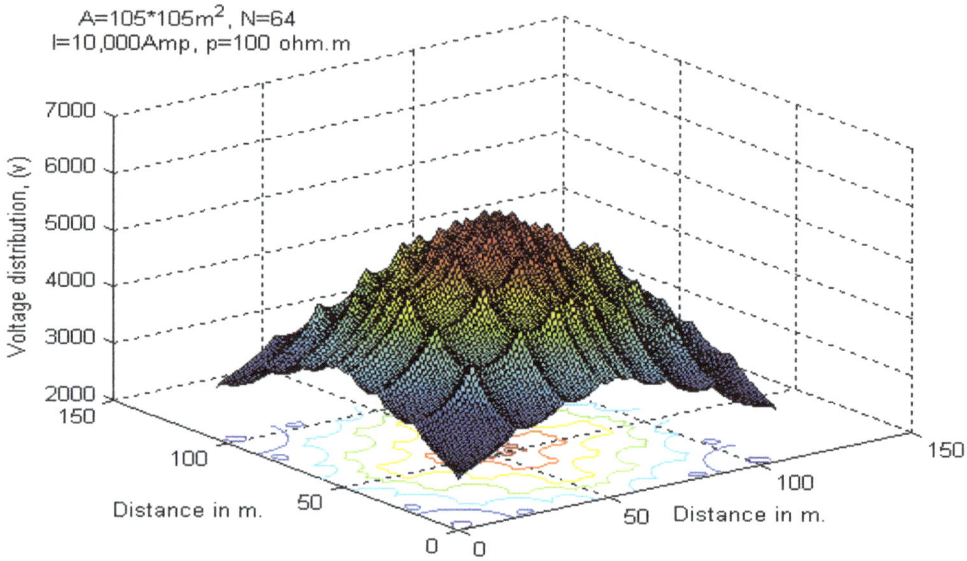

Figure 37c. Voltage profile along grid conductors in case of unequally spaced grid having denser conductors at the edges, the number of meshes = 64, short circuit current 10 kA, soil resistivity 100 Ω.m

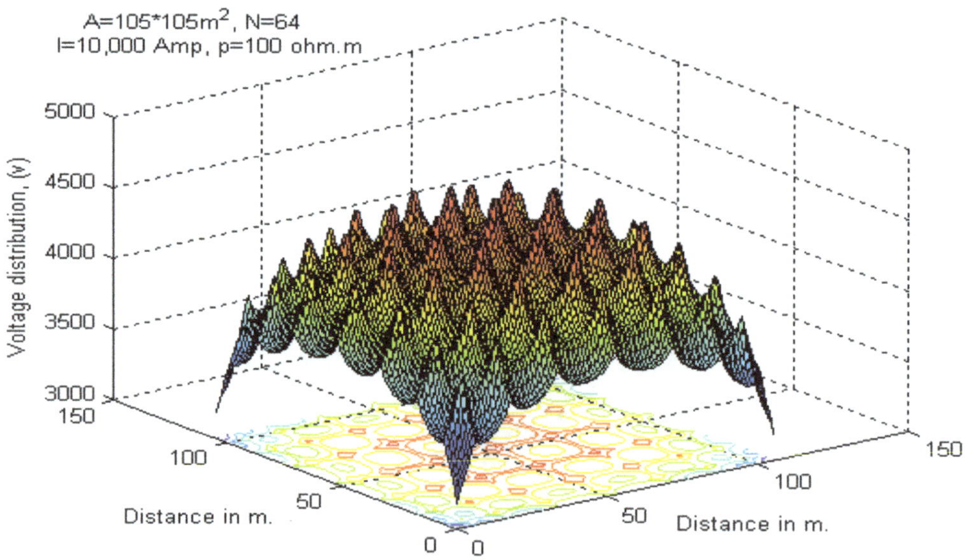

Table 9. GPR of meshes corners for different number of meshes, when the grid is loaded by 10 kA short circuit current and 100 Ω.m soil resistivity, grid area equals 105×105 m²

Grid Number of Meshes	Equally Space Grid Corner (A) GPR	Unequally GRID (B) CORNEr GPR	Unequally GRID (C) CORNER GPR
9×9	0.76	0.85	0.66
13×13	0.743	0.850	0.64
17×17	0.733	0.860	0.624

at the grid middle increases the GPR for this grid. While that having denser conductors at the edges has more flat earth voltage distribution.

As it is noticed from Table 9, the GPR whatever number of grid conductors and meshes for the same grid area, for unequally grid denser conductors at the grid edges has lowest GPR and the lowest potential differences between center and corner conductors compared with the other two grid configurations. It provides the most efficient grounding grid design. This is in agreement with according to IEEE 80, 2000

(Figure 38-a) gives the earth surface potential along the center of the square grounding grids of various meshes A, B and C. (Figures 38-b and 38-c) give the touch and step voltages for different shapes of the square grounding grids.

Figure 38. Earth surface potential along the center of the square grounding grids of various shapes of meshes, short circuit current 10 kA, soil resistivity 100 Ω.m

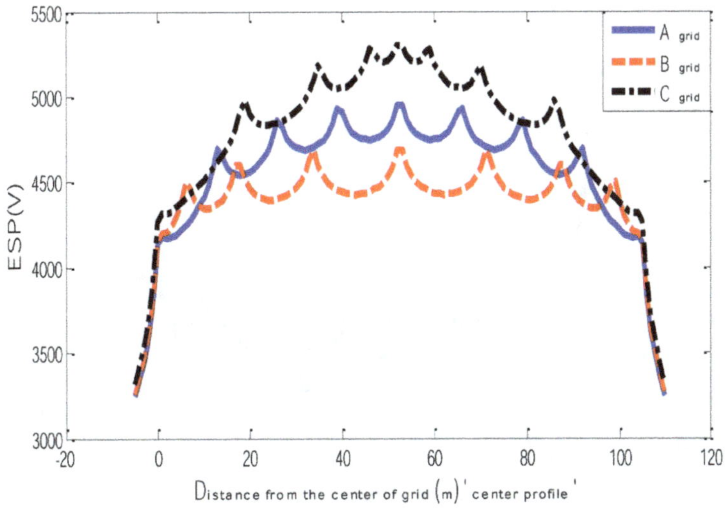

Optimal Compression Ratio (OCR)

In unequally spaced grid, when the conductors are arranged according to an exponent regularity, the span from the center can be calculated by equations 46 to 48. Empirical expression is used to calculate OCR as given in equations 49 to 55 and Table 7. To illustrate the effect of unequally space conductors' distribution on the earth surface potentials and touch voltage profiles for the grid buried in uniform soil and to satisfy the optimal grounding grid design, the grid module area,105×105 m^2, N=8×8, as an example is used. The grid is buried in a depth of 0.6 meter of uniform soil has 100 Ω.m resistivity. The earth surface potentials and touch voltages are calculated along the center conductor for this grid under different values of compression ratios (OCR) as presented in (Figure 41).

By the analysis of the calculated data represented in this figure, it is concluded that the touch voltage along the center conductor reaches its lowest value when the compression ratio for this grid reaches 0.538, it means that, the optimal compression ratio obtained gives the most save and economical grid design, and the best arrangement of grounding conductor span is achieved, so the minimum touch voltage and thus, the optimal design for grounding grid is satisfied. These findings are in agreement with results obtained by Hung, Chen, & Yan (1995). Figure 39 illustrates the relationship between the touch voltages over center conductor for the unequally spaced grid as a function

Figure 39. Effect of the different shapes of meshes on the touch voltage of the square grounding grids, short circuit current 10 kA, and soil resistivity100 Ω.m

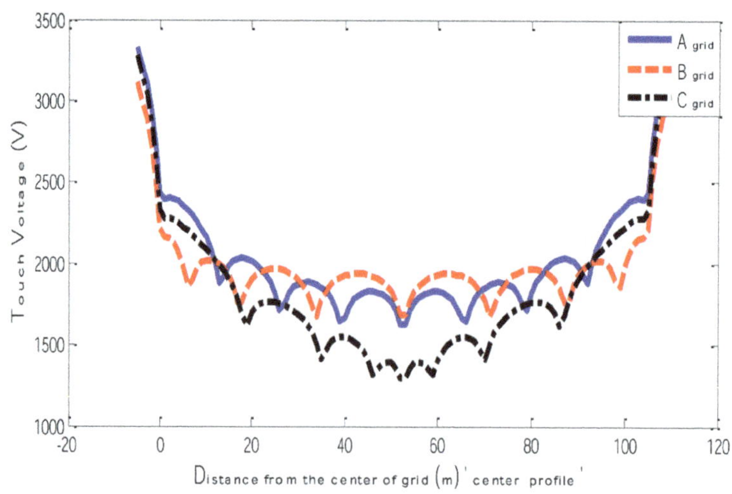

Figure 40. Effect of different shapes of meshes on the step voltage of the square grounding grids, short circuit current 10 kA, soil resistivity100 Ω.m

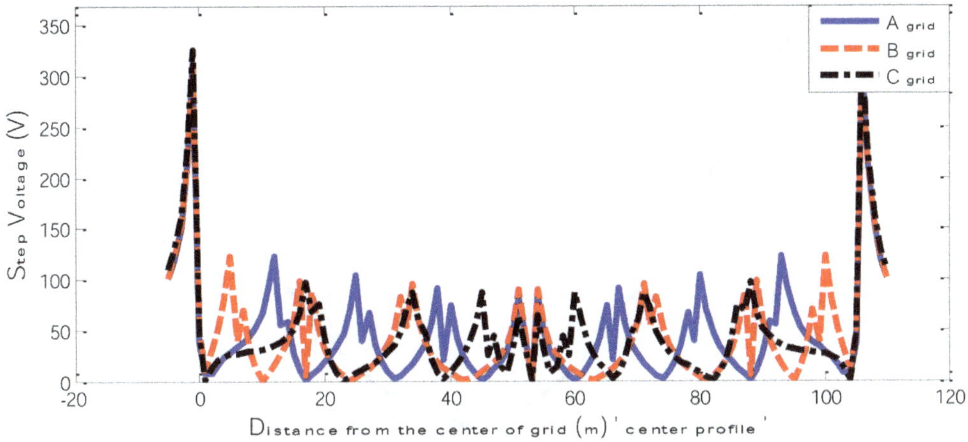

Figure 41. Earth surface potentials and touch voltages over the center conductor for A=105×105 m², N=8×8, grid having different compression ratios

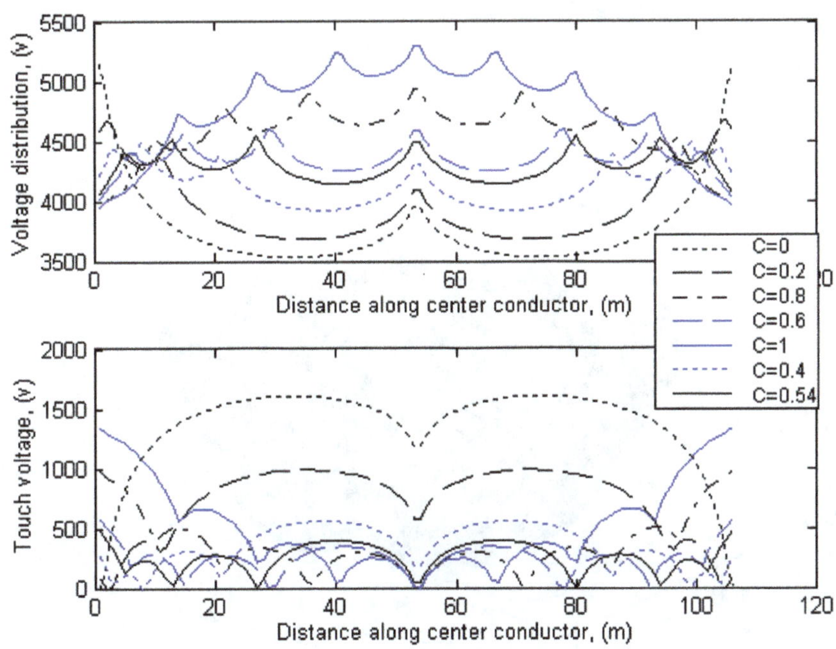

Figure 42. Touch voltages over center conductor for grounding grid module (A) as a function of conductor compression ratio in uniform soil of 100 Ω.m

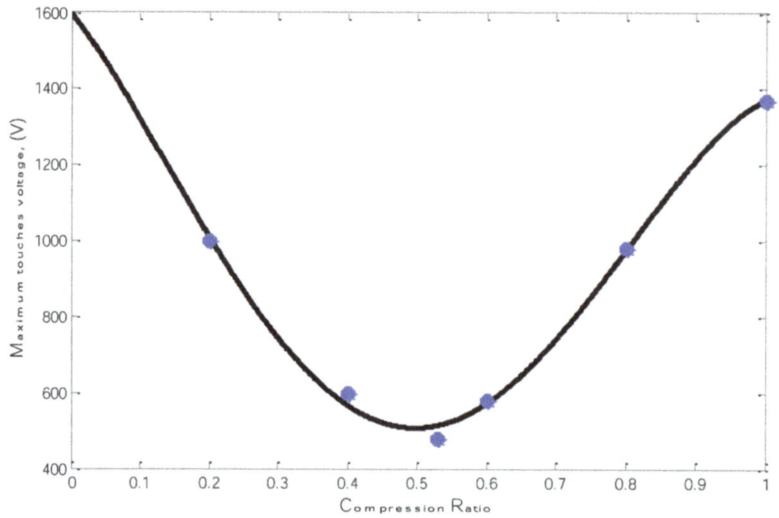

Figure 43. Earth surface potentials over all grid conductors for grounding grid of module (A) (A = 105 × 105 m², N = 8 × 8) in case of C = 0.2, short circuit current 10 kA, soil resistivity 100 Ω.m

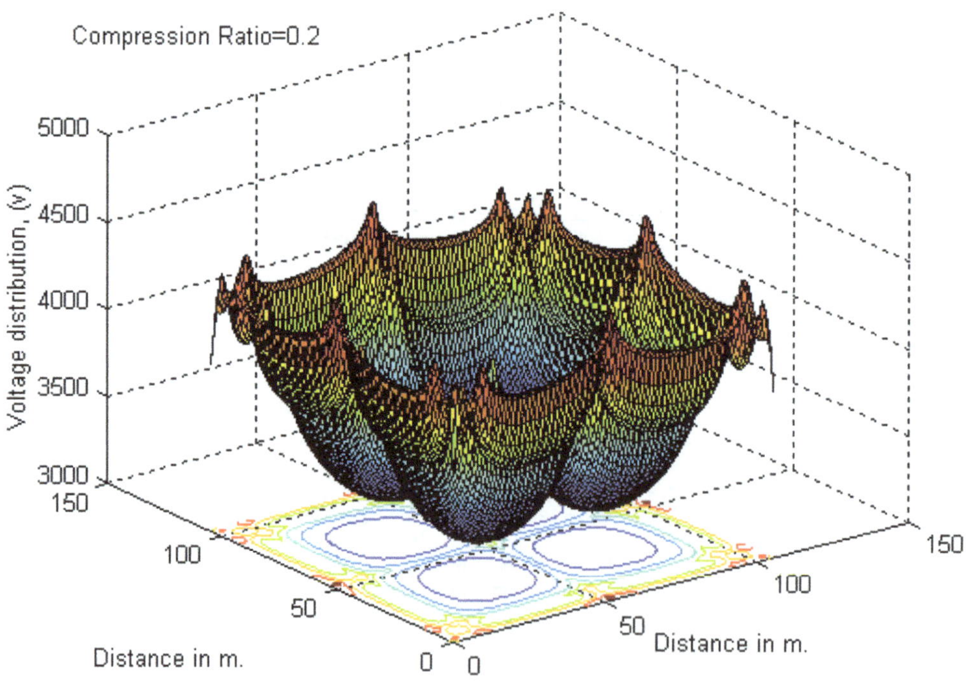

Figure 44. Earth surface potentials over all grid conductors for grid of (A = 105 × 105 m², N = 8 × 8) in case of C = 0.54, short circuit current 10 kA, soil resistivity 100 Ω.m

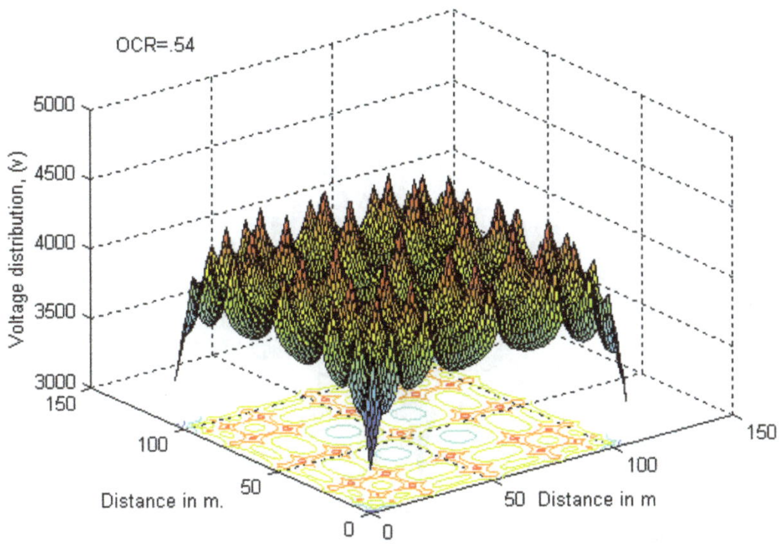

of conductor compression ratio. Remarkable, this relationship usually has an obvious ''U'' shape. As it is noticed from this figure the minimum touch voltage is obtained when the compressed ratio (OCR) is 0.538 in case of the grid is laid in uniform soil.

Samples of voltage distribution in 3D for different values of compressed ratios are given in (Figure 43). From this figure, it is clear that at C=0.54 the earth surface potential distribution is more uniform.

Voltage Profile of Grids Buried in Double Layers Soil

Often the installation of grounding grids of electric power stations and electrical networks is carrying in heterogeneous soil contains two or more layers. To illustrate the effect of changing the upper layer thickness in double layers soil structure with the variation of the upper and lower soil resistivity on the voltage profile and step and touch voltages, I.S.M method is used. The graphical representation for earth surface potentials is plotted along the center conductor for 105×105 m² grid area with eight equally spaces conductors buried in non-uniform soil having specifications that given in Table 10. The current emanating the grounding grid is 10 kA, and the grid laying depth

Figure 45. Earth surface potentials over all grid conductors for grid of (A = 105 × 105 m², N = 8 × 8) in case of C = 0.6, short circuit current 10 kA, soil resistivity 100 Ω.m

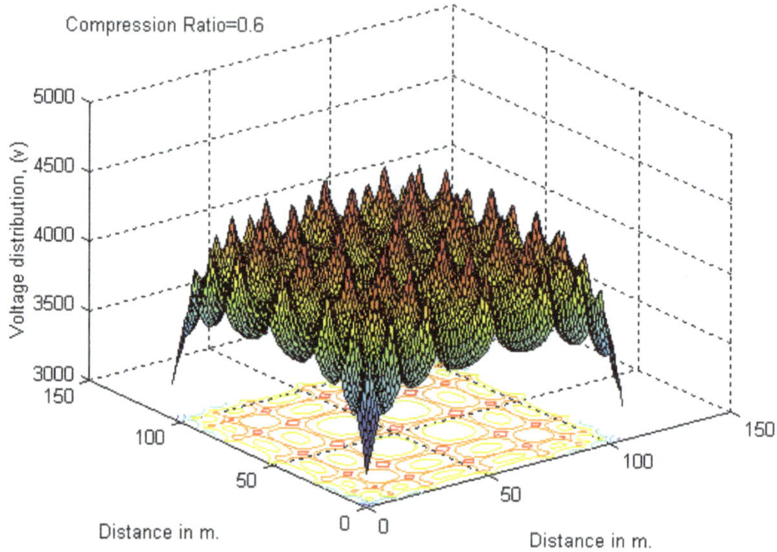

Figure 46. Earth surface potentials over all grid conductors for grid of (A = 105 × 105 m², N = 8 × 8) in case of C = 1, short circuit current 10 kA, soil resistivity 100 Ω.m

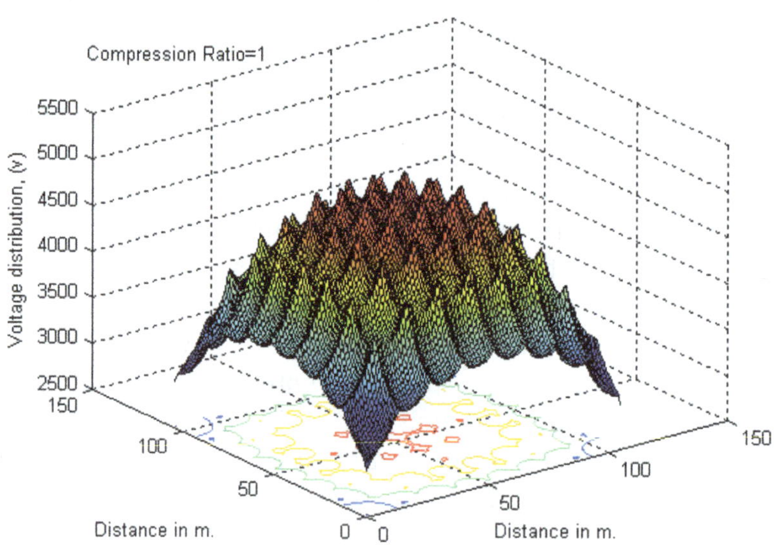

Figure 47. Earth surface potentials along center conductor for equally spaced grid (A = 105 × 105 m2, N = 8 × 8) with various soil structures

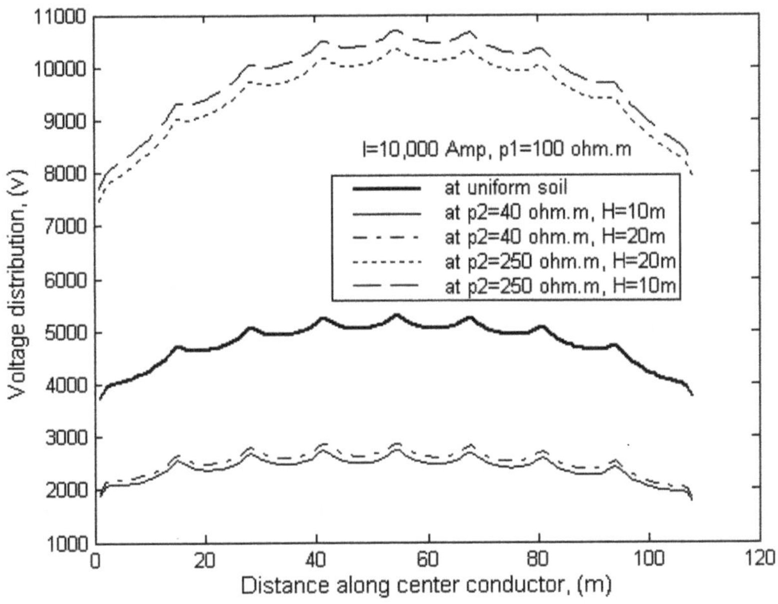

is 0.6 meter. (Figure 41-a) shows the earth surface potential along center conductors for equally spaced grid with 105×105 m^2, N = 8 × 8 (module A) with various soil structures. From this figure, it is noticed that the GPR voltage of the grid buried in upper layer of double layers soil structure in which the second layer has low resistivity, is lower than that having high soil resistivity of second layer. It is noticed also that at the same upper layer soil resistivity in which the grid is buried, the upper layer thickness has little effect on GPR.

To illustrate the effect of soil non-uniformity on the unequally spaced grounding grid system having different conductors' distribution, the maximum touch voltages along center conductor for the grounding grid module (A) which buried in laying depth of 0.6 meter of the upper layer of double layer soil having the specifications mentioned in Table 4 are plotted in (Figure 41-b).

Table 10. Specifications of used non-uniform soil

ρ_1 (upper layer resistivity), Ω.m	100	100	100	100	100
ρ_2 (lower layer resistivity), Ω.m	100	40	40	250	250
H (upper layer thickness) in meters	infinity	10	20	10	20

Figure 48. Maximum touch voltage over center conductor for mentioned grid buried in various soil structures as a function of compression ratio

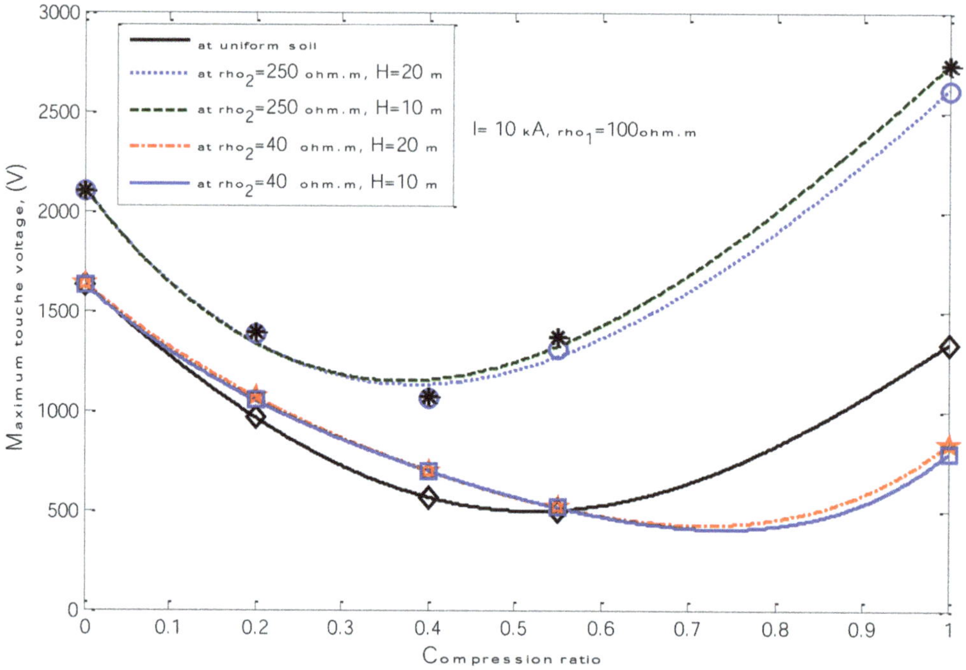

This figure shows that the compression ratio of grid conductors' arrangement satisfies the lowest touch voltage depends on the soil parameters in which the grid is buried in. From this figure, it is concluded that when $\rho_1 < \rho_2$, the compression ratio of lowest touch voltage takes value less than the optimal compression ratio (OCR) which determined by equation 49 for this grid, it's value reaches to 0.4 and this means that the grid conductors will be denser at the edges and having widely conductor spans at the middle. The results will be in the contrary if $\rho_1 > \rho_2$ in which the touch voltage reaches its lowest values at the compression ratio of 0.74. This is indication to denser conductors at the grid center. In case of uniform soil $\rho_1 = \rho_2$ the obtained results are in agreement with empirical OCR formula given in equation 49. The effect of upper layer thickness on OCR can be ignored as shown in (Figure 46, and (Figure 47).

Figure 49. Effect of vertical rods on the earth surface potential of square grounding grid (A), short circuit current 10 kA, soil resistivity 100 Ω.m

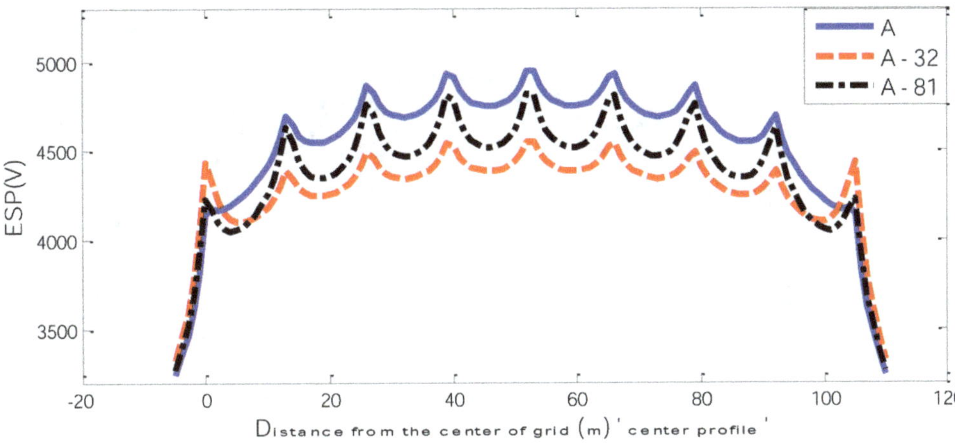

Effect of Grounding Rods on Voltage Profile of Grounding Grid

Each one of grounding grid modules A, B and C is provided with copper rods, each one has 3 meters length and 25 mm diameter. Thirty-two rods are distributed on the grid perimeter nodes in some calculations and eighty-one rods are distributed, one at each node of the grounding grid in other

Figure 50. Effect of vertical rods on the earth surface potential of square grounding grid (B), short circuit current 10 kA, and soil resistivity equals100 Ω.m

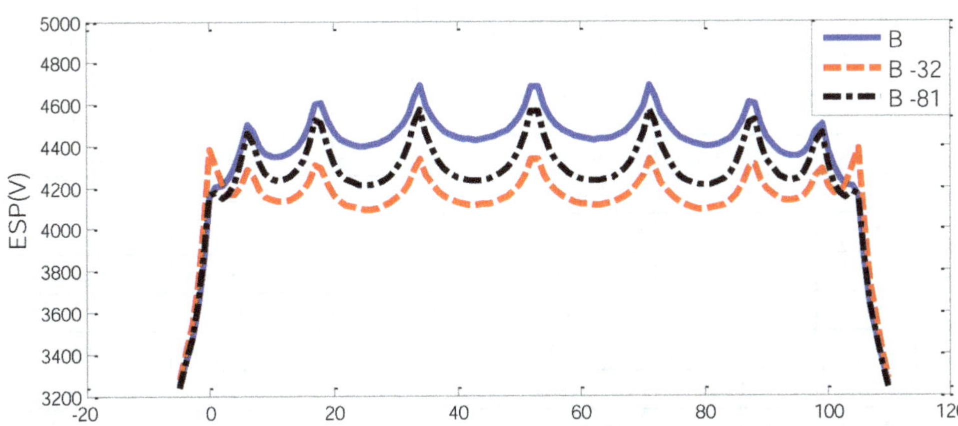

Figure 51. Effect of vertical rods on the earth surface potential of square grounding grid (C), short circuit current 10 kA, soil resistivity 100 Ω.m

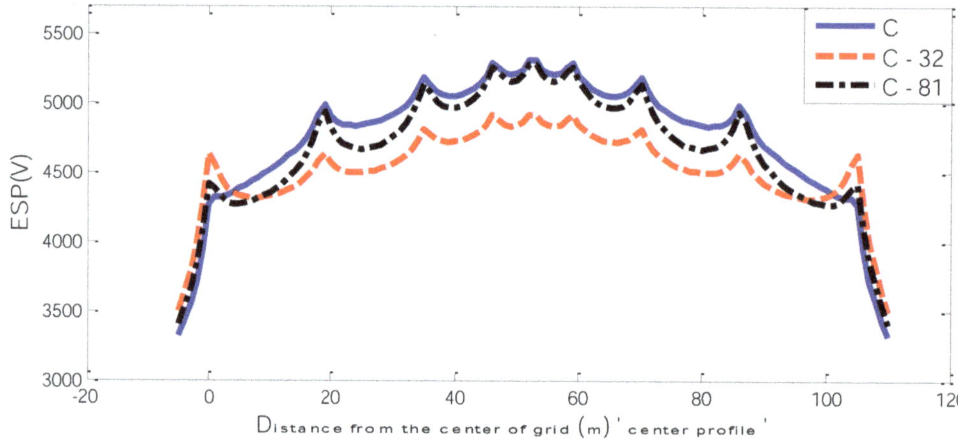

calculations. (Figures 42.a, 42.b and 42.c) show the vertical rods effect on the earth surface potential along the center conductor of the grounding grids A, B and C. Slightly decrease of the earth surface potential is noticed when the grids are provided with vertical rods. The range of this decrease reaches to 6.25% - 6.6% when installing rod at each node on the grid perimeter.

It is noticed also that as given in Table 11 the increase in grounding system length by installing more rods leads to decreasing in grounding resistance

Table 11. Total conductor length, ground resistance, ground potential raise, touch voltage and step voltage of different square grounding grid cases (center profile)

Case	L_t (m)	R_g (Ohm)	GPR (kV)	V_t Max % of GPR	V_s Max (kV)
A	1890	0.6578	6.578	37.03	4.76
B	1890	0.6367	6.367	34.75	4.89
C	1890	0.6600	6.600	35.30	4.94
A-32	1986	0.6287	6.287	30.22	5.63
B -32	1986	0.6142	6.142	30.18	5.77
C-32	1986	0.6308	6.308	26.58	5.45
A-81	2133	0.6036	6.036	29.92	5
B -81	2133	0.6118	6.118	31.93	4.80
C-81	2133	0.6182	6.182	28.69	4.911

Where (A), (B), and (C) are square grounding grids without vertical rods; (A-32), (B-32), and (C-32) are square grounding grids with 32 vertical rods; and (A-81), (B-81), and (C-81) are square grounding grids with 81 vertical rods.

and step potential. As well known the grounding system resistance, step and touch voltages are inversely proportional with the grounding system length according to IEEE 80 (2000).

Verifications of Calculated Results

Experimental Set-Up

For the verification of the calculated results an experimental model contains two identical tanks; each one has 1.5 m × 1.5 m × 0.5 m dimensions, filled with tap water. One is installed above the other, the top ends of the lower tank is coated with plastic to obtain electrical isolation between the metallic bodies of the two tanks. Only the electrical contact between the two mediums (representing the soil two layers) occurs through the metallic base of the upper tank. The lower tank is connected to earth. The restively of the upper and lower water layers can be controlled by adding an amount of sodium chloride.

A 220 volt alternating source provided with variac is connected with the grounding grid model which is suspended in the upper tank. Capacitor (10 Farad) is connected between the variac and the grid model to prevent any flow of D.C current due to the use of dissimilar materials. Two voltmeters are used; one for measuring the input voltage to the model and the other is connected to a probe for measuring the voltage distribution on the grid surface. One ammeter is used to measure the input grid current simulates the short circuit current, as shown in (Figure 43).

Grids Simulation

To simulate the grounding grids A, B and C scale factor of 1/100 is chosen, the dimensions of each grid are 105 ×105 cm. In opinion of several researchers, the effects of grid conductor diameter on grid resistance and the earth surface potentials are negligible. The simulated grid conductors are made from copper having 2.5 mm. The scale factor of the grid laying depth and the upper layer thickness is 1/100.

Water Resistivity Measurement

The two tanks are filled with water, to control the water resistivity of the two tanks (simulate the two soil layers) an amount of NaCL salt is used, and hence changing the reflection factor. The water conductivity is measured by means of DC probe.

Simulation of Short Circuit Current

The input voltage to the grid simulation is controlled by variac, until the current reaches 5 A with scaling factor 1: 2000 of short circuit current 10 kA.

Experimental Results

The step potential V, is measured using probe as shown in (Figure 43). The measurements are done on grid model 105×105 m^2, N= 64 meshes, for equal conductor distribution, unequal having denser conductors at the grid edges and unequal spaces having denser conductors at the grid middle. The resistivity of each layer is 100 Ω.m, (the soil is considered as uniform soil). The measured step potential V, is multiplied by the scaling factor of current and compared with the calculated values. The results are given in Table 12. From this table, it is noticed that the measured values are slightly high compared with the calculated values, the reason may be due to the boundary effects of the tank walls, as each grid model size is higher than 1/5 of the upper tank size.

Table 12. Earth surface potentials in volts and as percentages of GPR along center conductor for equally and unequally spaced grid of (A = 105 × 105 m^2, N = 64)

Equally Spaces Grid Conductors									
x, (m)	-52.5	-39.38	-26.25	-13.13	0	13.1	26.3	39.4	52.5
V, (volt)calculated	4003	4732	5076	5249	5297	5249	5076	4732	4003
V, (volt)measured	4010	4800	5100	5270	5300	5270	5100	4800	4010
Unequally Spaces Having Denser Conductors at the Grid Edges									
x, (m)	-52.4	-46.8	-35.1	-19.1	0.0	19.1	35.1	46.8	52.4
V, (volt)calculated	4080	4431	4629	4762	4799	4762	4629	4431	4080
V, (volt)measured	4100	4350	4650	4780	4800	4780	4650	4350	4100
Unequally Spaces Having Denser Conductors at the Grid Middle									
x, (m)	-52.4	-33.3	-17.3	-5.6	0.0	5.6	17.3	33.3	52.4
V, (volt)calculated	4156	5265	5785	6138	6197	6138	5785	5265	4156
V, (volt)measured	4160	5270	5780	6150	6200	6150	5780	5270	4160

Table 13. Earth surface potentials in volts along center conductor for equally spaced grid of A=105×105 m², N=64 in double layer soil

Upper and Lower Layers Resistivity	Upper Layer Thickness	Equally Spaces Grid Conductors									
Ohm. Meter	meter	x, (m)	-52.5	-39.38	-26.25	-13.13	0	13.1	26.3	39.4	52.5
100, 250	10	V, (volt) measured	7800	8200	9000	10000	10900	10000	9000	8200	7800
100, 250	20	V, (volt) measured	7500	8000	8600	9800	10600	0.99	0.96	0.89	0.76
100, 40	10	V, (volt) calculated	2000	2000	2100	2150	2300	2150	2100	2000	2000
100.40	20	V, (volt) measured	2010	2012	2112	2155	2310	2155	2112	2012	2010

Figure 52. Experimental setup

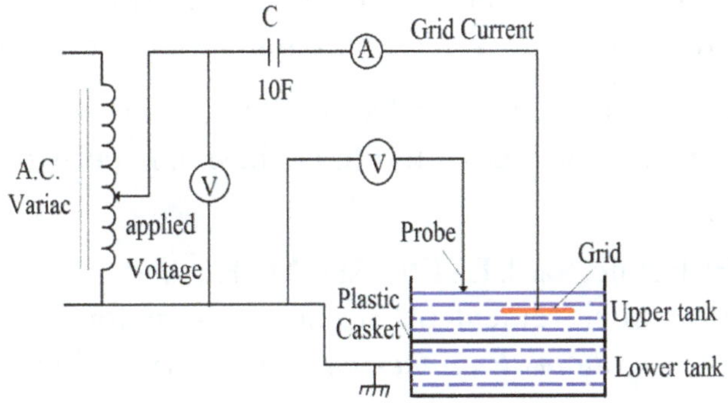

Similar results are obtained in Table 12 for equally spaces grid conductors for double layers simulation. The soil resistivity of upper and lower layers and the upper layer thickness are changed as given in Table 4. From this table it is noticed that the lower layer soil resistivity has significant change in the increase of step potential. On the contrary, the change in the thickness of the upper layer has little change in step potential.

REFERENCES

AIEE. (1961). Guide for Safety in A-Current Substation Grounding.

IEEE. (1986, August). ANSI/IEEE Std. 80-1986, Guide for Safety in AC substations Grounding. New York: John Wiley & Sons.

Bendito, E., Carmona, A., Encinas, A., M., Jimenez, M.J., (2000). The external charges method in grounding grid design. The commission interministerial de ciencia technology, Spanish research council.

Bharat Sanchar Nigam Limited. (2005). Measurement Earth Potential Rise (EPR) Zone (No. EPR/A-002). *CGM and Technical and Development Circle*, 1.

Dalziel, C. F., & Massogilia, F. P. (1956). Let-go currents and voltages. *AIEE Transactions*, *75*, 49–56.

Dawalibi, F., & Mukhedkar, D. (1975). Optimum design of substation grounding in a two layer earth structure. part I-Analytical study. *IEEE Trans.*, *94*(2), 252–261.

Du Zhongdong, Y. Z., Xishan, W., & Hua, X. (2005). The Optimum Design of Grounding Grid of Large Substation. In *Proceedings of the XIVth International Symposium on High Voltage Engineering*, Tsinghua University, Beijing, China.

El-Khodragy, T. M. (2007). Optimum Design of Earthing System of Substations by Using Artificial Neural Network [Ph. D. thesis]. Faculty of Engineering, Cairo University.

El-Sayed, M., El-Refaie, Said, E., Elmasry, M., K., Abd Elrahman, Mohamed H., Abdo, (2015) Achievement of the best design for unequally spaced grounding grids. *Ain Shams Engineering Journal*, *6*(1), 171–179

Heppe, R. J. (1979). Computation on potential at surface above an energized grid or other electrode, allowing for nonuniform current distribution. *IEEE Transactions on Power Apparatus and Systems*, *98*(6), 1978–1989. doi:10.1109/TPAS.1979.319377

Hung, L., Chen, X., & Yan, H. (1995). Study of Unequally Spaced Grounding Grid. *IEEE Transactions on Power Delivery*, *10*(2).

Ibrahim, E. A. (1993). Grounding of the Nuclear Power Plants Using Grounding Grids [Ph. D. thesis]. Faculty of Engineering, Cairo University.

IEC60 479-1(1984). Effect of current on human beings and livestock Engineering Standards Branch, Telecommunications Standards Division, (ESBTSD) (1994). Electrical Protection Grounding Fundamentals. United States department of agriculture, rural electrification administration.

IEEE. (1991). IEEE std 142 IEEE Recommended Practice for Grounding of Industrial and Commercial Power System.

IEEE. (2000). Guide for Safety in AC Substation Grounding.

Koch, W. (1950). Erdungmassnahmen fur Hoschtspannungsanlagen mit Geerdetem Stempunkt. Elektrotechnische Zeitschrifl, 71(4), 89-91.

Phithakwong, B., Kraisnachinda, N., Banjongjit, S., Chompoo-Inwai, C., & Kando, M. (2000). New Techniques the Computer-Aided Design for Substation Grounding. In *Power Engineering Society Winter Meeting* (Vol. 3, pp. 2011-2015).

Puttarach, A., Chakpitak, N., Kasirawat, T., & Pongsriwat, C. (2007). Ground Layer Depth and the Effect of GPR. In *Proceedings of the IEEE PES Power Africa 2007 Conference and Exposition.*

Rajotte, Y., De Sève, J., Fortin, J., Lehoux, R., & Simard, G. (2005). Earth Potential Rise Influence near HV Substations in Rural Areas. In Proceedings of CIRED, Turin.

Schon, J. H. (1998) Physical Properties of Rocks: Fundamentals and Principles of Petrophysics. In Handbook of Geophysical Exploration (Vol. 18). Pergamon.

IEEE Std. 80-1976, Guide for Safety in AC substations Grounding. (1976, June).

Sverak, J. G. (1984). Simplified analysis of electrical gradients above a ground grid. *IEEE Trans.*, *103*(1), 7–25.

Thabet, A. A. (2002). Grounding Systems of Electric Substations in Non-uniform Earth Structure with New Analysis [M. Sc. Thesis]. High Institute of Energy, Aswan.

Thapar, B., Gerez, V., Balakrishnan, K., & Blank, D. A. (1991). Simplified Equations for Mesh and Step Voltages in AC Substation IEEE Transaction on Power Delivery, *6*(2), 601–607.

Chapter 7
Cathodic Protection of Grounding Grids

ABSTRACT

This chapter contains the corrosion theory of grounding electrodes and basic electrochemistry in corrosion reactions. It contains also the forms of substation grounding grid corrosion (uniform corrosion, pitting corrosion, galvanic corrosion, microbial influenced corrosion), survey on corrosion rate of substation grounding grid, copper and steel corrosion rates, corrosion protection methods (coating, cathodic protection [CP]). The chapter contains also the methods of applying cathodic protection in grounding grids, anode selection, anode spacing, and impressed current in the grounding grid cathodic protection. Finally it contains the required information for design grounding system cathodic protection and sacrificial anode (galvanic) cathodic protection system design steps.

CORROSION THEORY AND BASIC ELECTROCHEMISTRY IN CORROSION REACTIONS

In this chapter, corrosion of substation grounding grid or any surround metallic structure due to current discharge buried in soil is fully introduced. Survey on calculating rate of corrosion and factors affecting the rate of corrosion are also introduced in this chapter.

Corrosion is defined as destruction or deterioration of material because of reaction with its environment. Some insists that the definition should be

DOI: 10.4018/978-1-5225-3853-0.ch007

Figure 1. Basic galvanic corrosion cell

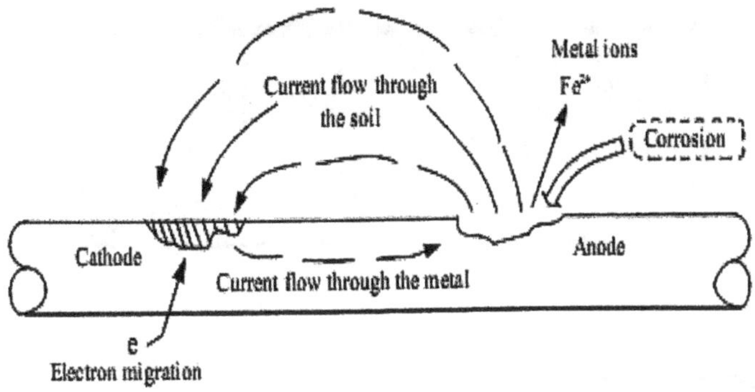

restricted to metals but often corrosion engineers must consider both metal and nonmetal for solution of a given problem as reported by Mars G. Fontana (1987). It is also considered the exothermic chemical transformation of a metal or metal alloy to a non-reactive covalent compound such as an oxide or silicate that is often similar or even identical to the mineral from which the metals were extracted as given by Howard (2004)

Most of the corrosion in soil is the result of an electrochemical reaction. Each galvanic corrosion cell comprises as shown in (Figure 1):

1. Anode and cathode areas on metal surface,
2. Soil electrolyte to enable occurrence of cathodic and anodic processes and current flow between anode and cathode and,
3. Conducting path for the flow of current from cathode to anode.

In the absence of any of these components, the corrosion cell would stop to operate as reported by Zhou, Wang, Liu et al. (2008) Mitolo, M., Pettinger, A., (2016), investigated the interactions between cathodically protected pipelines and grounding systems

FORMS OF SUBSTATION GROUNDING GRID CORROSION

There are number of corrosion forms of substation grounding grid. These forms can be summarized as follows:

Figure 2. Uniform corrosion

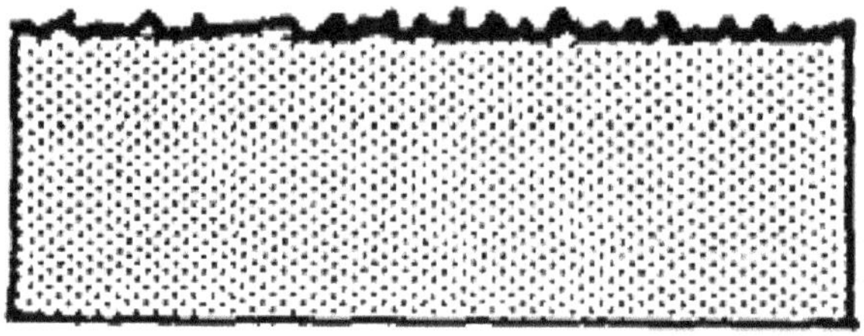

Uniform Corrosion

Uniform corrosion refers to corrosion damage occurring evenly over the metal/alloy surface as shown in (Figure 2), and the rate of corrosion is nearly the same over the entire surface. Although uniform corrosion is the most important form of corrosion, it is relatively easily measured and predicted, making disastrous failures relatively rare as reported by Tullmin and Roberge (1995)

Pitting Corrosion

Pitting corrosion is a form of localized corrosion that produces pits (holes) in the metal/alloy as shown in (Figure 3). It occurs when discrete areas of a material undergo rapid attack while the vast majority of the surface remains virtually unaffected. This is in sharp contrast to uniform corrosion in which all parts of the exposed surface recede at approximately the same rate. Pitting of a given material depends strongly upon the presence of an aggressive species in the environment as reported by Zhou, Wang, Yang et al. (2008)

Figure 3. Pitting corrosion

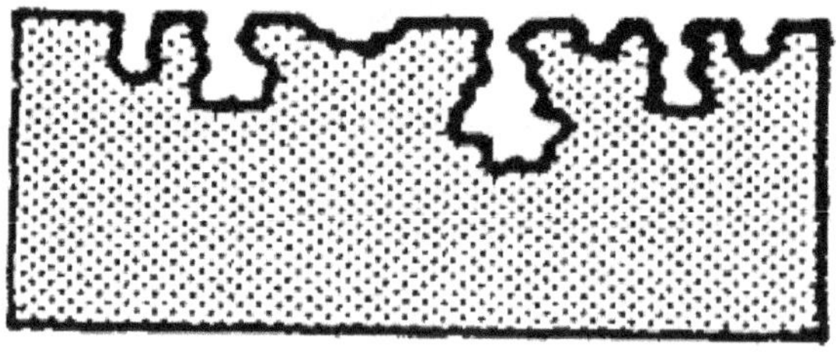

Figure 4. Galvanic corrosion

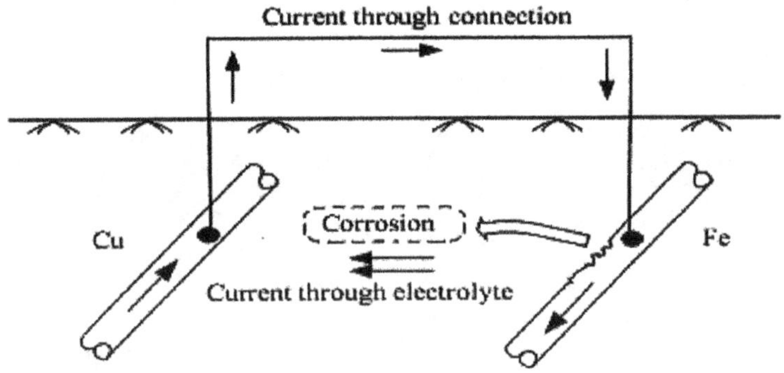

Galvanic Corrosion

When a difference in electric potential exists between two dissimilar metals, as an example, copper and iron, connected externally and buried in soil as shown in (Figure 4). The electrolyte (soil) allows some current to flow and causes corrosion in steel. Galvanic corrosion tends to be particularly severe if the anodic material's surface area is small compared to that of the cathode. In this case the galvanic current is concentrated on a small anodic area; it results in rapid corrosive penetration of the anode material as reported by Zhou, Wang, Liu et al. (2008).

Microbial Influenced Corrosion

In contaminated soils, another form of corrosion called microbial influenced corrosion (MIC) can occur. The most common form is where a range of bacteria called sulphate reducing bacteria (SRB) in anaerobic environments, can react to produce hydrogen sulphide (HS). The acidic conditions generated can lead to rapid, localized corrosion.

Pictures of different corrosion forms in substation grounding grid are presented in (Figure 5) as reported by Zhou, Wang, Liu et al. (2008).

SURVEY ON CORROSION RATE OF SUBSTATION GROUNDING GRID

Two ways are found to calculate the rate of corrosion. In the following sections, full details of them are introduced.

Figure 5. Corrosion forms of (a) uniform corrosion, (b) pitting corrosion and (c) microbial influenced corrosion in substation grounding grid

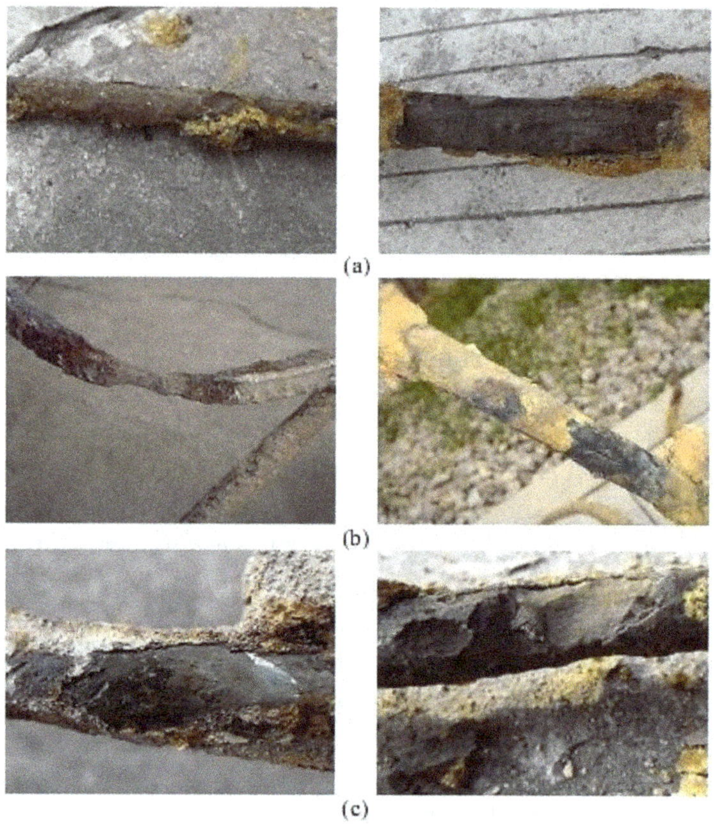

Steel Corrosion Rate

Experimental Formula

Formula for Calculating Expected Corrosion Rate

This relationship is obtained from the National Bureau of Standards Data to calculate the corrosion rate of steel (Bessemer Steel). These data were produced from actual experiments in 44 different soils conducted over a period of 12 years. As reported by as reported by Sen, Malmeda and Nelson (2002)

These tests used 1.5-inch and 3-inch diameter samples. The 3-inch samples corroded 13% more than the 1.5-inch sample with an error of 10%. It is found that:

$$Y = F\left(\rho, x_1, x_2, x_3\right) \tag{1}$$

where:

Y = Corrosion rate (mils/year),
ρ = Resistivity (ohm.cm),
x_1 = pH value,
x_2 = Moisture (%), and
x_3 = Aeration (%)

Finally, the following equation is obtained:

$$Y = 3.36 - 9.63(10^{-5})(\rho) + 0.29(x_1) + 0.034(x_2) + 0.012(x_3) \tag{2}$$

Equation 2 is obtained from experimental data and is limited by extreme corrosion conditions such as, extremely high resistivity (> 10,000 ohm.cm) or extremely low aeration quantities (< 3%). Corrosion reduces by these conditions to almost zero. This equation is applicable for all steel rods up to 3-inch diameter and is applicable for the first 12 years. It is concluded also experimentally that the average corrosion rate in the following 12 years reduces to half, that value in the first 12 years. The corrosion rate can be estimated from Table 1 for several material corrosion rates based on equation 2 as reported by Sen, Malmeda and Nelson (2002)

Table 1. Average corrosion rate of several materials

Materials	Corrosion Rate (mils/years)
Open earth steel	5.90
Wrought iron	5.00
Bessemer steel	5.30
Copper	1.25
Lead	3.00
Zinc	5.00

As reported by Sen, Malmeda and Nelson (2002).

Formula for Accounting the Weight Loss After Exposure to a Corrosive Environment Conditions

21 spot-checks by excavation were made on substations of different voltage levels distributed over foshan area, whose operation time varies from 1 to 30 years. Corrosion data computation was done on 70 samples from main grounding conductors (buried in soil), grounding conductors in cable trench and grounding down lead. The average and maximum corrosion rates are calculated as reported by Zhou, Wang, Liu et al. (2008) as follows:

The average corrosion rate (mm/y) = $10.\Delta W \,/\, (d_m.A.t)$ (3)

where:

ΔW = weight loss in grams,
d_m = metal density (g/cm),
A = area of metal (cm),
t = time of exposure in corrosive environment (years).

Thus,

maximum corrosion rate (mm/year) $= H/t$ (4)

where:

H = depth of the deepest hole in mm, and
t = time of exposure in corrosive environment (years).

Laboratory Formula

Faraday's first law, that given in (5), is also used for conversion of corrosion rates (A/cm^2) to penetration rates (millimeter per year) as given by Fan, Wang and Zhou et al. (2011).

$$\Delta m = \frac{A_t.I.s}{n.F_r} \tag{5}$$

where:

Δm = mass of metal corroded,
A_t = atomic weight of the metal,
I = current in amperes,
s = time in seconds,
n = number of equivalents transferred per mole of metal,
F_r = Faraday constant = 96,500.

Copper Corrosion Rate

Unlike steel, it is not possible to estimate the copper corrosion rates by oxygen and sulfur mass balance calculation since other species takes part in the copper corrosion process as well. Water molecules as moisture or in aqueous corrosion as hydroxide ions are probably the most frequent species that will react and form copper corrosion products in such a complex environment as expected in are positron with different trace gases and salts as reported by Szakalos and Seetharaman (2012)

In grounding grid design for AC substation, copper is used as the primary grid material. Steel-grounding system is widely used in China and readily accepted in many other countries worldwide, where copper is very expensive. However, it is still common in this kind of grounding system for grounding faults problems to occur due to corrosion, which might result in plenty of economic loss as reported by Zheng, He and Hu et al. (2002).

Reliable performance of steel grounding systems can be insured by selecting a conductor cross-section area as explained by Lawson (1988).

Factors Affecting Corrosion Rate

Different factors include factors connected with the metal, with the atmosphere, and with the corrosion product are introduced in the following subsections as reported by Chauhan (2008)

Factors Connected With the Metal

- Position of the metal in the E.M.F Series
- Purity of the metal
- Surface of the metal
- Stress corrosion

- Anode to cathode area ratio stress
- Physical state of a metal

Factors Connected With the Atmosphere

- Nature of the atmosphere
- Temperature of the atmosphere
- PH of the atmosphere
- Amount of moisture in the atmosphere
- Amount of oxygen in the atmosphere
- Amount of chemical fumes in the atmosphere, etc.

Factors Connected With the Corrosion Product

- In some cases, the corroded product sticks to the surface and absorbs more moisture. This induces further corrosion.
- In some cases, the corroded product acts as the protective coating which prevents further corrosion.
- In some other cases, the corroded product falls out of position exposing the fresh metal surface for further corrosion.

Corrosion Protection Methods

In this section; description of cathodic protection will be demonstrated. Forms, types of anode material, selection of their material, their life time, comparison between cathodic protection types, limitations that must be kept in mind and designing steps are also discussed in detail. (Figure 6) gives picture of galvanized rod removed after 10 years. (Figure7) gives picture of copper-bonded ground rod removed after 10 years (Pawnee Site). (Figure 8) gives picture of copper-bonded rod after 10 years- Electrode type H. Finally. In (Figure 9) Galvanized rod picture after 10 years -Electrode type I is presented.

Some of the most common corrosion prevention measures that are used to minimize corrosion of grounding material and/or steel piping are as reported by Sen, Malmeda and Nelson (2002):

- Using of tin coated bare copper cable or hot galvanized or corrosion-resistant steel as a grounding conductor to minimize the potential difference between the steel structures and the grounding system of the substation.

Figure 6. Galvanized rod removed after 10 years (Pawnee Site)

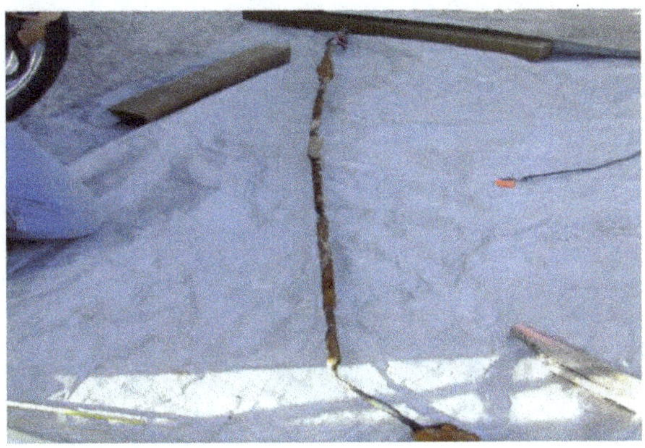

- Cathodic protection that achieved by applying a current to the structure to be protected from some outside source. When enough current is applied, the whole structure will be at one potential; thus, anode and cathode sites will not exist. Cathodic protection is commonly used on many types of structures, such as pipelines, underground storage tanks, locks, and ship hulls. Cathodic protection can be designed for 30 years of life expectancy and it can be used to protect the grounding system of electrical substation.
- Using of insulated copper ground cable in areas near pipes, which is a method to reduce corrosion by minimizing the difference in potential between anode and cathode.

Figure 7. Copper-bonded ground rod removed after 10 years (Pawnee Site)

Figure 8. Copper-bonded rod after 10 years: Electrode type H

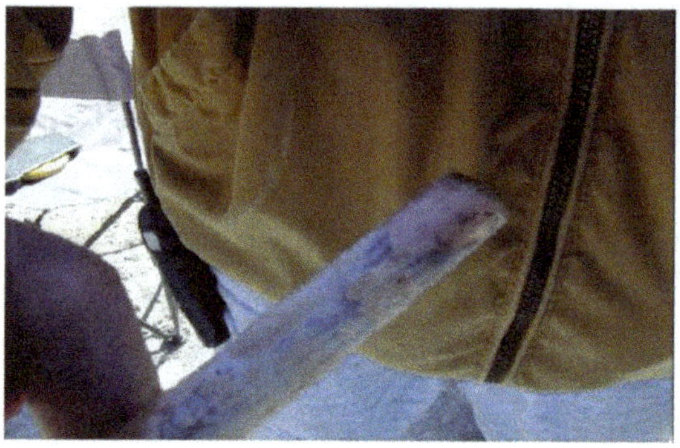

- Electrical isolation of piping from other plant structures and the grounding system.

The first two methods are the most commonly adopted in practical designs.

COATING

It has been used to control corrosion rate. It initially protects the underlying metal. When the continuity of the coating is destroyed, it is observed that the corrosion rate of the base metal is normal or above normal depending on the type of coating and the underlying metal. Zinc-coated (galvanized) steel is the most used type of coating. However, any coating as a sole means

Figure 9. Galvanized rod after 10 years: Electrode type I

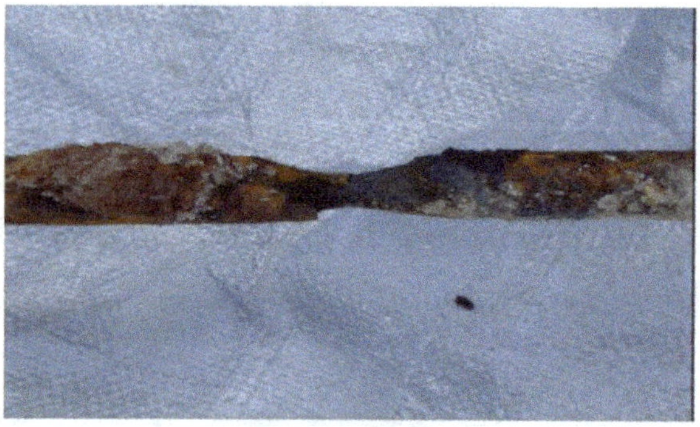

of protection is not very effective as reported by Sen, Malmeda and Nelson (2002).

This type of protection to control corrosion can be used in protection of grounding system of substation by using galvanized steel.

External coatings must be properly selected and applied, and the coated system must be carefully handled and installed to fulfill these functions. Various types of external coatings can accomplish the desired functions. Desirable characteristics of external coatings include: effective electrical insulator; effective moisture barrier; and applied to a pipe by a method that does not adversely affect the properties of system as reported by NACE Standard RP0169 (2002).

CATHODIC PROTECTION (CP)

One definition of cathodic protection is a technique to reduce the corrosion rate of a metal surface by making it the cathode of an electrochemical cell. It has inherent personnel protection by keeping grounding system protected from corrosion to maintain system performance.

CP systems are typically designed to drive the potential to about 1 V negative. Potentials more negative than 2.5 V can be damaging to coatings. Equipment properly connected to a CP system has a very low resistance path to ground ($< 2\ \Omega$). This provides an adequate path for dissipation of current in a fault condition as given by Durham and Durham (2005).

In the case of an impressed current system, a current is impressed on the structure by means of a power supply, referred to as a rectifier, and an anode buried in the ground. In the case of a sacrificial anode system, the galvanic relationship between a sacrificial anode material, such as zinc or magnesium, and the pipe steel is used to supply the required cathodic protection current as reported by Beavers (2001).

Methods of Applying Cathodic Protection

Cathodic protection may be achieved in either of two ways: sacrificial (or galvanic) anode cathodic protection (SACP) and impressed current cathodic protection (ICCP). The main difference between the two methods is that ICCP uses an external power source with inert anodes and SACP uses the naturally occurring electrochemical potential difference between different metallic elements to provide protection.

Sacrificial or Galvanic Anodes

They are applicable and effective where current requirements are low (less than 100 mA), and the structures to be protected are well-coated in a low resistivity soil. The anode consists of a metal, which is electro-negative to the protected structure. These anodes are self-energized and are connected directly to the protected structure as shown in (Figure 10). Sacrificial anodes, usually distributed along or around the protected metal parts, consume themselves in protecting the other metal. The consumption rate depends on the magnitude of current generated as villas the material of which the anode is made as reported by Bianchetti (2001).

The Main Advantages of Sacrificial Anodes

- Economically feasible when installed with the structure.
- Very little operation or maintenance requirements (very small chance of premature failure or breakdown).
- Extremely small possibility of overprotection (which may cause coating damage or hydrogen embitterment).
- Small likelihood of stray current causing interference damage to other metallic (foreign) structures.

Figure 10. Sacrificial or galvanic anode material

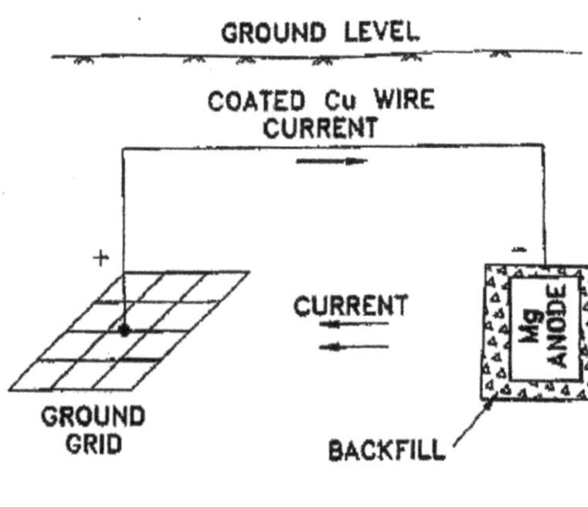

The Main Disadvantages of Sacrificial Anode

- Small driving voltage available (limited potential difference).
- Extremely small current available in higher resistivity electrolytes.
- Not economically feasible to install or replace anodes.
- Extensive existing structures.

Galvanic Anodes

The use of galvanic anodes for cathodic protection (CP) is a simple application of the dissimilar metal corrosion cell, where a steel grounding system is electrically connected to a metal higher in the electromotive force series and both are in a common conductive electrolyte such as the soil, the more active metal is corroded and discharges current in the process. Magnesium and zinc are such metals. If the amount of current needed for a given CP application is known, anode systems can be designed using sufficient anode material to produce the desired current output continuously over a desired number of years as reported by Bianchetti (2001).

Galvanic anodes consume themselves in protecting the other metal. The two galvanic anode metals commonly used for buried any metallic structure like grounding system for electrical substations are magnesium and zinc (A_t = 24.32 for Mg and 65.38 for Zn, $n = 2$ for Mg and 2 for Zn).

The consumption rate depends on the magnitude of current generated as well as the material of which the anode is made and is given by Farady's Second Law of Electrolysis as given in equation 5 as reported by Sen, Malmeda and Nelson (2002).

Magnesium Anode

Magnesium is the most widely used material for galvanic anodes. Typical characteristics are given in Table 2. Magnesium anodes are available in various shapes and weights from the manufacturers. Some of the sizes available, suitable for use in soil, are listed in Tables 3 and 4 as reported by Bianchetti (2001).

Zinc Anode

Zinc anodes are available in a number of sizes for use in earth anode beds. Typical characteristics of zinc used as an anode material are given in Table

Table 2. Characteristics of Magnesium anodes

Specific Gravity	1.94
Pounds per Cubic Foot	121
Theoretical Amp Hours per Pound	1000
Theoretical Pounds per Amp per Year	8.7
Current Efficiency - Percent	50
Actual Amp Hours per Pound	500
Actual Pounds per Amp per Year	17.4
Solution Potential - Volts to CSE	
Standard H -1 Alloy	-1.50 to 1.55
High Potential Alloy	-1.75 to 1.77
Driving Potential to Pipe Line	
Polarized to -0.90 Volt to CuSO4	
Standard Alloy - Volts	0.55
High Potential Alloy - Volts	0.80

As reported by Bianchetti (2001).

5. These are long slender shapes to achieve low resistances to earth and practical current output at the usual low driving voltage between anode and protected structure. Some available sizes as standard commercial items are shown in Table 6. Nonstandard sizes may be obtained on special order as reported by Bianchetti (2001).

Table 3. Standard alloy chemical composition

Weight Content %			
Element	**Grade A**	**Grade B**	**Grade C**
Al	5.3 - 6.7	5.3 - 6.7	5.0 - 7.0
Mn	0.15 min	0.15 min	0.15 min
Zn	5.3 - 6.7	5.3 - 6.7	5.3 - 6.7
Si	0.10 max	0.30 max	0.30 max
Cu	0.02 max	0.05 max	0.10 max
Ni	0.002 max	0.003 max	0.003 max
Fe	0.003 max	0.003 max	0.003 max
Other	0.30 max	0.30 max	0.30 max
Magnesium	Remainder	Remainder	Remainder

As reported by Bianchetti (2001).

Table 4. High potential alloy chemical composition

Element	Weight Content %
Al	0.010
Mn	0.50 to 1.30
Cu	0.02 max
Ni	0.001 max
Fe	0.03 max
Other	0.05 each or 0.3 max total
Magnesium	Remainder

As reported by Bianchetti (2001).

Table 5. Characteristics of zinc anodes

Specific Gravity	7
Pounds per Cubic Foot	440
Theoretical Amp Hours per Pound	372
Theoretical Pounds per Amp per Year	23.5
Current Efficiency - Percent	90
Actual Amp Hours per Pound	335
Actual Pounds per Amp per Year	26
Solution Potential - Volts to CSE	
Driving Potential to Pipeline	-1.1

As reported by Bianchetti (2001).

Table 6. Standard H-1 chemical composition

Element	Weight Content %	
	MLA – A – 18001(ASTM B – 418 Type I)	ASTM B-418 Type II
Al	0.1 – 0.5	0.05 max
Cd	0.02 – 0.07	0.003 max
Fe	0.005 max	0.0014 max
Pb	0.006 max	0.003 max
Cu	0.005 max	0.002 max
Zinc	Remainder	Remainder

As reported by Bianchetti (2001).

Anode Backfill

For reliable operation in earth installations, both zinc and magnesium anodes are used with a chemical backfill to surround the anode completely. Typical data on chemical backfill are given in Table 7. With the anode surrounded with a uniform material of known composition, anode current is more efficient. If the soil contacts the anode directly, variations in soil composition can set up local corrosion on the anode resulting in non-uniform consumption of the anode. By isolating the anode material from the native soil, the backfill material greatly reduces the possibility of adverse effect on anode performance. In the presence of phosphates, carbonates and bicarbonates, zinc anodes can develop passive films and cease to produce useful amounts of current. Carbonates and bicarbonates will influence magnesium the same way. Chlorides tend to increase self-corrosion of magnesium and reduce its current efficiency as reported by Bianchetti (2001).

Backfill serves some purposes such as providing uniform soil around the anode in order to assure uniform corrosion of the anode, minimizing the resistance between the anode and the earth, retaining moisture which aids in minimizing there distance, and aiding in depolarizing any film build up, which would reduce the flow of the galvanic currents as reported by Nelson and Holm (1985).

Anode Selection

After considering the available materials, one must make the suitable selection. The criterion for selection is as one would expect an analysis of performance versus cost, the performance of an anode is measured by the following criteria:

Table 7 Backfill data for magnesium and zinc anodes

Backfill	Hydrated	Bentonite Clay %	Sodium Sulfate %	Approximate Soil Resistivity in ohm. cm
(A)	50	50	--	250
(B)	75	20	5	50

- Backfill (A) is commonly used with zinc anodes.
- Backfill (B) with low resistivity is useful in high soil resistivity areas to reduce the anode resistance to earth.

Anode Life

Life is a function of three factors: weight, current output, and efficiency. Longer life is achieved through heavier weight, lower current output and high efficiency. The following expressions are used to calculate the estimated anode life time as reported by Bushman (2001).

$$\text{Anode life time} = \frac{F_c}{h_y} \times \frac{W_A.\eta_A.X}{I_A} \tag{6}$$

where:

F_c = Faradays consumption rate (amperes/hours/pound),
h_y = no. of hours per year,
W_A = anode weight (Ibs),
η_A = anode efficiency,
X = the utilization factor, and
I_A = anode current in Amperes.

The utilization factor may be taken as 85% meaning that when the anode is 85% consumed, it will require replacement because there is insufficient anode material remaining to maintain a reasonable percentage of its original current output as reported by Bushman (2001).

Magnesium Life Time as Given by Bianchetti (2001):

$$L_m = \frac{0.116 \times \text{Anode weight(pound)} \times \text{Efficiency} \times \text{Utilization Factor}}{\text{Design current(Amperes)}} \tag{7}$$

It can be reduced to the simpler form as follows:

$$L_m = \frac{48.5 \quad W}{I_m} \tag{8}$$

where:

L_m = the magnesium anode life in years,
W = the anode metal weight in pounds, and
I_m = the current output in mA as given by Bushman (2001).

Zinc Life Time as Given by Bianchetti (2001):

$$L_Z = \frac{0.0424 \times \text{Anode weight(pound)} \times \text{Efficiency} \times \text{Utilization Factor}}{\text{Design current(Amperes)}} \tag{9}$$

It can be reduced to the simpler form as follows:

$$L_Z = \frac{32.5.W}{I_m} \tag{10}$$

where:

L_Z = the zinc anode life in years,
W = the anode metal weight in pounds and
I_m = the current output in mA as given by Bushman (2001).

Anode Current

The current output of anode can be calculated in accordance with ohms law as follows

$$I_A = \frac{E_{A-}E_P}{R_A + R_{AP} + R_{Pg} + R_C + R_W} \tag{11}$$

where:

I_A = the anode output current in amperes,
EA = the potential of anode,
E_P = the backfill potential,
R_A = the anode film resistance,
R_{AP} = the backfill resistance,

R_{Pg} = the backfill to earth resistance,
R_C = the cathode film resistance, and
R_W = the resistance of connecting wire.

Correction factors are then applied to adjust the result of various shapes and structures potentials:

Output Current for Bare Structure as Given by Bushman (2001)

For Magnesium:

$$I_{mb} = 150{,}000.\frac{F.Y}{\rho} \tag{12}$$

For Zinc:

$$I_{Zb} = 50{,}000.\frac{F.Y}{\rho} \tag{13}$$

Output Current for Coated Structure as Reported by Bushman (2001)

For Magnesium:

$$I_{mc} = 110{,}000.\frac{F.Y}{\rho} \tag{14}$$

For Zinc:

$$I_{Zc} = 40{,}000.\frac{F.Y}{\rho} \tag{15}$$

where:

I_{mb} = the current output for magnesium anode on bare structure, mA
I_{Zb} = the current output for zinc anode on bare structure, mA
I_{mc} = the current output for magnesium anode on coated structure, mA

Table 8. Driving voltage correction (Y)

Structure Potential (Cu-Cuso4)	Std. Mag.	High Potential Mag.	Zinc
-0.7	1.21	2.14	1.60
-0.8	1.07	1.36	1.20
-0.85	1.00	1.29	1.00
-0.90	0.93	1.21	0.80
4.00	0.79	1.07	0.40
-1.10	0.64	0.93	n/a
-1.20	0.50	0.79	n/a

As given by Bushman (2001).

I_{Zc} = the current output for zinc anode on coated structure, mA
Y = the factor of driving voltage as given in Table 8.
F = the factor from anode shape as given in Table 9.
ρ = the soil resistivity, ohm.cm

Anode Spacing

Anodes may be connected in parallel in order to achieve a higher total current output at a given location, the closer together the anode are spaced the more the current output is restricted. To determine the approximate current output of a multiple anode ground bed, the single anode current previously calculated is multiplied by the appropriate adjusting factor that found in Table 10 as given by James B. Bushman PE. (2001).

For a more exact calculation an adjusting factor may be determined from Equation (16) which is given by Bushman (2001) as follows:

$$MA_{ADJ} = \frac{N_P}{1 + \frac{2.L\left(In0.656.N_P\right)}{S\left[\left(In\frac{8L_a}{d_a}\right) - 1\right]}} \tag{16}$$

where:

MA_{ADJ} = the multiple anode adjusting factor,

Table 9. Anode shape correction (F)

Alloy	Weight (Ibs)	Package Dimension	Anode Factor (F)
Mg	3	3" × 3" × 4.5"	0.53
Mg	5	3" × 3" × 7.5"	0.60
Mg	9	3" × 3" × 13.5"	0.71
Mg	9	2.75" × 2.75" × 26"	1.01
Mg	10	1.5" × 1.5" × 72" ingot 4" × 78" package	1.71
Mg	15	1.6" dia. × 10" extrusion 6" × 10" backfill	2.61
Mg	17	4" × 4"×17"	1.00
Mg	18	2" × 2" × 72" ingot 5" × 78" package	1.81
Mg	20	2.5" × 2.5" × 60" ingot 5" × 66" package	1.60
Mg	20	1.3" dia. × 20" extrusion 6" × 20" backfill	4.28
Mg	25	2" dia. × 10" extrusion 8" × 10" backfill	2.81
Mg	32	5" × 5" × 21"	1.06
Mg	40	3.75" × 3.75" × 60" ingot 6.5" × 66" package	1.72
Mg	42	3" × 3" × 72" ingot 6" × 78" package	1.90
Mg	50	8" dia. × 16"	1.09
Mg	50	5" × 5" × 31"	1.29
Zn	10	1.4" × 1.4" × 36" ingot 5" × 42" package	1.68
Zn	30	2" × 2" × 30" ingot 5" × 36" package	1.44
Zn	30	1.4" × 1.4" × 72" ingot 5" × 78"package	1.81
Zn	36	2" × 2" × 60" ingot 6.5" × 66" package	1.72

As given by Bushman (2001).

N_P = the No. of anodes in parallel,
L_a = the length of anode in feet,
S = the spacing center-to-center in feet,
d_a = the diameter of anode in feet.

Table 10. Multiple anode adjusting factor

No. of Anodes Bank	Anode Spacing in Feet				
	5'	10'	15'	20'	25'
2	1.84	1.92	1.95	1.97	2.03
3	2.46	2.71	2.80	2.85	3.02
4	3.04	3.46	3.63	3.71	4.01
5	3.59	4.19	4.43	4.56	4.98
6	4.13	4.90	5.22	5.31	5.96
7	4.65	5.60	6.00	6.23	6.91
8	5.15	6.28	6.77	7.04	7.85
9	5.67	6.96	7.54	7.88	8.82
10	6.16	7.64	8.30	8.68	9.75
11	6.76	8.41	9.14	9.56	10.75
12	7.30	9.12	9.93	10.40	11.71
13	7.83	9.83	10.72	11.23	12.68
14	8.37	10.54	11.51	12.07	13.64
15	8.91	11.25	12.30	12.91	14.61
16	9.44	11.96	13.09	13.75	15.57
17	9.98	12.68	13.89	14.58	16.54
18	10.51	13.39	14.68	15.45	17.50
19	11.05	14.10	15.47	16.26	18.47
20	11.59	14.81	16.26	17.10	19.43

As given by James B. Bushman PE. (2001).

Impressed Current

Impressed-current systems employ inert (zero or low dissolution) anodes and use an external source of DC power (rectified AC) to impress a current from an external anode onto the cathode surface as reported by Kean and Davies (1981).

An electrolytic cell is developed with the protected structure and the ground rod of the rectifier. The ground bed (anode) consists of a number of parallel graphite, carbon, duriron, or junk iron anodes, usually distributed along the protected structure and placed deep in the soil. The structure, connected to the negative of the rectifier, receives the current from the soil and is protected. The protective current from the rectifier is usually sufficient to overcome the galvanic currents leaving the anodic areas of the structure.

Figure 11. Impressed current or rectifier type

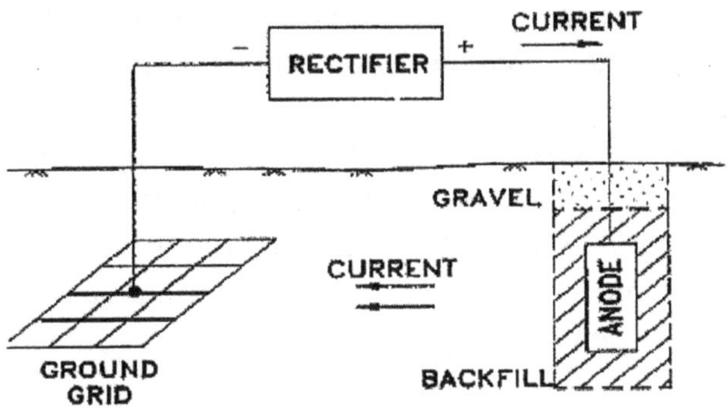

The general arrangement is shown in (Figure 11) as reported by Sen, Malmeda and Nelson (2002).

The Main Advantages of Impressed Current

- Unlimited current output capacity: the amount of current that can be designed into an impressed current system can be from a few amperes to as much as several hundred amperes. The amount of current available will be a function of the number of provided anodes, the rectifier voltage and amperage capacity and the soil (electrolyte) resistivity in which the anodes are installed.
- Adjustable output capacity: the output of the rectifier power supply is easily adjusted to accommodate either change in circuit resistance or current requirement. They can be provided with automatic control circuitry to eliminate the need for manual adjustment.
- Lower cost per ampere of cathodic protection current. Galvanic anode systems are significantly more expensive where amperes rather than milli-amperes of total current are required.

The Main Disadvantages of Impressed Current

They are more costly if only a few mill- amperes of current are required to protect a small or very well coated structure. Impressed current systems typically have a base cost of several thousand dollars. If only a few galvanic anodes are required for protection of a specific structure, this will often be the more economic choice.

- Impressed current systems are inherently more prone to failure. Switches can be turned off and fuses can blow. Thus, the system needs to be monitored more frequently and some repair may be required once every several years.
- Impressed current systems may create stray current corrosion on other nearby structures. This is an inherent potential problem with any impressed current system. This can sometimes be minimized through the use of distributed anode designs where the anodes are placed nearby the protected structures.

Impressed Current Anodes

Materials currently popular for use as anode material include graphite, high silicon cast iron, mixed metal oxide, platinum, and steel.

COMPARISON BETWEEN THE TWO TYPES OF CATHODIC PROTECTION

Table 11 gives a comparison between the two types of cathodic protection shown by Roberge (2014). The comparison includes the general uses, the benefits and features.

REQUIRED INFORMATION FOR DESIGN OF GROUNING GRID CTHODIC PROTECTION

Before deciding which type, galvanic or impressed current cathodic protection system will be used and before the system is designed, certain preliminary data must be gathered as reported by Headquarters Department of the army Washington (2005).

- **Physical Dimensions of Structure to be Protected:** One important element in designing a cathodic protection system is the structure's physical dimensions (for example, length, width, height). These data are used to calculate the surface area to be protected (A) as reported by Headquarters Department of the army Washington (2005).
- **Electrical Isolation:** If a structure is to be protected by the cathodic system, it must be electrically connected to the anode. Sometimes

Table 11. Comparison between the two types of cathodic protection

Sacrificial Anodes (Galvanic)	Impressed Current (ICCP)
Uses	
• Generally used for protection of well coated areas where protective current requirements and soil or water resistivities are low. • Where the surface area of a protected structure is relatively small due to economic restrictions.	• For structures where protective current requirements and life requirements are high. • Can be used over a wider range of soil and water resistivities. • For protection of large uncoated areas, where relatively few anodes are required.
Benefits and Features	
• No independent source of electric power required. • Limited effects on neighboring structures. • Extremely simple to install. May be directly fixed to the structure. • Simple additions can be made until the desired effect is achieved. • Anode connections are also protected. • Self-adjusting but the output generally cannot be controlled • Correct material selection ensures no over protection, thus avoiding metal embitterment and coating damage • No possibility of plant damage due to incorrect connections (i.e. reversed polarity). • Straight forward to install, operate and maintain.	• Requires external power source restrictions. • Can be applied to a wide range of structures in various states of coating condition. • May be adjusted manually or automatically to cater for changing conditions. • May be remotely adjusted, monitored and connected to Plant Alarm System. • Anodes are very compact, thus drag and water flow restriction are negligible. • Requires a small number of anodes compared to a galvanic system. Needs careful design and operation to ensure ongoing protection • Can affect other structures if not properly monitored • Installation needs to ensure all connections have a high integrity of insulation and that damage does not occur due to reversed polarity.

As reported by Roberge (2014).

parts of a structure or system are electrically isolated from each other by insulators as reported by Headquarters Department of the army Washington (2005).

- **Short Circuits:** All short circuits must be eliminated from existing and new cathodic protection systems. A short circuit can occur when one protected system contacts another, causing interference with the cathodic protection system. When updating existing systems, eliminating short circuits would be a necessary first step as reported by Headquarters Department of the army Washington (2005).
- **Corrosion History of Structures in the Area:** Studying the corrosion history in the area can prove very helpful when designing a cathodic protection system. The study should reinforce predictions for corrosivity of a given structure and its environment; in addition, it may reveal abnormal conditions not otherwise suspected as reported by Headquarters Department of the army Washington (2005).
- **Electrolyte Resistivity Survey:** A structure's corrosion rate is proportional to the electrolyte resistivity. Without cathodic protection,

Table 12. Soil resistivity and soil

Soil Resistivity Range (ohm.cm)	Corrosistivity
0 to 2000	Severe
2000 to 10,000	Moderate to severe
10,000 to 30,000	Mild
Above 30,000	Not likely

as electrolyte resistivity decreases, more current is allowed to flow from the structure into the electrolyte; thus, the structure corrodes more rapidly. As electrolyte resistivity increases, the corrosion rate decreases as shown in Table 12 Resistivity can be measured either in a laboratory or at the site with the proper instruments. The resistivity data will be used to calculate the sizes of anodes and rectifier required in designing the cathodic protection system as reported by Headquarters Department of the army Washington (2005).

- **Electrolyte pH Survey:** Corrosion is also proportional to electrolyte pH. The value of soil pH affects the corrosion process greatly. The more acidic the soil is, the higher the corrosion rate. The value ranges generally from 5 to 10 in soil; a value of 7 indicates neutrality (lower values, acidity; and higher values, alkalinity). The general relationship between the pH values and the corresponding corrosion is shown in Table 13 as reported by Sen, Malmeda and Nelson (2002).
- **Structure vs. Electrolyte Potential Survey:** For existing structures, the potential between the structure and the electrolyte will give a direct indication of the corrosivity. According to NACE Standard number RP-01, the potential requirement for cathodic protection is a negative (cathodic) potential of at least 0.85 volt as measured between the structure and a saturated copper-copper sulfate reference electrode in contact with the electrolyte. A potential which is less than negative -0.85 volt would probably be corrosive, with corrosivity increasing as the negative value decreases (becomes more positive) as reported by Headquarters Department of the army Washington (2005).
- **Current Requirement:** A critical part of design calculations for cathodic protection systems on existing structures is the amount of current required per square foot (called current density) to change the structure's potential to -0.85 volt. The current density required to shift the potential indicates the structure's surface condition. The amount of

Table 13. pH values and corrosion

Soil Characteristics	pH Values	Corrosion Rate
Extremely Acid	Below 4.5	Highest Corrosion
Very Strongly Acid	4.5 - 5.0	
Strongly Acid	5.1 – 5.5	
Medium Acid	5.6 – 6.0	
Slightly Acid	6.1 – 6.5	
Mildly Alkaline	7.4 – 7.8	Higher Corrosion
Moderately Alkaline	7.9 – 8.4	
Strongly Alkaline	8.5 – 9.0	
Very Strongly Alkaline	9.1-Higher	
Neutral	6.6 – 7.3	Least Corrosion

current required for complete cathodic protection can be determined by three ways:

- ◦ An actual test on existing structures using a temporary cathodic protection setup.
- ◦ A theoretical calculation based on coating efficiency. The efficiency of the coating as supplied will have a direct effect on the total current requirement, as given in equation 12 as reported by Headquarters Department of the army Washington (2005).

$$I_T = \left(A_s\right)\left(I_r\right)\left(1.0 - CE\right) \tag{17}$$

where:

I = the total protective current,
A_s = the total structure surface area in square feet,
$\left(I_r\right)$ = the required current density as given in Table 14, and
CE = the coating efficiency.

- ◦ An estimate of current requirements using tables based on field experience.

- **Coating Resistance:** A coating's resistance decreases greatly with age and directly affects structure to-electrolyte resistance for design

Table 14. Typical current density requirements for cathodic protection of uncoated steel

Environment	(I_p) (mA/sq.ft) AFM 88 – 9	(I_p) (mA/sq.ft) I.S Gerrard
Neutral soil	0.4 to 1.5	0.4 to 1.5
Well aerated neutral soil	2 to 3	2 to 3
Wet soil	1 t 6	2.5 t 6
Highly acidic soil	3 to 15	5 to 15
Soil supporting active sulfate reducing bacteria	6 to 42	Up to 42
Head soil	3 to 25	5 to 25
Stationary freshwater	1 to 6	5
Moving freshwater containing dissolved	5 to 15	5 to 15
Seawater	3 to 10	5 to 25

As reported by Headquarters Department of the army Washington (2005).

calculations. The coating manufacturers supply coating resistance values are given in reference as reported by Headquarters Department of the army Washington (2005).

- **Protective Current Required:** By knowing the physical dimensions of the structure to be protected, the surface area can be calculated. The product of the surface area multiplied by current density obtained previously in Table (14) gives the total current required as reported by Headquarters Department of the army Washington (2005).
- **The Need for Cathodic Protection:** For existing structures, the current requirement survey will verify the need for a cathodic protection system. For new systems, standard practice is to assume a current density of at least 2 mill amperes per square foot of bare area will be needed to protect the structure. However, local corrosion history may demand a different current density. In addition, cathodic protection is mandatory for underground gas distribution lines and for water storage tanks with a 250,000-gallon capacity or greater. Cathodic protection also is required for underground piping systems located within 10 feet of steel reinforced concrete because galvanic corrosion will occur between the steel rebar and the pipeline. It can be used also for protecting the underground metallic structure of grounding system as reported by Headquarters Department of the army Washington (2005).

DETERMINING TYPE AND DESIGN OF CATHODIC PROTECTION SYSTEM

Before the design work starts, a design basis shall be prepared and accepted by the customer/ management. The design basis shall include the following information as reported by Johnsen (2006)

- Which standard or recommended practice to base the design upon.
- Type of protection; sacrificial anodes or impressed current.
- Combined protection with coating.
- Current densities to be used (if not directly in accordance with the standard).
- Protection potentials.
- If coating, type of coating and degradation rate.
- Documentation to base the design upon.
- Is the structure in electrical contact with other structures permanently or from time to time?

If the soil resistivity is low (less than 5000 ohm-centimeters) and the current density requirement is low (less than 1 mill ampere per square foot), a galvanic system can be used. However, if the soil resistivity and/or current density requirement exceed the above values, an impressed current system should be used as reported by Headquarters Department of the army Washington (2005).

Sacrificial Anode (Galvanic) Cathodic Protection System Design Steps

Step 1: Review soil resistivity (p) in Ω. cm.
Step 2: Select anode material

If $\rho < 2000$, Zinc Anode is selected

If $\rho > 2000$, Magnesium Anode is selected

Each anode specification will include anode weight, anode dimensions, and package dimensions (anode plus backfill). Table 15 shows size and weight of high potential magnesium alloy anodes.

Table 15. Weights and dimensions of selected high potential magnesium-alloy anodes for use on soil or water

Weight (Ib)	Size (in.)	Packaged Weight (Ib)	Packaged Size (in.)
3	3.75 × 3.75 × 5	12	6×10
5	3.75 × 3.75 × 7.5	17	6×12
9	2.75 × 2.75 × 26	35	6×31
9	3.75 × 3.75 × 13.25	27	6×17
12	3.75 × 3.75 × 18	36	6 ×23
14	2.75 × 2.75 × 4	50	6 ×46
14	13.75 × 3.75 × 21	42	6.5 × 26
17	2.75 × 2.75 × 50	60	6×55
17	3.75 × 3.75 × 26	45	6.5 × 29
20	2.5 × 2.5 × 59.25	70	5 ×66
24	4.5×4.5 ×23	60	7×30
32	5.5 × 5.5 × 21	74	8×28
40	3.75 × 3.75 ×59.25	105	6.5 × 66
48	5.5 × 5.5 × 30	100	8×38
48	8×16	100	12×25
60	4.5 ×4.5 ×60	---	---

Step 3: Calculate the net driving potential for anodes (E) Volt.

Step 4: Calculate the number of anodes needed to meet grounded resistance limitations. The total resistance (R_T) of the galvanic circuit is given by the following equations [21]:

$$R_T = R_A + R_{Wa} + R \tag{18}$$

$$R_T = E/I \tag{19}$$

$$R_c = R_{av}/A_s \tag{20}$$

where:

R_A = the anode-to-electrolyte resistance,
R = the structure-to-electrolyte resistance,
R_{Wa} = the anode lead wire resistance (negligible),
R_c = the coating resistance in ohms per square feet,
R_{av} = the average coating resistance (specified by the supplier) Ω/sq. ft, and
A_s = the structure's surface area in square feet.

$$N = \frac{0.0052.\rho}{R_A.L_b}\left[1n\frac{8L_b}{d_b} - 1\right] \quad (21)$$

where:

N = the number of anodes,
d_b = the diameter of the backfill column in feet (specified by supplier), and
L_b = the length of the backfill column in feet (specified by supplier).

Step 5: Calculate the number of anodes for systems by the following equation

$$N = \frac{L_f.I}{49.3 \times W_a} \quad (22)$$

where:

W_a = the weight of one anode (in pound) and
L_f = the expected lifetime in years.

Step 6: Select number of anodes to be used. The greater value of the two equations.

Step 7: Select ground bed layout. When the required number of anodes has been calculated, the area to be protected by one anode in square feet is calculated by the following equation as reported by Headquarters Department of the army Washington (2005).

$$A_a = \frac{A_s}{N} \quad (23)$$

Step 8: Calculate the life-cycle cost for proposed design according to NACE Standard RP-02. The design process should be done for several different anode choices to find the one with minimal life-cycle cost.

Step 9: Prepare plans and specifications. When the design procedure has been done for several different anodes and the final anode has been chosen, plans and specifications can be completed as reported by Headquarters Department of the army Washington (2005).

Impressed Current Cathodic Protection System Design Steps

The steps of designing ICCP are discussed below in full details.

Step 1: Review soil resistivity ρ in Ω. cm.

Step 2: Review current requirement test, it can be obtained from Table 14

Step 3: The used anodes are most often made of high silicon chromium-bearing cast-iron (HSCBCI). When impressed current-type cathodic protection systems are used to mitigate corrosion on an underground steel structure, Table 16 gives (HSCBCI) anode sizes and specifications.

Step 4: Calculate the number of anodes needed to satisfy manufacture's current density limitations. Impressed current anodes are supplied with a recommended maximum current density as reported by Headquarters Department of the army Washington (2005).

$$N = \frac{I}{A_1 . I_1} \tag{24}$$

where:

A_1 = the anode surface area per square feet per anode,

I_1 = the recommended maximum current density output in mA.

Step 5: Calculate number of anodes needed to meet design life requirement as reported by Headquarters Department of the army Washington (2005).

$$N = \frac{L_f . I}{1000 W_a} \tag{25}$$

Table 16. Weights and dimensions of selected circular high silicon chromium bearing cast iron

Anode Weight (lb)	Anode Dimensions (in.)	Anodes Surface Size (in.)	Package Area (sq.ft)
12	1×60	1.4	10×84
44	2×60	2.6	10×84
60	2×60	2.8	10×84
110	3×60	4.0	10×84

where:

W_a = the weight of one anode (in pound),
L_f = the expected lifetime in years.

Step 6: Calculate the number of anodes needed to meet maximum anode ground bed resistance requirements as reported by Headquarters Department of the army Washington (2005).

$$R_a = \frac{\rho.K}{N.L_b} \cdot \frac{\rho.P}{S_b} \tag{26}$$

where:

K = the anode shape factor it can be obtained from Table 17,
S_b = the center-to-center spacing between anode backfill columns in feet,
L_b = the length of the anode backfill column in feet, and
P = the paralleling factor that can be obtained from Table 18.

Step 7: Select number of anodes to be used. The highest number calculated by equation will be the number of anodes used.
Step 8: Select area for placement of anode bed. The area with the lowest soil resistivity will be chosen to minimize anode-to-electrolyte resistance.
Step 9: Determine total circuit resistance. The total circuit resistance will be used to calculate the rectifier size needed.

Calculate anode ground bed resistance using Equation 26 as reported by Headquarters Department of the army Washington (2005).

Table 17. Shape function (K) for impressed current cathodic protection anodes where L_a is effective anode length and d_a is anode/backfill diameter

L/d	K	L/d	K
5	0.0140	20	0.0213
6	0.0150	25	0.0224
7	0.0158	30	0.0234
8	0.0165	35	0.0242
9	0.0171	40	0.0249
10	0.0177	45	0.0255
12	0.0186	50	0.0261
14	0.0194	55	0.0266
16	0.0201	60	0.0270
28	0.0207	20	0.0213

Table 18. Anode paralleling factor (P) of various numbers of anodes (N) installed in parallel

N	P	N	P
2	0.00261	14	0.00168
3	0.00289	16	0.00155
4	0.00283	18	0.00145
5	0.00268	20	0.00135
6	0.00252	22	0.00128
7	0.00237	24	0.00121
8	0.00224	26	0.00168
9	0.00212	28	0.00155
10	0.00201	30	0.00145
12	0.00182	----	0.00135

As reported by Bryan (1975).

As reported by Bryan (1975).

Calculate ground bed header cable resistance. The cable is typically supplied with a specified resistance in ohms per 100 feet. The wire resistance then is calculated as reported by Headquarters Department of the army Washington (2005)

$$R_W = \frac{R_m.L_w}{100ft} \tag{27}$$

where:

R_m = the wire resistance specified by the manufacturer,
L_w = the structure's length in feet.

Calculate structure-to-electrolyte resistance:

$$R_C = \frac{R}{N_c} \tag{28}$$

where:

R = the structure-to-electrolyte resistance,
R_C = the coating resistance in ohms per square feet,
N_c = the coated grounding system or pipe area in square feet.

Calculate total circuit resistance as reported by Headquarters Department of the army Washington (2005):

$$R_T = R_A + R_W + R_C \tag{29}$$

Step 10: Calculation of rectifier voltage. Equation 30 is used to determine voltage output (V_{rec}) of the rectifier as reported by Headquarters Department of the army Washington (2005)

$$V_{rec} = I * R_T * 150\% \tag{30}$$

where:

V_{rec} = the voltage output of rectifier,
I = the total protection current in amperes, and
R_T = the total circuit resistance.

Step 11: Selection of a rectifier. A rectifier must be chosen based on the results of Equation 30). Many rectifiers are available commercially; one that satisfies the minimum requirements of (I) and (V_{rec}) in Equation 25 should be chosen.

Step 12: Calculation of system cost. As with the galvanic cathodic protection system, the choice of anode for design calculation is arbitrary. When several anodes have been used in the design calculations, an economic analysis should be done as recommended in NACE Standard RP-02.

Step 13: Prepare plans and specifications as reported by Headquarters Department of the army Washington (2005)

REFERENCES

Nelson, J. P., & Holm, W. K. (1985). A cathodically protected electrical substation ground. *IEEE Transactions on Industry Applications*, *21*(2), 357–361. doi:10.1109/TIA.1985.349656

US Air Force. (1962). Corrosion Control. Air Force Manual AFM 88-9 (p. 203).

Beavers, J. A., (2001). Cathodic Protection - How It Works, Control of Pipeline Corrosion. In *Proceedings of NACE International* (Ch. 3, pp. 21-48).

Bianchetti, R.L. (2001). Cathodic Protection with Galvanic Anodes. Control of Pipeline Corrosion. In *Proceedings of NACE International* (Ch. 3, pp. 21-48).

Bryan, W. T. (1970). *Designing Impressed Current Cathodic Protection Systems With Durco Anodes*. The Duriron Company.

Chauhan, B. S. (2008). *Engineering Chemistry*. Laxmi Publications Pvt Limited.

Corrichem Company for Cathodic protection. (n. d.). info@corrichem.com.

Fan, X., Wang, J., Zhou, M., Huang, S., Wang, Y., & Dou, P. (2011). Laboratory Soil Corrosion Test of Steel Grounding Materials. In *Proceedings of the 7th Asia-Pacific International Conference on Lightning*, Chengdu, China.

Gerrard, J. S. (1976). Practical Applications of Cathodic Protection. Corrosion, 2, 11-65.

Headquarters Department of the army Washington (2003). Operation and Maintenance Cathodic Protection Systems, Unified Facilities Criteria (UFC), UFC 3-570-06, January 31, 2003.

Headquarters Department of the army Washington. (2005). Electrical Design Cathodic Protection, Unified Facilities Criteria (UFC), UFC 3-570-02A.

Howard, A. P. (2004) Introduction to Corrosion and Galvanizing of Helix Foundations. Presented at Helical Foundations and Tiebacks Committee, Deep Foundation Institute, Tampa, FL.

IEEE Guide for Safety in AC Substation 80. (1986). IEEE. Std. No. 80-1986.

IEEE Guide for Safety in AC Substation 80. (2000). IEEE Std. No. 80-2000.

James, B., & Bushman, P. E. (2001). *Galvanic Anode System Design. Bushman, Associates, Inc. Corrosion Consultants.*

Johnsen, R. (2006). Cathodic Protection. Institutt Produktutvikling og materialer (NTNU).

Kean, R. L., & Davies, K. G. (1981). Cathodic Protection. Retrieved from www.npl.co.uk/upload/pdf/cathodic_ororection.pdf

Lawson, V. R. (1988). Problems and detection of line anchor and substation ground grid corrosion. *IEEE Transactions on Industry Applications*, *24*(1), 25–32. doi:10.1109/28.87245

Durham, M.O., & Durham, R.A. (2005). Consequences and standards from using CP systems to prevent corrosion. IEEE Industry Applications Magazine.

Mars, G. (1987). *Fontana* (pp. 1–11). New York: Corrosion Engineering, Mc-Graw Hill Books Company.

Mi, Z., Jianguo, W., Yang, L., Nianwen, X., Zhen, S., Junjie, C., & Chunhua, F. (2008) Causes, forms and remedies of substation grounding grid corrosion. In Proceedings of the *International Conference on High Voltage Engineering and Application*, Chongqing, China. doi:10.1109/ICHVE.2008.4773904

Mitolo, M., & Pettinger, A. (2016). Interactions between cathodically protected pipelines and grounding systems. *IEEE Transactions on Industry Applications*, *52*(5), 3694–3698. doi:10.1109/TIA.2016.2582795

NACE. (2002). Standard RP0169-2002 Control of external corrosion on underground or submerged metallic piping systems.

Roberge, P.R. (2017). Cathodic Protection Part 1 – Code of Practice for Land and Marine Applications. *Cathodic Protection Co. Limited.* Retrieved from www.cathodic.co.uk/files/1246442149cathodic%20protection%20overview.pdf

Sen, P.K., Malmeda, K., & Nelson, J.P. (2002). Steel Grounding Design Gide and Application Notes. In *Proceedings of the IEEE Rural Electric Power Conference,* Chengdu, China (pp. 6-7).

Szakalos, P., & Seetharaman, S. (2012). Corrosion of copper canister (Technical Note 2012:17).

Tullmin, M., & Roberge, P. R. (1995). Tutorial corrosion of metallic. *IEEE Transactions on Materials Reliability*, *44*, 271–278. doi:10.1109/24.387383

Zheng, R., He, J., Hu, J., Lu, G., & Luo, B. (2002). The theory and implementation of corrosion diagnosis for grounding system. In *Proceedings of the Industry Applications Conference* (Vol. 2, pp. 1120 - 1126). doi:10.1109/IAS.2002.1042700

Chapter 8
Transient Response of Grounding Systems

ABSTRACT

This chapter describes a model used to simulate lightning current. Comparison of different approaches used for lighting simulation has been mentioned. Transmission line approach has been chosen to simulate the grounding system behavior under lightning condition. Two different types of the grounding system have been studied under lightning conditions vertical driven electrodes and buried grids. This chapter draws attention also to the following points: equivalent circuit of a vertical ground rod on lightning condition, sensitivity analysis of soil parameters and grounding rod dimensions, equivalent circuit of grounding grid on lightning conditions and sensitivity analysis of soil parameters and grounding grid dimensions.

LIGHTNING CURRENT IMPULSE MODEL

A typical lightning is an electrical discharge between:

1. Cloud and the earth (cloud-to-ground flash)
2. Within the cloud (intra-cloud flash)
3. Different clouds (inter-cloud flash).

DOI: 10.4018/978-1-5225-3853-0.ch008

Out of all the flashes occurring in the natures, only 10% of flashes are cloud-to-ground flashes. A typical cloud-to-ground lightning flash starts with a descending stepped leader after a preliminary discharge in the cloud. As the stepped leader propagates towards the ground, the electric field at the ground or at the tip of grounded objects is increased. At a certain instant, the field would be sufficiently high to launch upward connecting leaders from the ground or grounded structures. The upward connecting leaders have a polarity opposite to that of the downward leader. The incepted upward leader propagates in such a way that it seeks the tip of the descending leader. When the gradient between the two leader tips is sufficiently high (500 kV/m), then a final jump between the two occurs. The length of this final jump depends upon the charges in the above said two leaders. Because of the interception between the two leaders, the gap between the cloud and ground is bridged. At this instant, a ground potential wave called the return stroke propagates upward, discharging the leader channel. The return stroke current is the most important parameter for the transient behavior study of the grounding system. The properties of it have been summarized by Thottappillil R., (2002). Usually, the discharge current of return stroke increases from zero to a maximum in few µs (from 0.1 to 10 µs), then declines to half the peak value in about 20 to 1000 µs. The typical value of the peak current derivative *di/dt* is about 110 kA/µs. The peak value of the stroke current is about 15-30 kA (median value), and some stroke current could be about 250 kA (probability of occurrence less than 0.1%) as reported by by Thottappillil R., (2002). The above said lightning return stroke current parameters forms the basis of the impulse current sources which will be used as an input for the transient analysis of grounding system in the present study Table 1 gives two equations simulate the lightning current, (Figure 1) shows fast and slow impulses simulate the lightning current of 12 kA peak current.

Table 1. Equations simulate the lightning current

Type	Equation
Fast impulse	$I(t) = 1 \cdot \left(e^{-27000.t} - e^{-5600000.t}\right)$
Slow impulse	$I(t) = 1 \cdot \left(e^{-7924.t} - e^{-400109.t}\right)$

Figure 1. Fast and slow lightning current

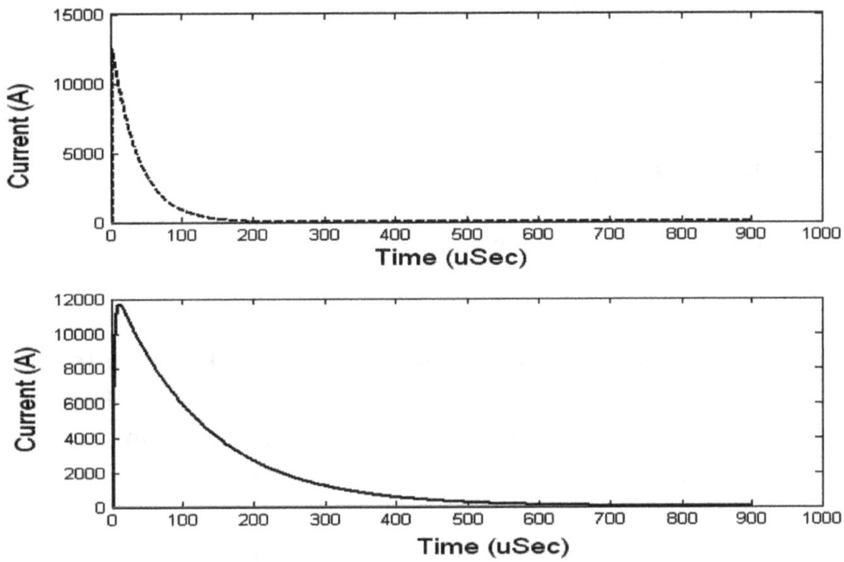

MODELING OF GROUNDING SYSTEM FOR LIGHTNING STUDIES

The modeling of complex transient behavior of grounding system under lightning strikes had a better future specifically because of the following reasons as reported by Mitra (1973)

- The earlier models had several assumptions in order to arrive at simple equations. But by using numerical methods, most of complex equations can be solved.
- Practical complex grounding systems can be modeled easily because of the large memory and speed of the computer.

The various numerical modeling methods for grounding systems under lightning studies, developed since 1980's until now can be classified as reported by Liu (2004) as follows

1. Circuit approach
2. Electromagnetic field approach
    a. Method of moment
    b. Finite element method
3. Hybrid approach
4. Transmission line approach

Summary of comparison between the different approaches is given in Table 2.

Based on the above comparisons in Table 2, it is found that the transmission line approach is a better choice for the present study. Unfortunately, the development of this approach was not as fast as that of circuit and electromagnetic field approaches. As of now, all the transmission line approaches were limited to model simple grounding systems, i.e., counterpoise wires or single grounding rods.

Consequently, from scope of this study, it becomes necessary to improve transmission line approach for modeling transient behavior of grounding systems for engineering application. The conventional transmission line approach with parameters (*L, C,* and *R*) will be extended from a single grounding wire/rod to grounding grids.

Table 2. Comparison of different approaches for modeling the transient behavior of grounding system

Approaches	Mathematical Expressions	Visualization	Solution Procedure	Requirement of Computer Power	Accuracy
Electromagnetic field approach	Complicated	Difficult to understand	Complicated	Very powerful computer, large computation time	Believed to be most accurate
Circuit approach	Simple	Easy to understand	simple	Normal computer, small computation time	Reasonably accurate
Hybrid approach	Complicated	Not very easy to understand	Slightly complicated	Normal computer, small computation time	Reasonably accurate
Transmission line approach	The simplest	Very easy to understand	very simple can predict wave propagation delay	Normal computer, small computation time	Reasonably accurate

IMPROVED TRANSMISSION LINE APPROACH

The grounding systems are structured with good conductors. Every conductor is assumed to be a lossy transmission line. It is also assumed that the radius of each conductor is much smaller than the buried depth and the length of the wire. The conductor is characterized by its electrical properties and dimensions. The soil is modeled as a linear and homogeneous half space characterized by resistivity, relative permittivity and permeability, thus making the soil a lossy medium. The current is partly flowing along the conductors, and partly dissipating from its surface into the soil in a radial direction. For the moment, the ionization of the soil is not considered. The skin effect (internal loss) of the conductor is neglected, as the losses in the soil are much larger than the internal loss. An infinite uniform transmission line approach, similar to the one used by Mazzetti et al. (1983), is used in the simulation. Also, the transmission line parameters are frequency independent. It means the parameters are constant for a given lightning stroke. On the other hand, practical grounding systems have some conducting structure above the ground (e.g., air-termination, and down conductors of the lightning protection system), in addition to the buried conductors. Any electromagnetic coupling between the over ground system and underground system are not considered. Besides, the vertical conductor that connects the buried grounding system with the over ground system is also not considered. The surge current is assumed to be injected directly into the grounding system.

Equivalent Circuit of a Vertical Ground Rod on Lightning Condition

Vertical ground rods are one of the simplest and most commonly used means for earth termination of electrical and lightning protection systems. Their behavior at 50 or 60 Hz is well understood using a simplified analysis based on static approximation. However, their transient performance is also of interest when, for example, they are subjected to lightning or other high-frequency excitations considered in electromagnetic-compatibility (EMC) studies or in telecommunications. The property that is of primary interest is their input impedance. Traditionally at low frequencies, this impedance is represented by a single resistor (Figure 2-b) and at high frequencies by a lumped R-L-C circuit (Figure 2-c).

Figure 2. Low-current models of a vertical ground rod

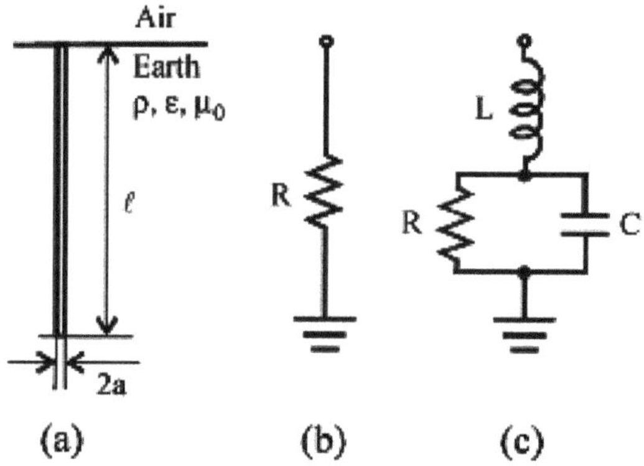

In (Figure 2) the current impinges on the rod electrode and enters the ground, which in addition to its resistivity has a dielectric constant ε. Thus, in parallel to the conductive current in the ground there develops a capacitive or displacement current in case the electrode voltage changes with time. The displacement current follows exactly the lines of the conductive current, and thus the ground electrode will have a capacitance, as reciprocal to the resistance as shown in the following equations:

$$R = \frac{\rho}{2\pi l}.\ln\frac{2l}{a}\,ohm \tag{1}$$

$$C = \frac{\varepsilon.l}{2\ln\frac{2l}{a}}.\frac{10^{-9}}{9}\,farad \tag{2}$$

The inductance of such a rod is

$$L = 2l.\ln\frac{2l}{a}.10^{-7}\,henry \tag{3}$$

where

l = the rod length in meter,
a = the radius of the driven rod and

ρ = the soil resistivity in $\Omega.m$

The capacitive time constant of any circuit element is given by the product of capacitance and resistance, and thus a rod electrode has a time constant

$$\tau_c = RC = \frac{\rho.\varepsilon}{4\pi \cdot 9 \cdot 10^9} \tag{4}$$

It depends only on resistivity and dielectric constant ε, the value of which for ordinary soil may be of the order of 9.

The inductive time constant as quotient of self-inductance and resistance is

$$\tau_L = \frac{L}{R} = 4\pi \frac{l^2}{\rho} \cdot 10^{-7} \tag{5}$$

It is proportional to the square of the length l of the electrode, and therefore greatly dependent on this dimension.

Since the inductive and capacitive time constant (10^{-9},10^{-7}) are less than the time usually assumed for the shortest possible lightning impulse (10^{-6}) impinging on a driven rod, current and voltage in the rod will follow such impressed impulse of the order of 10^{-9} second without any significant time delay and will behave as in the steady state. Hence capacitance and inductance of a driven rod of moderate length play no significant part even with rapid lightning phenomena.

Transient behavior of grounding systems can be simulated based on conventional transmission line approach since a lumped distributed circuit can be used to represent a transmission line if the line is divided into many electrically small sections. The transient behavior of grounding systems can be simulated in ATP-EMTP, based on the above said distributed circuit. In ATP-EMTP, every electrically small section of the transmission line is represented by lumped resistance, inductance, capacitance and other coupling elements. The number of the sections required for an accurate simulation is dependent on the highest frequency of the injected impulse, the higher the frequency, the larger the number of sections required.

An example of application of this model is modeling transient behavior of different grounding rod and grids, which have dimensions as given in tables 3, 4 and 5.

Table 3. Parameters of different soil resistivity, $l = 6m$, a = 0.02 m, ε = 9

Case Study	m	R (Ω)	L (H)	C (F)
R01	1000	73.7	7.67×10^{-6}	4.6×10^{-10}
R02	100	7.3	7.67×10^{-6}	4.6×10^{-10}
R03	50	3.6	7.67×10^{-6}	4.6×10^{-10}
R04	10	0.73	7.67×10^{-6}	4.6×10^{-10}

All the grids are buried at 0.6 m depth in a homogeneous soil, double-exponential current impulse,

$$I(t) = 1 \cdot \left(e^{-27000.t} - e^{-5600000.t}\right) \text{A},$$

is injected at the corner of the grounding grids.

Sensitivity Analysis of Soil Parameters and Grounding Rod Dimensions

Parameters of the grounding system, such as the soil resistivity and rod length are related to the electric circuit elements through their parameters. Therefore, the influence of these parameters can be explained based on circuit theory.

Figure 3. Influence of soil resistivity variation

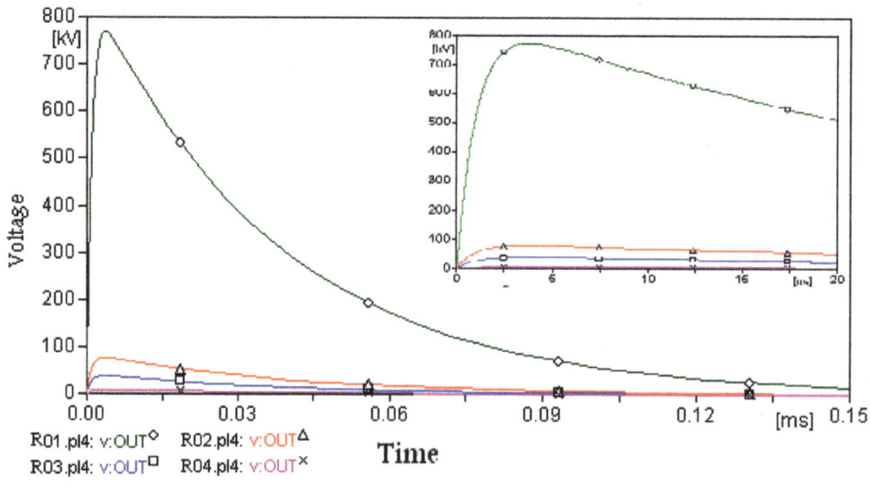

Figure 4. Transient impedance for R01, R02, R03 and R04 rods

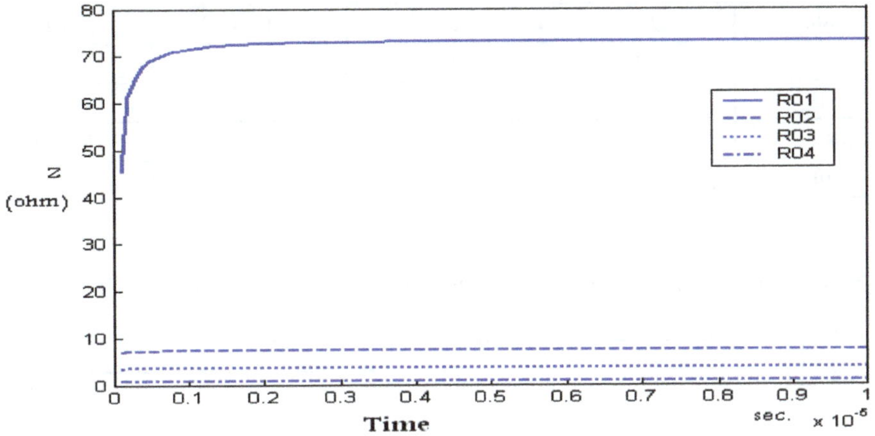

The Influence of the Soil Resistivity

To study the effect of soil resistivity variation on the ground impedance, the soil resistivity of the ground system changed between $10\,\Omega.m$ to $1000\,\Omega.m$. Results of fast lightning current impulse applied to the grounding system with parameters mentioned in table 3 are shown in figure (3-a, b)

It is noticed that changing the soil resistivity from $10\,\Omega.m$ to $1000\,\Omega.m$ has great effect on the resistance of the ground system as given in Table 3. It varies from $0.73\,\Omega$ to $73.7\,\Omega$, also as shown in (Figure 3) the maximum value of transient voltage magnitude increases from about 7.5 kV to 770 kV.

From (Figure 4) it is noticed that the transient impedance remains constant, inductance and capacitance part have no effect.

The Influence of the Rod Length

To study the effect of rod length variation on the ground impedance and transient voltages, the rod length of the ground system changed between 3 m to 15 m.

Result of fast lightning current impulse applied to the grounding system grids with parameters mentioned in table 4 are shown in (Figures (4-a) and (4-b))

Table 4 presents different rod length to study the transient response.

From (Figure 5) it is noticed that changing the rod length affect the peak value of the transient voltage magnitude of the ground system, when the rod length changes from 3 m to 15 m, the maximum voltage magnitude decreases from about 1.4 MV to 400 kV.

Table 4. Parameters of different rod length, $\rho = 1000\,\Omega.m$, $a = 0.02$ m, $\varepsilon = 9$

Case Study	*L*, m	R (Ω)	L (H)	C (F)
Rp04	3	131	3.4×10^{-6}	2.6×10^{-10}
Rp03	5	85	6.2×10^{-6}	4×10^{-10}
Rp02	9	52.2	1.22×10^{-5}	6.6×10^{-10}
Rp01	15	33	2.1×10^{-5}	1×10^{-9}

Figure 5. Influence of rod length variation on the rod transient voltage

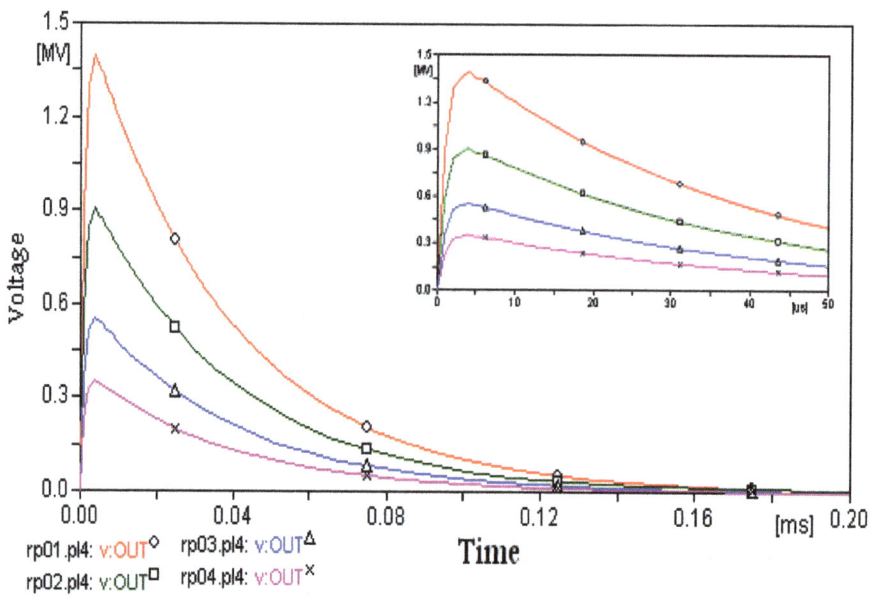

From (Figure 6) it is notice that the transient impedance remains constant, inductance and capacitance part have small effect at the beginning of the current impulse.

EQUIVALENT CIRCUIT OF A HORIZONTAL GROUNDING GRID ON LIGHTNING CONDITION

Ground grids are considered complex arrangement and many research efforts have been made to explain the performance of grounding impedance of the ground grid under lightning

Figure 6. Transient impedance for Rp01, Rp02, Rp03 and Rp04 rods

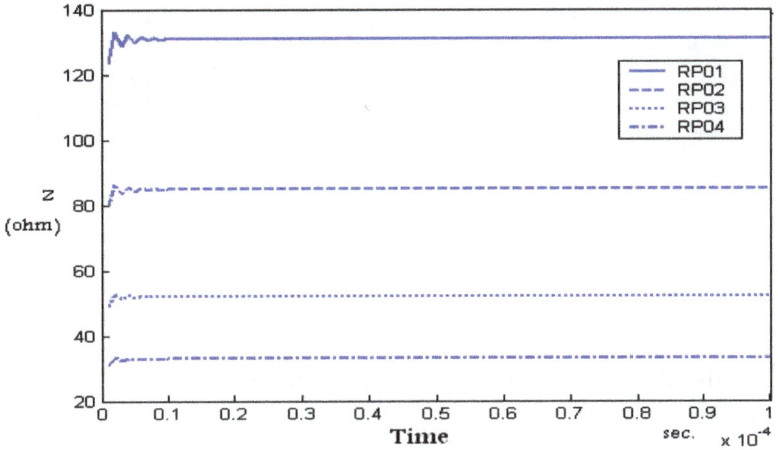

The program ATP-EMTP is used, the software widely used by power engineer for transient analysis.

Using the following formula to calculate the grid parameters:

$$R = \rho\left[\frac{1}{l} + \frac{1}{\sqrt{20A}}\left(1 + \frac{1}{1 + h\sqrt{20\,/\,A}}\right)\right]\Omega \tag{6}$$

where R is the resistance of the grid, ρ is the soil resistivity in $\Omega.m$, A is the area of the grid, and l is the total length of the grid.

$$C = \frac{\varepsilon \cdot \rho \cdot 10^{-9}}{36\pi R}\,\Omega \tag{7}$$

where C is the capacitance of the grid, ε is the permittivity of soil.

$$L_1 = \frac{2l''}{3 \cdot 10^{-7}} \ln \frac{2l''}{r''} H \tag{8}$$

where L_1 is the inductance of the main wire, l'' is the length of the main wire, and r'' is the main wire radius

Figure 7. Different number of meshes for square grids

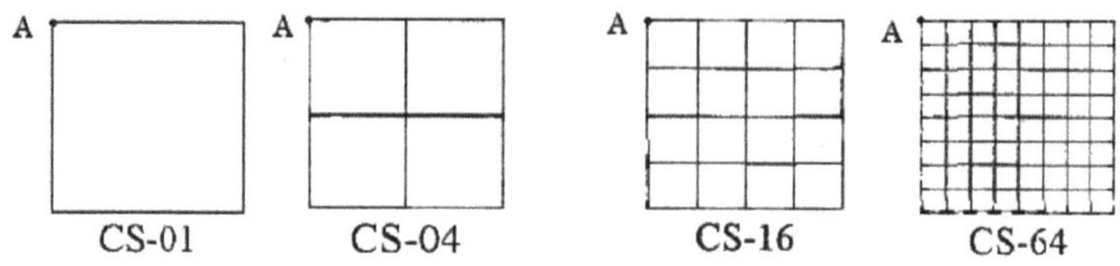

$$L_2 = \frac{2l'''}{3 \cdot 10^{-7}} \ln \frac{4l'''}{r'''} H \tag{9}$$

where L_2 is the inductance of the grid conductors, *l'''* is the side length of the grid, and *r'''* is the grid wire radius.

Different grid sizes are tested CS01, CS04, CS16 and CS64 mentioned in chapter three, shown in (Figure 7) and tabulated in Table 5.

The radius of the horizontal grounding grid conductor is 10 mm, the grid is buried at 0.6m depth in the soil with ε =50 and ρ=100 $\Omega.m$, the injected current impulse at point A in each grid has the formula

$$I(t) = 12000 \cdot (e^{-27000.t} - e^{-5600000.t}),$$

Table 5. Parameters of the four square grids given in Figure 5

Grid Model	Aream × m	n (Number of Parallel Conductors)	N (Number of Meshes)	D	Total Grid Length, m
CS01	10×10	2	1	10	40
CS04	10×10	3	4	5	60
CS16	50×50	5	16	12.5	500
CS64	100×100	9	64	12.5	1800

Table 6. RLC values for each grid configuration for $\rho = 100\,\Omega.m$

Grid Model	R (Ω)	L_1 (H)	C (F)	L_2 (H)
CS01	6.499	2.3966e-005	6.8025e-09	5.5294e-006
CS04	5.6657	2.3966e-005	7.8030e-09	5.5294e-006
CS16	1.0716	1.4129e-004	4.1254e-08	3.3012e-005
CS64	0.4969	3.0106e-004	8.8966e-08	7.0644e-005

the peak value of the current impulse is about 12 kA. The RLC parameters of the grids under study tabulated in Table 5 are given in Table 6.

(Figures 8, 9) give the transient impedance of different grids under testing. (Figures 10, 11) present simulation results of the transient voltages at injection point for different grids as response to the fast lightning current impulse.

Figure 8. Transient impedance for CS01 and CS04 grids

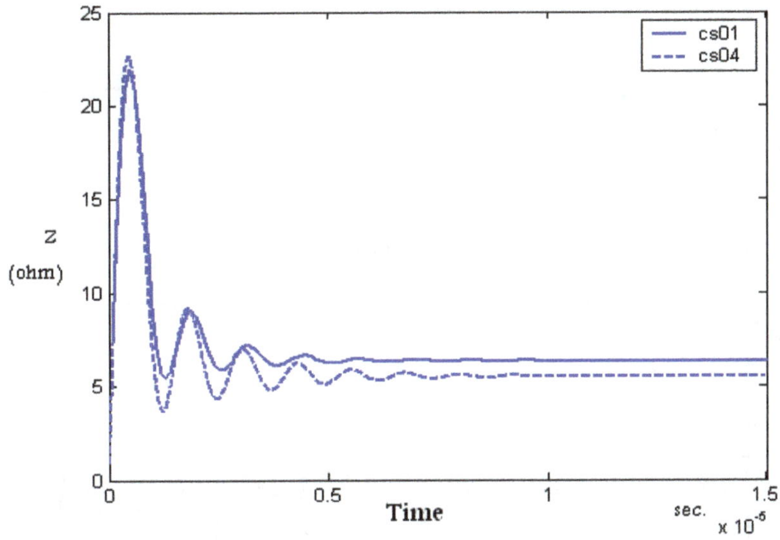

Figure 9. Transient impedance for CS16 and CS64 grids

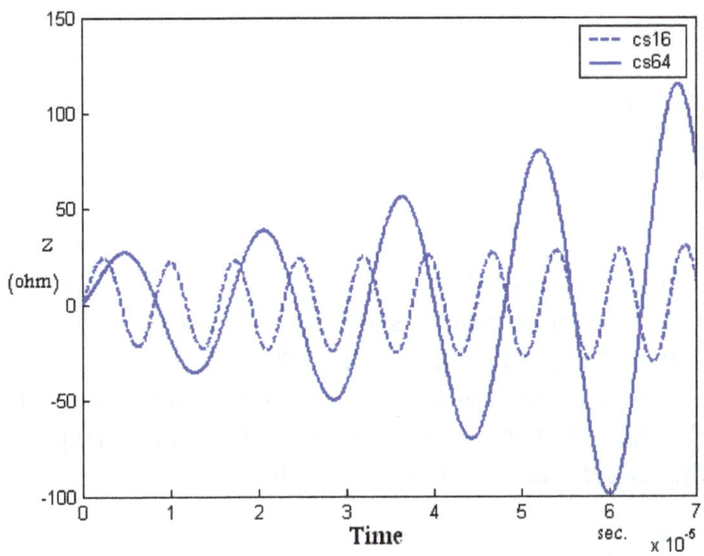

Figure 10. Voltage wave form for CS01 and CS04 grids

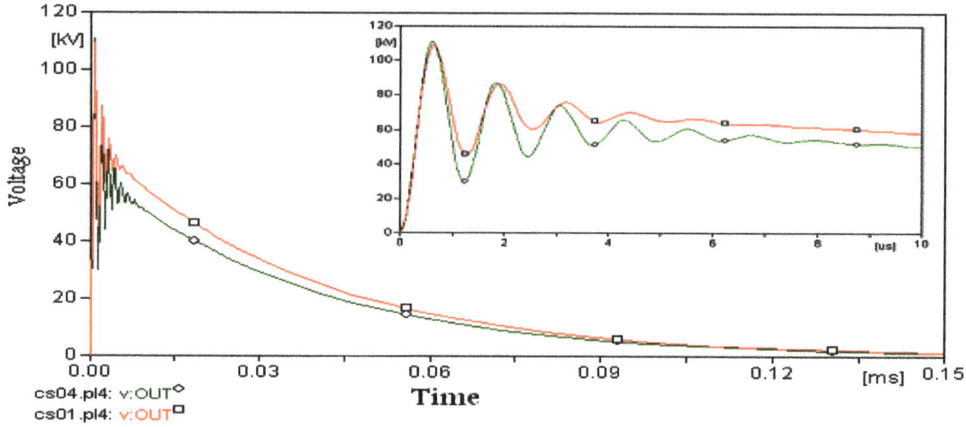

Figure 11. Voltage wave form for CS16 and CS64 grids

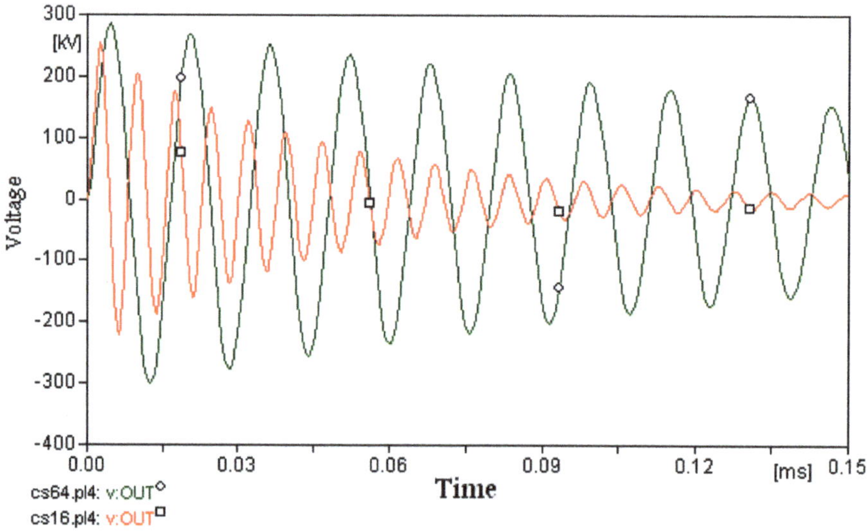

The maximum values of the transient voltages of CS01 and CS04 in the proposed simulation are almost the same as reported by Grcev and Heimbach, (1997) and Liu, Zitnik and Thottappillil (2001).

The transient impedance of CS01 and CS04 are shown in (Figure 5.6-a) the peak value for CS01 is 23Ω then settle to 6.5Ω after 10 μsec, for CS04 the peak value is 22Ω then settle to 5.5Ω after 10 μsec.

From (Figure 6-b) it is noticed that the oscillation of the ground system transient impedance increases due to the increase of inductance and capacitance of the ground system and decrease of the ground system resistance, i.e. the increase of grid dimensions increase the capacitance and inductance of the ground system and reduce the ground system resistance and this lead to oscillation of the transient impedance of the ground system.

During the 70 μsec period shown in (Figure 6-b) the impedance of CS16 oscillate for nine cycle, while the impedance of CS64 oscillate for four cycle only.

As seen in (Figure 6-c) it is noticed that as the area of the grid increase the voltage magnitude decrease, after 6μsec from the beginning of the lightning impulse, peak voltage appear over CS01 is about 78 kV and over CS04 is about 60 kV.

SENSITIVITY ANALYSIS OF SOIL PARAMETERS AND GROUNDING GRID DIMENSIONS

All the parameters of the grounding system, such as the soil relative permittivity, soil resistivity, and the conductors' diameter, are related to the electric circuit elements parameters, so the influence of these parameters can be explained from the point of electric circuit theory.

The Influence of the Soil Resistivity

Result of fast lightning current impulse applied to the grounding grids with radius of the horizontal grounding grid conductor is 10 mm, the grid is buried at 0.6m depth in the soil with $\varepsilon = 50$.

(Figure 7-a) shows the transient impedance for tested grids CS01 and CS04 for $\rho = 100\,\Omega.m$ and $400\,\Omega.m$, (Figure 7-b) shows the transient impedance for tested grids CS16 and CS64 for $\rho = 100\,\Omega.m$ and $400\,\Omega.m$. The transient voltages response for CS01, CS04, CS16 and CS64 are shown in (Figure 8-a) for $\rho = 100\,\Omega.m$, in case of $\rho = 400\,\Omega.m$ results are shown in (Figure 8-b).

Table 7 shows the values of Lumped RLC circuit for each grid configuration for $\rho = 400\,\Omega.m$.

As shown in (Figure 7-a) for the grid CS01 when the soil resistivity equal $100\,\Omega.m$ the peak value of the transient impedance was (23Ω) 350% of the transient impedance after 10 μsec (6.5Ω) while in case of soil resistivity equal

Table 7. RLC values for $A = 400\,\Omega.m$

Grid Model	R (Ω)	L (H)	C (F)	L_2 (H)
CS01	25.9963	2.3966e-005	6.8025e-09	5.5294e-006
CS04	22.6630	2.3966e-005	7.8030e-09	5.5294e-006
CS16	4.2866	1.4129e-004	4.1254e-08	3.3012e-005
CS64	1.9877	3.0106e-004	8.8966e-08	7.0644e-005

Figure 12. Transient impedance for CS0 and CS04 grids for ρ=100,400

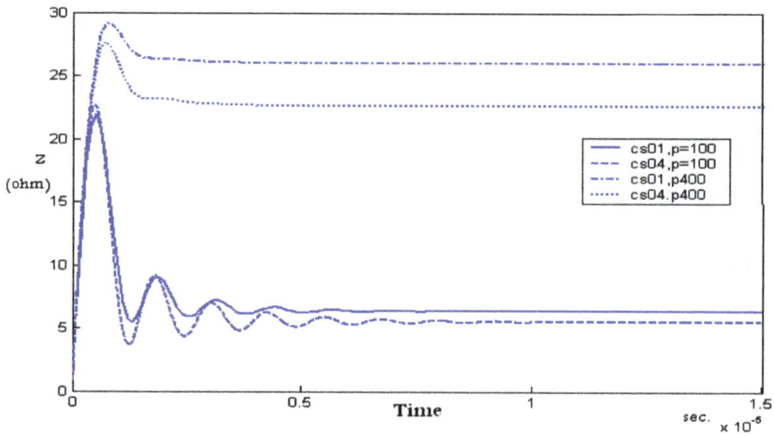

Figure 13. Transient impedance for CS16 and CS64 grids for ρ=100,400 Ω.m

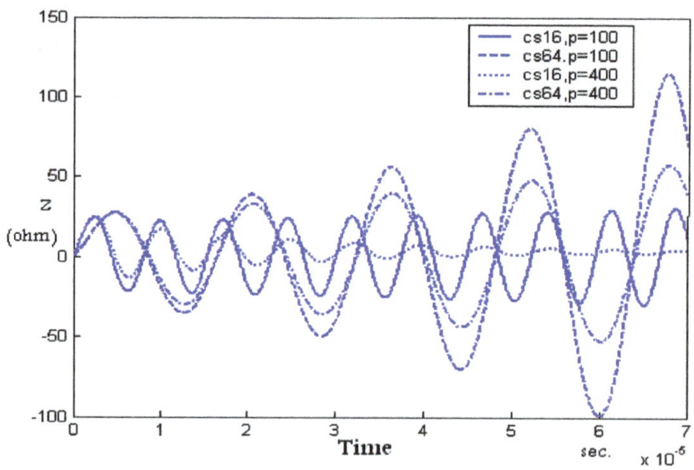

Figure 14. Voltage wave form for ρ=100 Ω.m

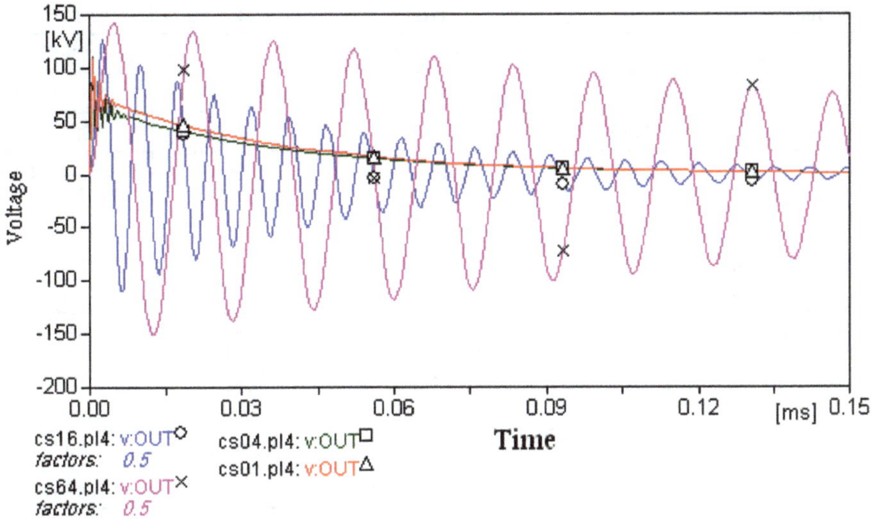

Figure 15. Voltage wave form for ρ=400 Ω.m

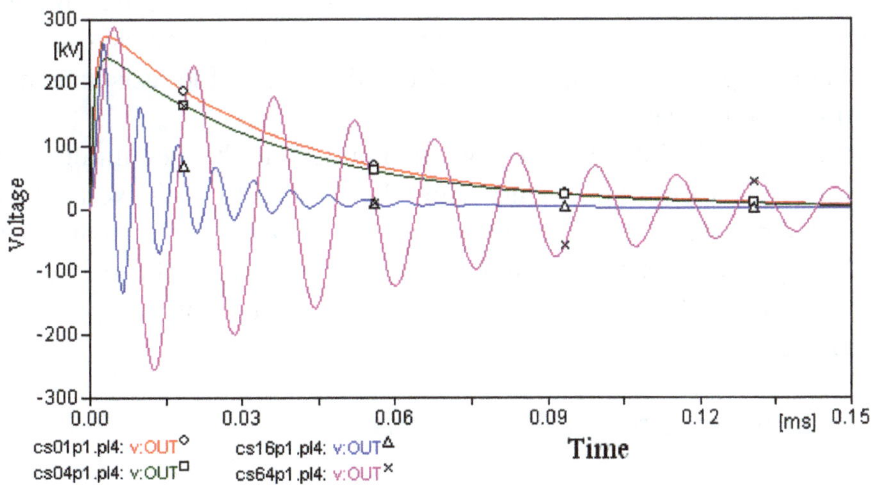

400 Ω.m the peak value of the transient impedance was (28Ω) 107% of the transient impedance after 10 µsec (26Ω).

Comparing Table 6 and Table 7 together gives the following conclusion, when the soil resistivity change from 100 Ω.m to 400 Ω.m only the resistive part of the grounding system RLC model change while capacitive and inductive parts does not change and due to this changes the results of the simulation

Table 8. RLC values for ε = 9

Grid Model	R (Ω)	L_1 (H)	C (F)	L_2 (H)
CS01	6.4991	2.3966e-005	1.2244e-09	5.5294e-006
CS04	5.6657	2.3966e-005	1.4045e-09	5.5294e-006
CS16	1.0716	1.4129e-004	7.4257e-09	3.3012e-005
CS64	0.4969	3.0106e-004	1.6014e-08	7.0644e-005

Table 9. RLC values for ε = 36

Grid Model	R (Ω)	L_1 (H)	C (F)	L_2 (H)
CS01	6.4991	2.3966e-005	4.8978e-09	5.5294e-006
CS04	5.6657	2.3966e-005	5.6182e-09	5.5294e-006
CS16	1.0716	1.4129e-004	2.9703e-08	3.3012e-005
CS64	0.4969	3.0106e-004	6.4056e-08	7.0644e-005

change as shown in (Figure 8-a,b) the voltage magnitude for CS01 and CS04 after 6μsec from the beginning of the lightning impulse change from 65 kV and 54.5 kV to 264 kV and 230 kV respectively also the voltage oscillation at the beginning of the lightning impulse is damped as the resistance of the ground system increase.

For CS16 and CS64 the voltage magnitude at the beginning of the lightning impulse does not change while the oscillation of the voltage wave form is damp faster due to the increase of the resistive part, the number of complete cycles oscillation during the simulation time 150 μsec for CS64 does not change (9 cycles) while for CS16 the oscillation damped after 9 cycles instead of 20 cycles.

The Influence of the Soil Permittivity

Result of fast lightning current impulse applied to the grounding grids with radius of the horizontal grounding grid conductor is 10mm, the grid is buried at 0.6m depth in the soil with $\varepsilon = 9$ and $\varepsilon = 36$ respectively and $\rho = 100\ \Omega.m$ shown in (Figure 9-a, b), (Figure 10-a,b), (Figure 11-a,b) and (Figure 12-a,b). Table 8 gives the values of Lumped RLC circuit for each grid configuration for $\varepsilon = 9$.

Table 9 gives the values of Lumped RLC circuit for each grid configuration for $\varepsilon = 36$.

Figure 16. Voltage wave form for CS01, ε = 9 and 36

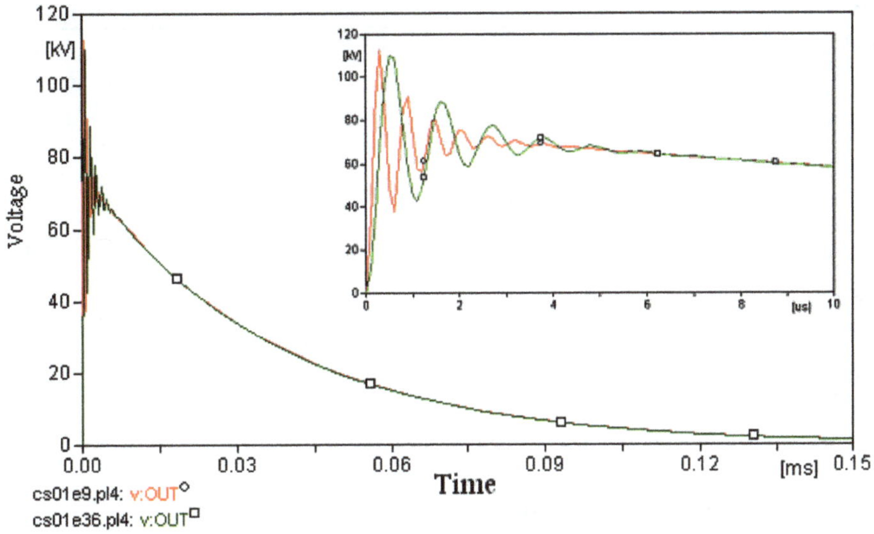

It is noticed that increasing the soil relative permittivity increases the capacitance of the ground system.

From (Figure 17) it is notice that the change in the soil permittivity from ε = 9 to 36 led to small phase shift in the transient voltage response for CS01 grid (about 0.5 μsec.).

Figure 17. Transient impedance for CS01, ε = 9 and 36

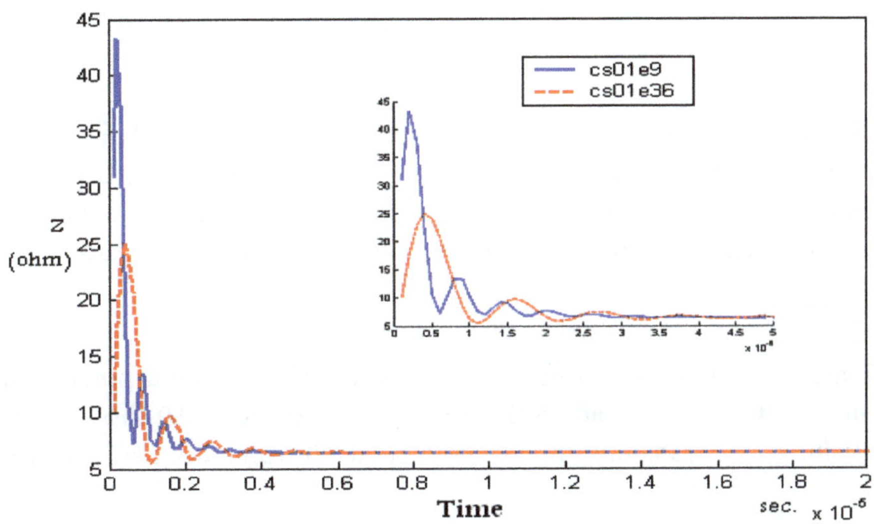

Figure 18. Voltage wave form for CS04, ε =9 and 36

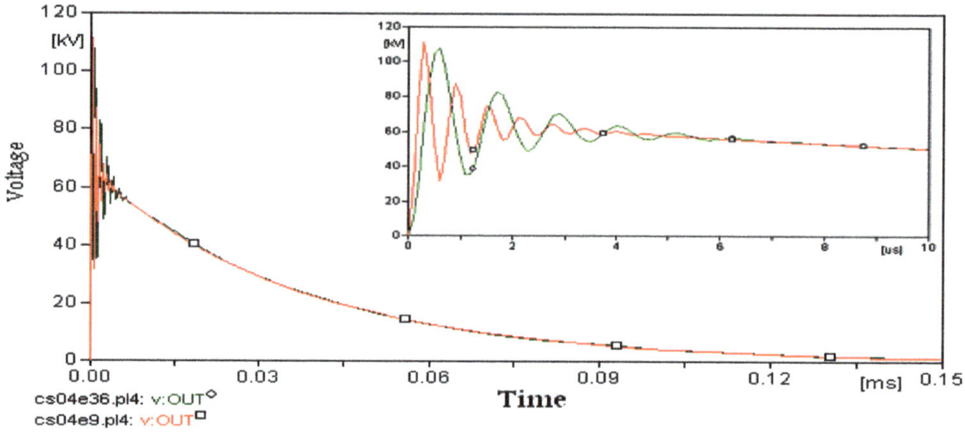

Figure 19. Transient impedance for CS04, ε =9 and 36

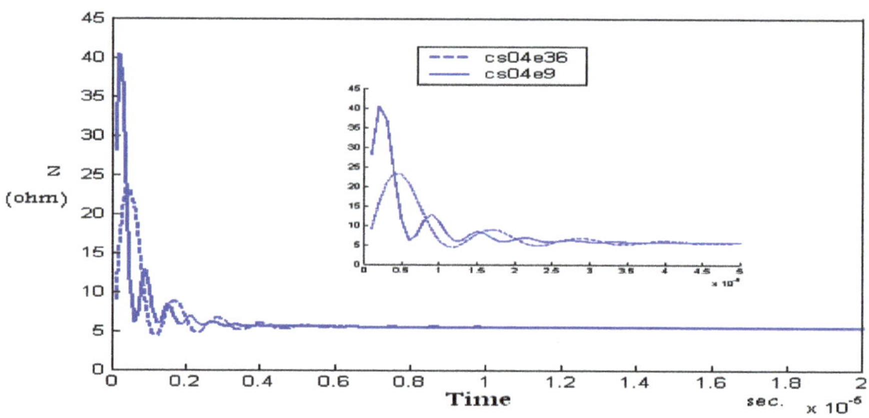

Also from (Figure 18) the peal value of the transient impedance reduced from about 43 Ω to 25Ω and this is due to the increase of the capacitance value of the ground system, which gives low resistance in transient and high frequency cases as the impedance of the capacitance is inverse proportional to its capacitance $\left(x_c = \frac{1}{Jwc}\right)$.

Figure 18 shows small time shift in the transient ground potential rise response for CS04 grid, (about 0.5 μ sec.). Also (Figure 19) shows that the peak value of the transient impedance is reduced from about 40 Ω to 24Ω.

Figure 20. Voltage wave form for CS16, ε =9 and 36

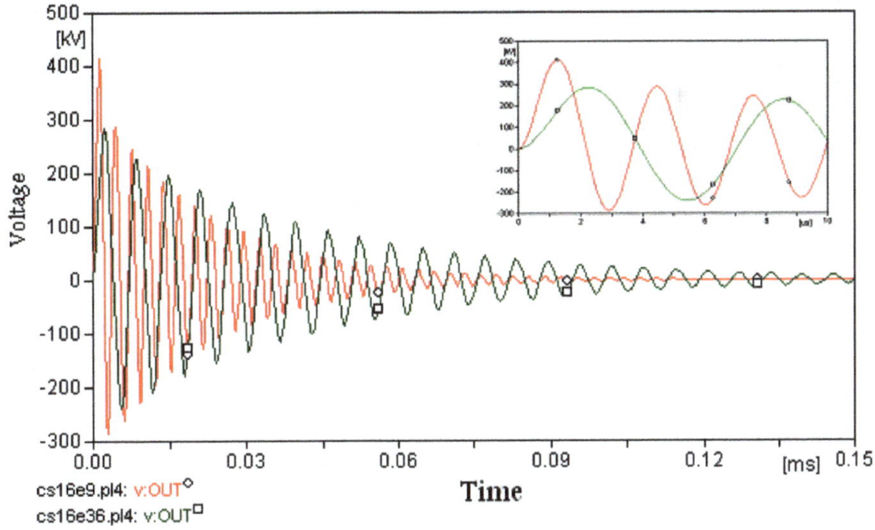

Figure 21. Transient impedance for CS16, ε =9 and 36

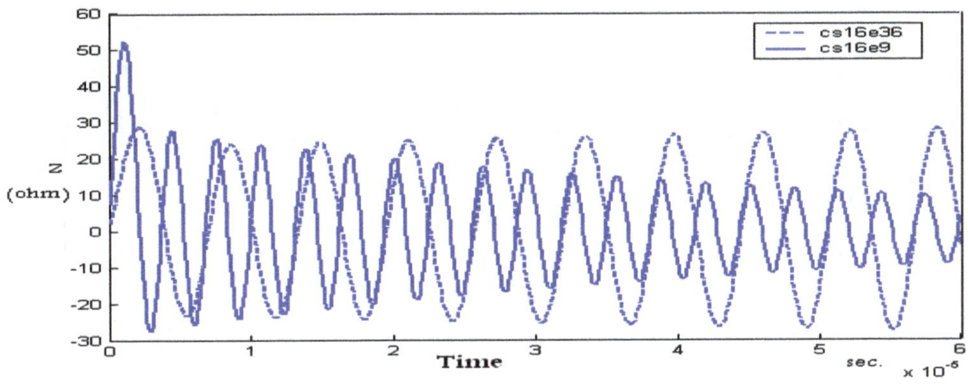

Figure 20 shows small time shift in the transient voltage response for CS16 grid (about 0.5 μsec.).

Also (Figure 21) shows that the peal value of the transient impedance reduced from about 52Ω to 30Ω.

Figure 23 shows small phase shift in the transient voltage response for CS64 grid (about 0.5 μsec.). Also (Figure 12-b) gives the peal value of the transient impedance reduced from about 60Ω to 33Ω.

Figure 22. Voltage wave form for CS64, ε =9 and 36

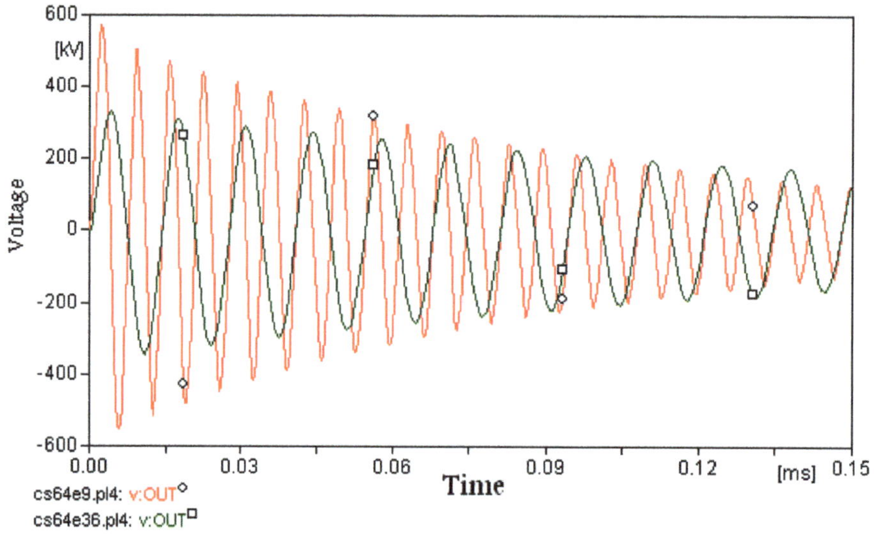

Figure 23. Transient impedance for CS64, ε =9 and 36

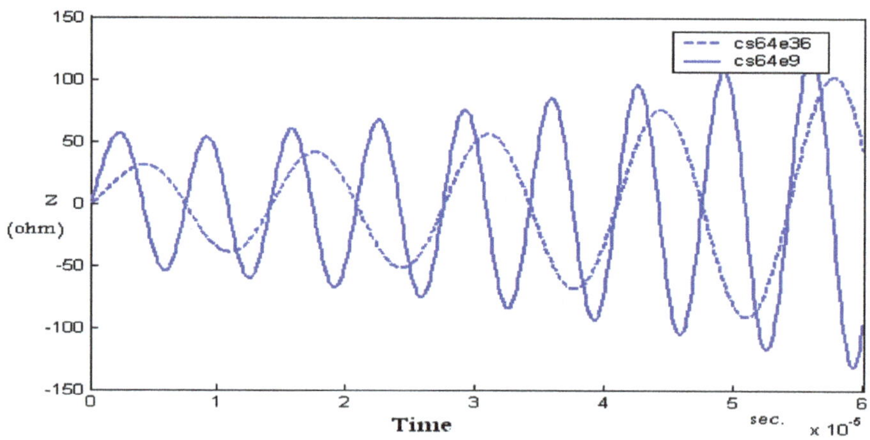

The Influence of the Conductors' Diameter

Result of fast lightning current impulse applied to the grounding grids with radius of the horizontal grounding grid conductor is 7mm and 20mm, the grid is buried at 0.6m depth in the soil with $\varepsilon = 50$ and $\rho = 100\ \Omega.M$ shown in (Figure 13-a, b), (Figure 14-a, b), (Figure 15-a, b) and (Figure 16-a, b).

Table 10. RLC values for r = 7mm

Grid Model	R (Ω)	L_1 (H)	C (F)	L_2 (H)
CS01	6.4991	2.4917e-005	6.8025e-09	5.7671e-006
CS04	5.6657	2.4917e-005	7.8030e-09	5.7671e-006
CS16	1.0716	1.4604e-004	4.1254e-08	3.4201e-005
CS64	0.4969	3.1057e-004	8.8966e-08	7.3022e-005

Table 11. RLC values for r = 20mm

Grid Model	R (Ω)	L_1 (H)	C (F)	L_2 (H)
CS01	6.4991	2.2117e-005	6.8025e-09	5.0673e-006
CS04	5.6657	2.2117e-005	7.8030e-09	5.0673e-006
CS16	1.0716	1.3205e-004	4.1254e-08	3.0701e-005
CS64	0.4969	2.8258e-004	8.8966e-08	6.6023e-005

Table 10 shows the values of Lumped RLC circuit for each grid configuration for r = 7mm.

Table 11 shows the values of Lumped RLC circuit for each grid configuration for r =20mm.

Figure 24. Voltage wave form for CS01, a = 7 and 20mm

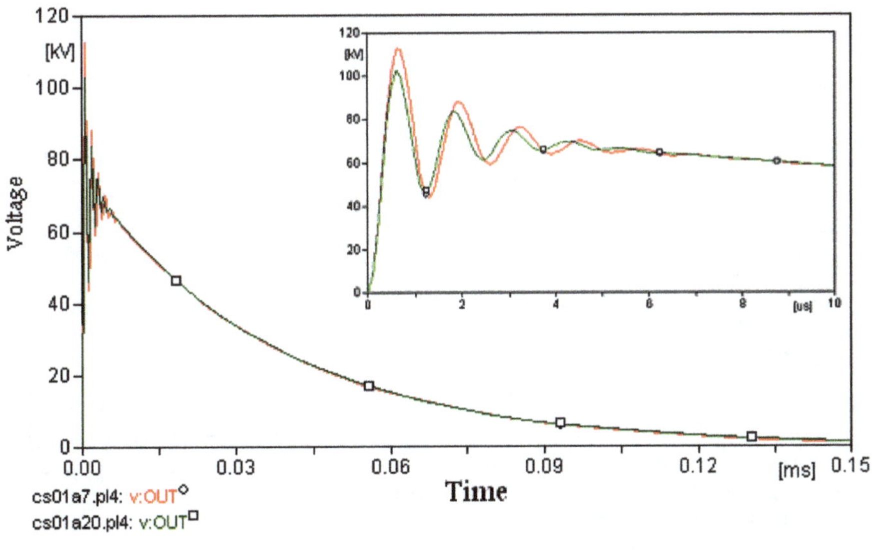

Figure 25. Transient impedance for CS01, a = 7 and 20mm

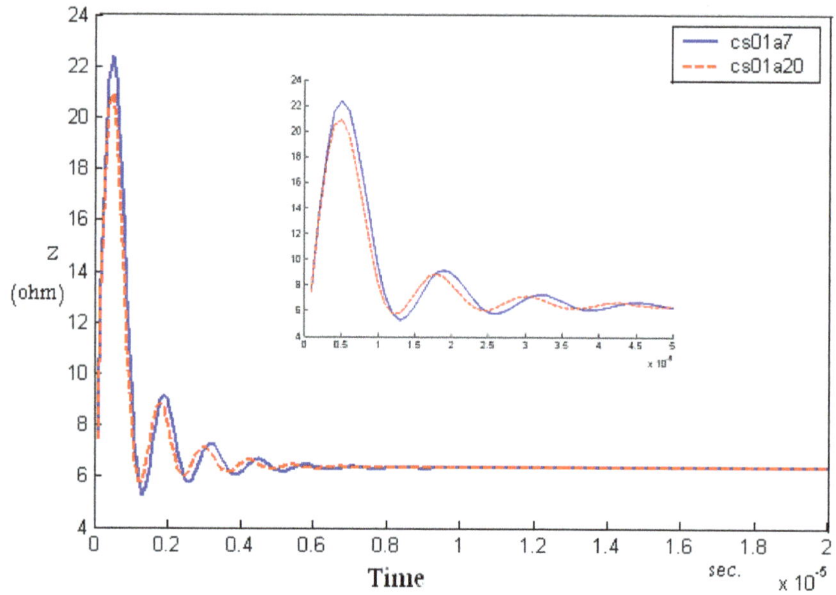

From (Figure 23) it is noticed that changing the radius of the conductors reduce the peak value of the transient voltages from 118kv to 100kv for CS01, for CS04 from 116kv to 99kv as shown in (Figure 24), for CS16 from 250

Figure 26. Voltage wave form for CS04, a = 7 and 20mm

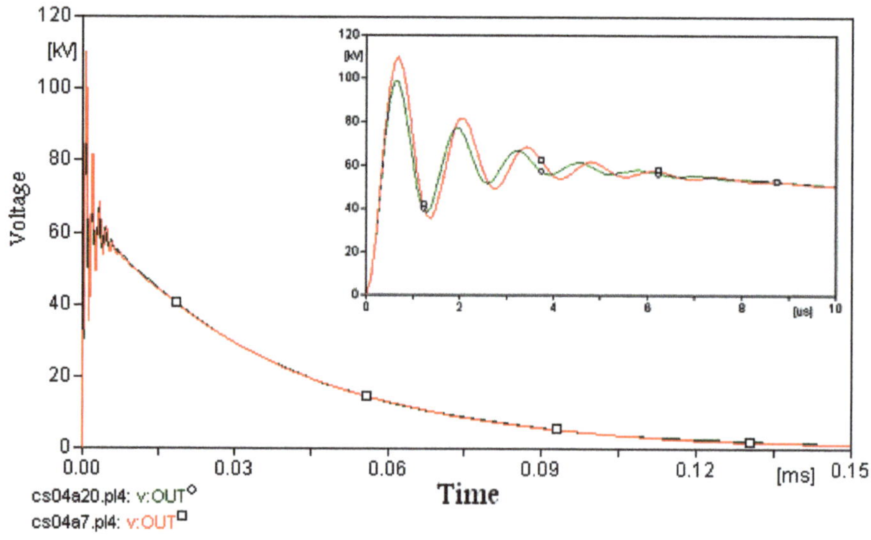

Figure 27. Transient impedance for CS04, a = 7 and 20mm

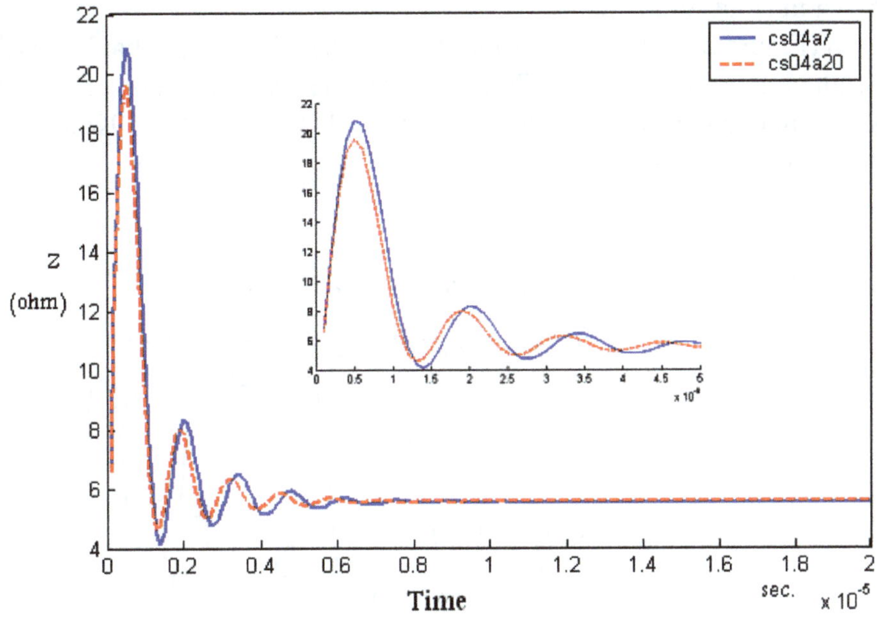

Figure 28. Voltage wave form for CS16, a = 7 and 20mm

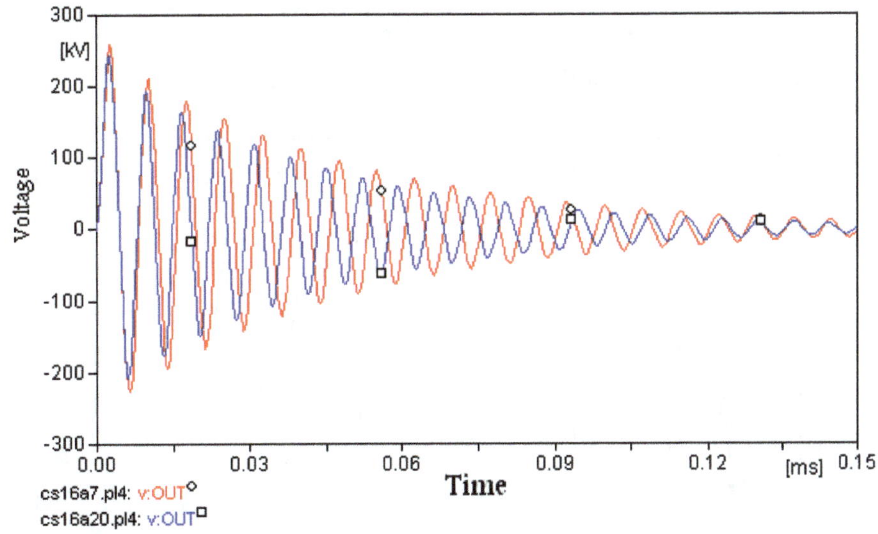

kV to 246 kV as shown in (Figure 27) and for CS46 from 292 kV to 287 kV as shown in (Figure 26).

From (Figure 23), (Figure 24) and (Figure 25) it is noticed that increasing the conductor's diameter from 7mm to 20mm reduce the peak value of the transient impedance from 23Ω to 21Ω for CS01, from 21Ω to 19Ω for CS04 and from 26Ω to 24.5Ω for CS16.

Figure 29. Transient impedance for CS16, a = 7 and 20mm

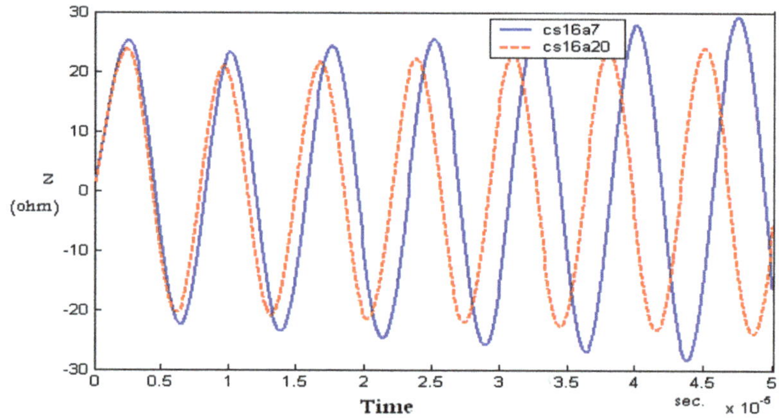

Figure 30. Voltage wave form for CS64, a = 7 and 20mm

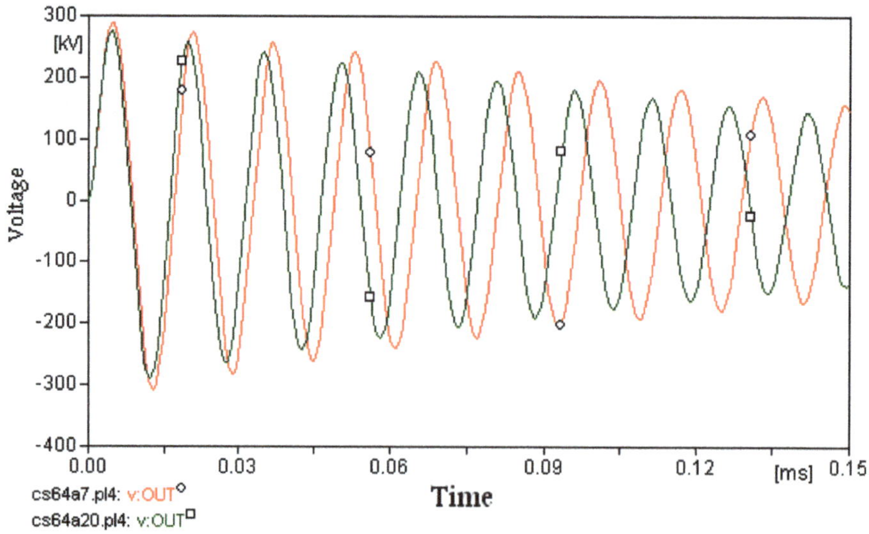

Figure 31. Transient impedance for CS64, a = 7 and 20mm

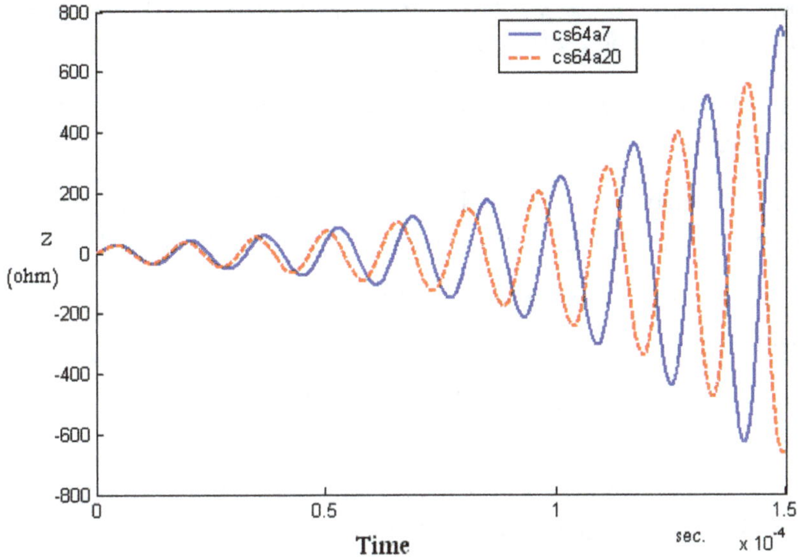

As shown in (Figure 26) when the conductor radius changes from 7mm to 20mm the maximum voltage magnitude for CS04 grid is reduced from about 114 kV to 98 kV.

REFERENCES

Bourg, S., Sacepe, B., & Debu, T. (1995). Deep earth electrodes in highly resistive ground: Frequency behavior. In *Proc. IEEE Int. Symp. Electromagnetic Compatibility* (pp. 584–589). doi:10.1109/ISEMC.1995.523625

Grcev, L. D., & Heimbach, M. (1997). Frequency dependent and transient characteristics of substation grounding system. *IEEE Transactions on Power Delivery*, *12*(1), 172–178. doi:10.1109/61.568238

Harrington, R. F. (1986). *Field computation by moment of methods*. New York: Macmillan.

Mata, C. T., Fernandez, M. I., Rakov, V. A., & Uman, M. A. (2000). EMTP modeling of a triggered-lightning strike to the phase conductor of an overhead distribution line. *IEEE Transactions on Power Delivery*, *15*(4), 1175–1181. doi:10.1109/61.891499

Mazzetti, C., & Veca, G. M. (1983). Impulse behavior of grounding electrodes. *IEEE Transactions on Power Apparatus and Systems*, *-102*(9), 3148–3156. doi:10.1109/TPAS.1983.318122

Mitra, R. (1973). *Computer techniques for electromagnetic*. New York: Pergamon Press.

Imece, A. F., Durbak, D. W., & Elahi, H. (1996). Modeling guidelines for fast front transients. *IEEE Transactions on Power Delivery*, *11*(1), 493–506.

Paul, C. R. (1992). *Introduction to Electromagnetic Compatibility. John Wiley & Sons.*

Paul, C. R. (1994). Analysis of multi-conductor transmission lines. John Wiley & Sons, Inc.

Ramahi, O. M., & Brench, C. (2001). *EMI/EMC computational modeling handbook*. Kluwer Academic Publishers.

Thottappillil R., (2002). Electromagnetic pulse environment of cloud-to ground lightning for EMC studies, *IEEE transactions on EMC*, *44*(1), 203-213.

Volakis, J. L., Chatterjee, A., & Kempel, L. C. (1998). Finite element method for electromagnetic. New York: IEEE Inc.

Yaqing, L. (2004). Transient Response of Grounding Systems caused by Lightning: Modeling and Experiments [Phd thesis]. Acta Universitatis Upsalinesis Uppsala.

Yaqing, L., Mihael, Z., & Rajeev, T. (2001), An improved transmission-line model of grounding system. IEEE Trans. on electromagnetic compatibility, 43(3).

Index

N

P

R

S

T

U

V

www.ingramcontent.com/pod-product-compliance
Lightning Source LLC
LaVergne TN
LVHW081529300926
849161LV00006B/1361